T0320384

HIGHER SPECIAL FUNCTIONS

Higher special functions emerge from boundary eigenvalue problems of Fuchsian differential equations with more than three singularities.

This detailed reference provides solutions for singular boundary eigenvalue problems of linear ordinary differential equations of second order, exploring previously unknown methods for finding higher special functions.

Starting from the fact that it is the singularities of a differential equation that determine the local, as well as the global, behaviour of its solutions, the author develops methods that are both new and efficient and lead to functional relationships that were previously unknown.

All the developments discussed are placed within their historical context, allowing the reader to trace the roots of the theory back through the work of many generations of great mathematicians. Particular attention is given to the work of George Cecil Jaffé, who laid the foundation with the calculation of the quantum mechanical energy levels of the hydrogen molecule ion.

Encyclopedia of Mathematics and Its Applications

This series is devoted to significant topics or themes that have wide application in mathematics or mathematical science and for which a detailed development of the abstract theory is less important than a thorough and concrete exploration of the implications and applications.

Books in the **Encyclopedia of Mathematics and Its Applications** cover their subjects comprehensively. Less important results may be summarised as exercises at the ends of chapters. For technicalities, readers can be referred to the bibliography, which is expected to be comprehensive. As a result, volumes are encyclopaedic references or manageable guides to major subjects.

ENCYCLOPEDIA OF MATHEMATICS AND ITS APPLICATIONS

All the titles listed below can be obtained from good booksellers or from Cambridge University Press. For a complete series listing visit

www.cambridge.org/mathematics.

138 B. Courcelle and J. Engelfriet *Graph Structure and Monadic Second-Order Logic*
139 M. Fiedler *Matrices and Graphs in Geometry*
140 N. Vakil *Real Analysis through Modern Infinitesimals*
141 R. B. Paris *Hadamard Expansions and Hyperasymptotic Evaluation*
142 Y. Crama and P. L. Hammer *Boolean Functions*
143 A. Arapostathis, V. S. Borkar and M. K. Ghosh *Ergodic Control of Diffusion Processes*
144 N. Caspard, B. Leclerc and B. Monjardet *Finite Ordered Sets*
145 D. Z. Arov and H. Dym *Bitangential Direct and Inverse Problems for Systems of Integral and Differential Equations*
146 G. Dassios *Ellipsoidal Harmonics*
147 L. W. Beineke and R. J. Wilson (eds.) with O. R. Oellermann *Topics in Structural Graph Theory*
148 L. Berlyand, A. G. Kolpakov and A. Novikov *Introduction to the Network Approximation Method for Materials Modeling*
149 M. Baake and U. Grimm *Aperiodic Order I: A Mathematical Invitation*
150 J. Borwein *et al. Lattice Sums Then and Now*
151 R. Schneider *Convex Bodies: The Brunn–Minkowski Theory (Second Edition)*
152 G. Da Prato and J. Zabczyk *Stochastic Equations in Infinite Dimensions (Second Edition)*
153 D. Hofmann, G. J. Seal and W. Tholen (eds.) *Monoidal Topology*
154 M. Cabrera García and Á. Rodríguez Palacios *Non-Associative Normed Algebras I: The Vidav–Palmer and Gelfand–Naimark Theorems*
155 C. F. Dunkl and Y. Xu *Orthogonal Polynomials of Several Variables (Second Edition)*
156 L. W. Beineke and R. J. Wilson (eds.) with B. Toft *Topics in Chromatic Graph Theory*
157 T. Mora *Solving Polynomial Equation Systems III: Algebraic Solving*
158 T. Mora *Solving Polynomial Equation Systems IV: Buchberger Theory and Beyond*
159 V. Berthé and M. Rigo (eds.) *Combinatorics, Words and Symbolic Dynamics*
160 B. Rubin *Introduction to Radon Transforms: With Elements of Fractional Calculus and Harmonic Analysis*
161 M. Ghergu and S. D. Taliaferro *Isolated Singularities in Partial Differential Inequalities*
162 G. Molica Bisci, V. D. Radulescu and R. Servadei *Variational Methods for Nonlocal Fractional Problems*
163 S. Wagon *The Banach–Tarski Paradox (Second Edition)*
164 K. Broughan *Equivalents of the Riemann Hypothesis I: Arithmetic Equivalents*
165 K. Broughan *Equivalents of the Riemann Hypothesis II: Analytic Equivalents*
166 M. Baake and U. Grimm (eds.) *Aperiodic Order II: Crystallography and Almost Periodicity*
167 M. Cabrera García and Á. Rodríguez Palacios *Non-Associative Normed Algebras II: Representation Theory and the Zel'manov Approach*
168 A. Yu. Khrennikov, S. V. Kozyrev and W. A. Zúñiga-Galindo *Ultrametric Pseudodifferential Equations and Applications*
169 S. R. Finch *Mathematical Constants II*
170 J. Krajíček *Proof Complexity*
171 D. Bulacu, S. Caenepeel, F. Panaite and F. Van Oystaeyen *Quasi-Hopf Algebras*
172 P. McMullen *Geometric Regular Polytopes*
173 M. Aguiar and S. Mahajan *Bimonoids for Hyperplane Arrangements*
174 M. Barski and J. Zabczyk *Mathematics of the Bond Market: A Lévy Processes Approach*
175 T. R. Bielecki, J. Jakubowski and M. Nięwęgłowski *Structured Dependence between Stochastic Processes*
176 A. A. Borovkov, V. V. Ulyanov and Mikhail Zhitlukhin *Asymptotic Analysis of Random Walks: Light-Tailed Distributions*
177 Y.-K. Chan *Foundations of Constructive Probability Theory*
178 L. W. Beineke, M. C. Golumbic and R. J. Wilson (eds.) *Topics in Algorithmic Graph Theory*
179 H.-L. Gau and P. Y. Wu *Numerical Ranges of Hilbert Space Operators*
180 P. A. Martin *Time-Domain Scattering*
181 M. D. de la Iglesia *Orthogonal Polynomials in the Spectral Analysis of Markov Processes*
182 A. E. Brouwer and H. Van Maldeghem *Strongly Regular Graphs*
183 D. Z. Arov and O. J. Staffans *Linear State/Signal Systems*
184 A. A. Borovkov *Compound Renewal Processes*
185 D. Bridges, H. Ishihara, M. Rathjen and H. Schwichtenberg (eds.) *Handbook of Constructive Mathematics*
186 M. Aguiar and S. Mahajan *Coxeter Bialgebras*
187 K. Broughan *Equivalents of the Riemann Hypothesis III: Further Steps towards Resolving the Riemann Hypothesis*

ENCYCLOPEDIA OF MATHEMATICS AND ITS APPLICATIONS

Higher Special Functions

A Theory of the Central Two-Point Connection Problem Based on a Singularity Approach

WOLFGANG LAY

Universität Stuttgart

Shaftesbury Road, Cambridge CB2 8EA, United Kingdom

One Liberty Plaza, 20th Floor, New York, NY 10006, USA

477 Williamstown Road, Port Melbourne, VIC 3207, Australia

314–321, 3rd Floor, Plot 3, Splendor Forum, Jasola District Centre, New Delhi - 110025, India

103 Penang Road, #05–06/07, Visioncrest Commercial, Singapore 238467

Cambridge University Press is part of Cambridge University Press & Assessment, a department of the University of Cambridge.

We share the University's mission to contribute to society through the pursuit of education, learning and research at the highest international levels of excellence.

www.cambridge.org
Information on this title: www.cambridge.org/9781009123198

DOI: 10.1017/9781009128414

When citing this work, please include a reference to the DOI 10.1017/9781009128414

First published 2024

A catalogue record for this publication is available from the British Library

A Cataloging-in-Publication data record for this book is available from the Library of Congress

ISBN 978-1-009-12319-8 Hardback

To
Cornelia Charlotte
Ann-Sophie
Clara Ruth Emilia

Contents

Preface

This book presents a mathematical method for calculating previously unknown functions. These functions have specific properties; therefore, they are called **special functions**. The most outstanding of their properties is that they are particular solutions of certain differential equations – linear, ordinary, homogeneous differential equations of second order with polynomial coefficients:

$$P_0(z)\,\frac{\mathrm{d}^2y}{\mathrm{d}z^2} + P_1(z)\,\frac{\mathrm{d}y}{\mathrm{d}z} + P_2(z)\,y = 0, \quad z \in \mathbb{C}; \tag{1}$$

here $P_i(z)$, $i = 0, 1, 2$ are polynomials in z.

As is well known, such differential equations determine – as a consequence of their order – a two-dimensional set of functions. That is, among those functions which satisfy the equation, only after two quantities have been determined is one of them uniquely determined. Usually, these two quantities are the value $y(z_0)$ and the derivative $\left.\frac{\mathrm{d}y}{\mathrm{d}z}\right|_{z=z_0}$ of the function $y(z)$ at a certain point $z = z_0$ of the differential equation (1).

Now, for the differential equation (1) there are inevitably points z_1 in the domain of definition $z \in \mathbb{C}$ where the function $P_0(z_1)$ has the value zero. Such points are called **singularities of the differential equation (1)**, in contrast to the **ordinary points of the differential equation** at which $P_0(z)$ is not zero. The special thing about the singularities $z = z_1$ of the differential equation (1) is that – under specific conditions that will be discussed – already the determination of the function value $y(z_1) = y_1$ is sufficient to uniquely determine the solution among all possible functions determined by the differential equation (1). One does not have the possibility to choose both the function value and the derivative of its solution $y(z)$ at the singularity $z = z_1$ of the differential equation (1).

However, it can also happen that solutions of the differential equation (1) become singular at their singularities $z = z_1$, i.e., that the function value increases over all limits when approaching the singularity. However, it is a characteristic of linear differential equations that their solutions can only become singular at their singularities: at ordinary points of the differential equation (1), all of its solutions are holomorphic.

In this singular case, a specific behaviour of the function value when approaching the singularity has to be taken, rather than the function value itself.

As is well known, two functions $y_1(z)$ and $y_2(z)$ are called **linearly independent** if there is no constant $C \in \mathbb{C}$ such that

$$y_2(z) = C\, y_1(z) \tag{2}$$

hold; otherwise, these two functions are called **linearly dependent**. It is also known that the general solution $y^{(g)}(z)$ of the differential equation (1) is given by

$$y^{(g)}(z) = C_1\, y_1(z) + C_2\, y_2(z), \tag{3}$$

where the two particular solutions $y_1(z)$ and $y_2(z)$ of the differential equation (1) are supposed to be linearly independent and the coefficients $C_1 \in \mathbb{C}$ and $C_2 \in \mathbb{C}$, which are independent of z, take all values of their domain of definition $\mathbb{C}$. Such a pair of solutions is called a **fundamental system**.

What is so special about the special functions? Let us consider two singularities of the differential equation (1), which for the sake of simplicity should lie on the positive real axis, i.e., at $z = z_1 = 0$ and at $z = z_2 = 1$. The interval in between we call the **relevant interval**. Let us now consider a solution $y(z)$ of the differential equation (1), which is supposed to have a certain function value $y(z_1) = y_1$ at the singularity at $z = z_1$. Thus, the function, and therefore the solution $y(z)$, is uniquely determined. We can no longer demand that this function should assume a certain value at z_2. This is only possible again if at least one of its coefficients $P_i(z), i = 0, 1, 2$, also depends on a parameter E in addition to the independent variable z. Only then may there be certain values $E = E_i, i = 0, 1, 2, \ldots$, of this parameter E for which there are partial solutions $y_{E_i}(z)$ of the differential equation (1), which take a given value at both points z_1 and z_2. The computation of this parameter $E = E_i, i = 0, 1, 2, \ldots$, is called the boundary eigenvalue problem, and the values $E = E_i, i = 0, 1, 2, \ldots$, are called eigenvalues. Because at least one of the boundary points of the relevant interval is a singularity of the differential equation (1), it is not only a boundary eigenvalue problem but a **singular boundary eigenvalue problem**. If we are dealing with singular solutions of the differential equation (1), then – as already mentioned above – the function value is replaced by the behaviour when approaching the singularity radially. This is especially true if the singularity z_2 is an improper point of the equation (1), i.e., if it is placed at infinity, which can be symbolised by $z_2 = \infty$.

It becomes obvious that the key to the calculation of special functions is a method to calculate their eigenvalues. This book explains under what conditions classical methods for the determination of eigenvalues exist, and where this is not yet the case. Then, a mathematical principle is formulated and a mathematical method is developed with the help of which one can calculate the eigenvalues and, thus, the special functions for all differential equations (1). It is immediately obvious that this considerably expands the hitherto rather limited number of special functions.

Now, one could suggest that the singularities of the differential equation (1) have only a finite number of singular points of the differential equation and are, therefore, negligible, while all other points (and thus the vast majority) are ordinary points of the differential equation. However, this is not true: singularities, although they occur only once in the differential equation (1), or at least only sporadically, essentially determine the solutions in the entire definition area of the equation: the singularities of the differential equation (1) are the cornerstones of its solution. So it is not surprising that the following applies: every differential equation (1) has at least one singularity.

Special functions have enjoyed great popularity among mathematicians as well as physicists. The following anecdote has been passed down, showing the tremendous popularity of probably the most-cited book in mathematics ever. **Sir Michael V. Berry**, the English theoretical physicist, was once invited to contemplate being marooned on the proverbial desert island. He was asked what book he would most like to have there, in addition to the Bible and (as an English person) the complete works of Shakespeare. His immediate answer was: Abramowitz and Stegun's *Handbook of Mathematical Functions* (which first appeared in 1964), perhaps the most successful work of mathematical reference ever published. This answer is by no means surprising: as soon as principles dominate in a field of science, special functions become important. Hereby, special functions are standardised functions that first originate from mathematically formulated rules and second are able to describe a variety of scientific phenomena.

This book is not designed as a reference book, nor to raise any claim of completeness, but it first displays the exact solution of the central two-point connection problem and second founds the hope that in the near future the topic of higher special functions will experience a revival in science and in lecturing, yielding lots of new insights, surprising scientific results and not-yet-seen phenomena.

Chapter 1 uses the example of classical special functions to show the aspect from which they must be considered, so that mathematical methods can be derived which are so general that they can also be used to calculate the higher special functions. In addition, the underlying mathematical principle is presented.

In Chapter 2, a concept for the treatment of singularities of linear differential equations and their local solutions is presented, which goes back to Henri Poincaré, but deviates from it in one fundamental respect.

In Chapter 3, the differential equations of the Fuchs class are presented as fundamental equations from which many others – so-called confluent cases – can be derived. Based on the methodology of Chapter 2, the Fuchs equation with four singularities is presented as an example. Finally, the totality of the differential equation is presented in a schematic.

In Chapter 4, the singular boundary eigenvalue problem is formulated and the methods developed in previous chapters are calculated and solved using the differential equations presented in Chapter 3.

In Chapter 5, the previously developed method is applied concretely to selected examples. On the one hand, this shows the usefulness of the basic method as developed in previous chapters; on the other hand, it shows how much more knowledge of the concrete problem is needed in order to move on from the general level of theory to the solution of such concrete problems.

Acknowledgements

A book like the present one cannot be produced without the direct and indirect participation of a whole series of people – even and especially if it is a monograph. Such is the case here. I would, therefore, like to take this opportunity to mention a group of people who, in one way or another, have made a significant contribution to the success of this project.

I would like to thank **Professor Dr Sir Michael Victor Berry** of the University of Bristol for a research stay at the H. H. Wills Physics Laboratory as a fellow of the Max Planck Society, which brought me into contact with the topic of special functions and just at a time when the first thoughts on the DLMF (Digital Library of Mathematical Functions) project were taking shape.

I would like to thank **Professor Dr Felix Medland Arscott**(†) of the University of Winnipeg, who accompanied and guided me on my way to becoming a scientist, with a sensitivity and respect that I have rarely experienced. He was a great teacher of science and I will always honour his memory. That I have remained true to the field of differential equations is, to a large extent, due to him.

I owe special thanks to **Professor Dr Serguei Yuriewitsch Slavyanov**(†) of St. Petersburg State University. As mentor he advised and shaped me over a 30-year period of scientific cooperation. Without his scientific supervision, the creation of this book would not have been possible.

I would like to thank **Dipl. Phys. Karlheinz Bay** for scientific cooperation, and carrying out the necessary numerical calculations over several years of research on the subject.

I would like to thank **Dr Achim Dannecker** for his support over the period of writing this book, and for valuable discussions on various topics concerning the drafting of a standard work in mathematics. His mental support and active interest in the development of the work over many years have encouraged and supported me, especially during difficult phases.

Special thanks go to **Dipl. Math. Hardy Wagner**, who accompanied me on the difficult and long journey from typescript to book. With his patience and empathy, he is a real friend and an excellent mathematician.

I would like to thank **Dr Tino Lukaschek** for a critical review of the typescript and for numerous valuable suggestions in connection with the writing of the book.

I would like to thank **Sandra and Stefan Mercamp** for their mental guidance over the years, and for their interest and participation in the creation of this book. Discussions with them in a relaxed atmosphere have given me courage when the process of writing was at a standstill.

I would like to thank **Professor Dr Charles William Clark** of the National Institute of Standards and Technology (NIST) in Gaithersburg (Maryland) for his recommendation to have the book published by Cambridge University Press and to have it reviewed by Professors Peter John Olver and Mark Jay Ablowitz. The large-scale DLMF project has contributed significantly to the writing of this book.

I would like to thank **Professor Dr Peter John Olver** of the University of Minnesota (Minneapolis) and **Professor Dr Mark Jay Ablowitz** of the University of Colorado (Boulder) for their scholarly reading, for numerous valuable suggestions and – above all – for recommending the work to Cambridge University Press for publication.

I would also like to thank my wife **Dr Cornelia Charlotte Matz** and my daughters **Ann-Sophie Lay** and **Clara Ruth Emilia Lay** for their critical comments and constant consideration when it came to supporting me in my research efforts. I thank my wife for her support in compiling the reference list, and Ann-Sophie for various discussions in connection with the writing and publication of the work, and for always listening to me when I needed her advice and opinion.

1
Introduction

Classical special functions are a traditional field of mathematics. As particular solutions of singular boundary eigenvalue problems of linear ordinary differential equations of second order, they are by definition functions that can be represented as the product of an asymptotic factor and a (finite or infinite) Taylor series. The coefficients of these series are by definition solutions of two-term recurrence relations, from which an algebraic boundary eigenvalue criterion can be formulated. This method is called the *Sommerfeld polynomial method* (Rubinowicz, 1972); thus, one can say that the boundary eigenvalue condition is by definition algebraic in nature. It is the central message in this book that one can resolve this restriction and it is shown how to do this methodically, and what the fundamental mathematical principle underlying this method is.

Now, of course, the method developed for this also applies to problems that can be solved with classical methods. So, in order to present the newly developed methods in the light of what is known, and to be able to understand the new perspective more easily (and also measure the results obtained against what is already known), this new method is applied in this chapter to the already known solutions. This makes it possible to classify the new according to the well known. Accordingly, it is a 'phenomenological' introduction, where the focus is not on definitions, theorems and their proofs, but on the ad hoc introduction of the relevant quantities. The systematic introduction based on definitions follows in the subsequent chapters, this time with respect to differential equations, which are no longer accessible by classical methods.

1.1 Historical Remarks

The history of the subject treated here goes back to the middle of the nineteenth century, to a time when the German mathematician **Lazarus Fuchs** (1833–1902) wrote down the local solution of a linear, ordinary differential equation with polynomial coefficients in the vicinity of a singularity. This solution consists of a product of an (in general) irrational power of the independent variable and an (in general) infinite Taylor, thus one-sided, series. Yet, the explicit form of the solution was only one aspect of the great importance of Fuchs' approach. It was just as important to have recognised the significance of the singularity of a differential equation in the first place. Singularities are something ubiquitous; they cannot be avoided, not in the equation and certainly not in the solution. And not only that: singularities of differential equations determine the behaviour of their solutions everywhere, even where the equations are holomorphic. This is perhaps the most significant peculiarity of differential equations. Therefore, it is more than justified to bring singularities to the centre of consideration if one wants to deal with differential equations in depth. This insight, which is now about 150 years old, is the basis of this work.

It is well known that the basic approach to solving a differential equation is based on Weierstrass' approximation theorem for integer powers, according to which powers can be used to approximate any holomorphic function. Based on this, the French mathematician **Paul Painlevé** (1863–1933) developed a basic existence theorem for local solutions, the so-called 'calcul des limites' or majorant method. In turn, all fundamental questions – such as those concerning analytical continuations or the uniqueness of solutions – are based on this. One can say that this complex of fundamental questions was largely settled at the beginning of the twentieth century.

The full significance of Fuchs' work was revealed in the fact that there are two types of singularities in differential equations: those that are now called *regular* and those that are called *irregular*. This distinction is triggered by the nature of their local solutions. For the irregular singularities of differential equations, the Fuchsian approach did not provide local solutions, only for regular ones. As the German mathematician **Meyer Hamburger** (1838–1903) showed a little later, this requires generalised, two-sided infinite power series, i.e., Laurent series. This makes for a significant technical complication. So, it remained to search for a one-sided series approach for the case of irregular singularities. Although the Fuchsian approach could not do this, it showed the way: a one-sided replacement for the Laurent series was then concretely worked out about 20 years later by the French mathematician **Achille Marie Gaston Floquét** (1847–1920) and, at about the same time, by the German mathematician **Ludwig Wilhelm Thomé** (1841–1910). (That is why these approaches are now called Thomé solutions in Germany and Floquét solutions in France.) The crucial idea was to write down the asymptotic factors, explicitly. The problem with Thomé's solutions, however, was that they did not converge in all cases. Thomé put a lot of effort into showing under what conditions his series approaches converged. It was finally the French mathematician **Henri Poincaré** (1854–1912) who showed the significance of Thomé's series: they are asymptotic solutions of the differential equation that represent their solutions within sectors, at the top of which is placed the singularity. The lateral lines delimiting these sectors are called *Stokes lines*, after the Irish mathematician and physicist **Sir George Gabriel Stokes** (1819–1903).

Another important insight into linear differential equations was that it was understood that differential equations with exclusively regular singularities could serve as the cornerstone of a mathematical theory, because the irregular singularities could be generated by merging regular singularities. Equations with exclusively regular singularities were given the name *Fuchsian differential equations* and the simplest forms among them were in turn given their own names. Thus, the simplest form of the Fuchsian differential equation with only one singularity is called *Laplace's equation*, the one with two singularities is called *Euler's equation* and the one with three singularities is called *Gauss' equation*.

The Fuchsian approach always leads to difference equations for its coefficients and thus to asymptotic questions of their solution for large values of the index. It has been found that significant differences in the asymptotic behaviour of the solutions can exist, although the ratio of two successive terms of the solutions tends towards one and the same value for large values of the index. Depending on whether this is the case with a difference equation or not, it is called *regular* or *irregular*.

As far as the investigation of regular difference equations is concerned, Henri Poincaré and the German mathematician **Oskar Perron** (1880–1975) excelled. For the study of irregular difference equations, in turn, the Canadian mathematician **George David Birkhoff** (1884–1944), the Russian mathematician **Waldemar Juliet Trjitzinsky** (1901–1973) and the American mathematician **Clarence Raymond Adams** (1898–1965) made lasting contributions in the 1920s (see, e.g., Birkhoff–Trjitzinsky theory and the Birkhoff–Adams theorem in Aulbach et al., 2004). Thus, while Henri Poincaré recognised the importance of Thomé's approaches to linear differential equations by proving their asymptotic character, Birkhoff and Trjitzinsky carried out the analogous investigations for linear difference equations.

The importance of all this work for the development of modern physics at the beginning of the twentieth century should be undisputed, namely relativity and quantum theory, just as this development in turn had an effect on mathematics. The realisation that nature is essentially linear on small scales, as evidenced by the *Schrödinger equation*, has given rise to this retroactive influence. Singular boundary eigenvalue problems of linear differential equations of second order were henceforth an important object of research. It turned out that it is not the order of the differential equation that determines the order of the difference equation resulting from a Fuchsian approach, but the number of its singularities.

In order for the solution of a linear differential equation to behave in a prescribed manner not only at one point, but at two, this solution must have a parameter. In quantum theory, this is the energy parameter. In order for the solution to behave in a prescribed manner at two different points, this energy parameter must assume certain values. Calculating these means solving the boundary eigenvalue problem. The condition on which this determination is based is called the boundary eigenvalue condition.

The mathematical form of the boundary eigenvalue condition is of central importance for the solution of the problem. It turns out that this condition is generally only algebraic in nature if the underlying differential equation is either of the Laplacian, Eulerian or Gaussian type, i.e., originates from a Fuchsian equation that has at most three singularities. If this number of singularities is larger than three, then the boundary eigenvalue condition becomes transcendental. And for such equations, there was only one method to write the boundary eigenvalue condition at all, and this is only if the order of the difference equation which the coefficients of Fuchs' solution approaches is at most two: the method of infinite continued fractions.

Transcendental boundary eigenvalue conditions thus occur systematically for Fuchsian differential equations that have more than three singularities. The simplest of these equations is called *Heun's differential equation*, because the German mathematician **Karl Heun** (1859–1929) studied it systematically for the first time in a paper from 1889.

Today, a distinction is made between algebraic and transcendental boundary eigenvalue conditions in two respects: the singular boundary eigenvalue problems that lead to transcendental eigenvalue conditions are called *central two-point connection problems* (CTCPs) and the resulting solution functions are called *higher special functions*; all others are called *classical special functions*.

To be precise, local solution approaches such as those of Lazarus Fuchs are no longer sufficient for solving singular boundary eigenvalue problems. This required a large-scale scientific development that began when **Arnold Sommerfeld** (1868–1951) had his great era as a professor at the Ludwig Maximilian University of Munich. In 1919, **Wolfgang Ernst Pauli** (1900–1958) asked Sommerfeld for a topic for a doctoral thesis. Pauli received from Sommerfeld the task of applying the rules of quantum mechanics established by **Niels Bohr** (1885–1962) in 1913 in order to show that the ionised hydrogen molecule is stable under the conditions of the then new quantum mechanics. Since the quantisation rules of Bohr and Sommerfeld (for which Bohr received the Nobel Prize) were not quite correct (they were what are now called semiclassical quantisation rules), he did not quite succeed, but obtained the result that the ionised hydrogen molecule was at least metastable. Thus the problem of calculating quantum mechanical energy levels became the most important in quantum physics, as it was supposed to explain the chemical bonding between two protons by a single electron.

In 1933 **George Cecil Jaffé** (1888–1965) took up the problem anew, then having available Schrödinger's differential equation and thus the correct quantisation rules. So, Jaffé was eventually able to show the stability of the ionised hydrogen molecule. However, what became much more important was that Jaffé applied a marvellous transformation of the underlying differential equation that enabled him to apply a solution ansatz that solved the problem exactly. This idea is the basis of the method of solution of the specific singular boundary eigenvalue problem, called the *central two-point connection problem* in this book. It consists of a series ansatz and a transformation that I call, in honour and commemoration of George Jaffé, the *Jaffé ansatz* and the *Jaffé transformation*, respectively. The particular solution of the underlying differential equation represented by the Jaffé ansatz is called the *Jaffé solution* (see Jaffé, 1933).

Unfortunately, George Jaffé, as a Jewish professor at the Universität Gießen, was dismissed in 1933 and in 1934 had to abandon his position as a professor, eventually going into exile in 1939, leaving Germany for ever. He went to Bâton Rouge in the US state of Louisiana but did not pick up his ideas there any more.

Sixty years after this great publication, in spring 1993, I visited the elderly **Friedrich Hund** (1896–1997) – one of the founders of quantum theory – in Göttingen. He told me the sad story of George Jaffé's fate. So this book was also written as a protest against oblivion, against the forgetting of a great scientist and his difficult life.

In 1989, Karl Heun's publication about the differential equation that nowadays bears his name celebrated its centennial appearance. In order to mark this date, an international conference was organised that brought together experts from all over the world who worked in the field. The result of this conference was several commitments to promote the topic, and even several notations that have been agreed. It was common opinion that the step from classical special functions to higher special functions was ripe to be made, and the participants were in agreement that this was a fundamental step.

Subsequently, following this conference, an unpublished conference booklet was written (Seeger and Lay, 1990), then a book written by several participants of the conference (Ronveaux, 1995), then a monograph on the topic by the Russian expert **Serguei Yuriewitsch Slavyanov** (1942–2019) and the present author (Slavyanov and Lay, 2000), and eventually a successor to the famous *Handbook of Mathematical Functions* by **Milton Abramowitz** (1915–1958) and **Irene Stegun** (1919–2008) that was edited by **Frank William John Olver** (1924–2013), entitled *The NIST Handbook of Mathematical Functions*, published (within the DLMF) by the National Institute of Standards and Technologies (NIST; Olver et al., 2010). In this book, the Heun differential equation has been dealt with as a new differential equation, not incorporated within the set of differential equations, the solutions of which belong to the classical special functions of mathematical physics (cf. Chapter 31). The authors of this part on Heun's differential equation were **Brian D. Sleeman** (1939–2021), a British professor and participant of the famous conference introduced above, and the Russian mathematician **Vadim B. Kuznetsov** (*1963).

However, the Heun equation has not been treated in an exhaustive manner, since several important topics had not been tackled or settled up to that time. These circumstances were the impetus for me to undertake a large-scale investigation in order to treat the main remaining problem: the central two-point connection problem for differential equations beyond Gaussian type. The solution of this problem in a satisfactory manner, i.e., theoretically as well as calculatory, yields the possibility of raising the whole field on a new level: singular boundary eigenvalue problems of differential equations beyond Gaussian type, stemming from applications that are solvable. Now, these mathematical problems lead to new functions (higher special functions) and show up new phenomena not yet seen, based on the fact that the differential equations have a sort of parameter that differential equations yielding classical special functions do not have.

1.2 Classical Special Functions: Testing the New in a Well-Known Area

In this section, those differential equations are described whose singular boundary eigenvalue problems lead to the *classical special functions*. It may seem superfluous to rewrite what has already been sufficiently described in books such as Whittaker and Watson (1927), Coddington and Levinson (1955), Ince (1956), Bieberbach (1965), Olver (1974) or Hille (1997). However, this is an inaccurate impression. Here, in order to get beyond the results of classical theory, one must adopt a somewhat different viewpoint than the one which developed the classical theory. This slightly different viewpoint, which allows a generalisation, must of course also be applicable to the classical theory. Thus, the known terrain is useful because it allows the new theory to be proven on the basis of known results. Anyone who picks up pencil and paper and starts calculating will appreciate being able to direct the new along the lines of the old.

1.2.1 Differential Equations

The Gauss Equation

Basically, there is one differential equation whose singular eigenvalue problems produce classical special functions: that is, the *Gauss differential equation*. It is a differential equation (1), where $P_0(z)$ is the fourth-order polynomial

$$P_0(z) = z^2\,(z-1)^2,$$

$P_1(z)$ is the third-order polynomial

$$P_1(z) = (A_0 + A_1)\,z^3 - (2\,A_0 + A_1)\,z^2 + A_0\,z$$

and $P_2(z)$ is the second-order polynomial

$$P_2(z) = (B_0 + B_1 - C)\,z^2 - (2\,B_0 - C)\,z + B_0.$$

By means of

$$P(z) = \frac{P_1(z)}{P_0(z)}, \quad Q(z) = \frac{P_2(z)}{P_0(z)}, \tag{1.2.1}$$

equation (1) on page ix becomes

$$\frac{d^2y}{dz^2} + P(z)\,\frac{dy}{dz} + Q(z)\,y(z) = 0, \quad z \in \mathbb{C} \tag{1.2.2}$$

with

$$
\begin{aligned}
P(z) &= \frac{P_1(z)}{P_0(z)} = \frac{A_0}{z} + \frac{A_1}{z-1}, \\
Q(z) &= \frac{P_2(z)}{P_0(z)} = \frac{B_0}{z^2} + \frac{B_1}{(z-1)^2} + \frac{C}{z} - \frac{C}{z-1}.
\end{aligned} \tag{1.2.3}
$$

As is seen, the coefficient $P(z)$ has a first-order pole at the singularities of the differential equation, located at $z = 0$ and at $z = 1$, and the coefficient $Q(z)$ has a second-order pole there. This is the crucial significance of the singularity of the differential equation at hand. Astonishingly, this differential equation has not just two but three singularities, two of which are placed at finite points, being called *finite singularities* for short, namely at $z = 0$ and at $z = 1$; the third one is located at infinity, being called *infinite singularity* for short. Since infinity is neither a proper point nor a number, this improper point of the differential equation (1.2.2) is dealt with by means of inverting it, viz. applying the transformation

$$
\zeta = \frac{1}{z}, \tag{1.2.4}
$$

thus getting

$$
\frac{d^2y}{d\zeta^2} + \left[\frac{2}{\zeta} - \frac{1}{\zeta^2} P(\zeta)\right] \frac{dy}{d\zeta} + \frac{1}{\zeta^4} Q(\zeta)\, y = 0, \quad \zeta \in \mathbb{C}, \tag{1.2.5}
$$

with $P(\zeta) = P(1/z)$ and $Q(\zeta) = Q(1/z)$, and then considering the point $\zeta = 0$. With equation (1.2.3) this means

$$
\begin{aligned}
\tilde{P}(\zeta) &= \frac{2}{\zeta} - \frac{1}{\zeta^2} P(\zeta) = \frac{\tilde{A}_0}{\zeta} + \frac{A_1}{\zeta - 1}, \\
\tilde{Q}(\zeta) &= \frac{1}{\zeta^4} Q(\zeta) = \frac{\tilde{B}_0}{\zeta^2} + \frac{B_1}{(\zeta-1)^2} + \frac{\tilde{C}}{\zeta} - \frac{\tilde{C}}{\zeta - 1}.
\end{aligned} \tag{1.2.6}
$$

As may be seen, this differential equation is just the same as (1.2.2), (1.2.3), except for the coefficients A_0, B_0, C. These do have other values, namely

$$
\begin{aligned}
\tilde{A}_0 &= 2 - A_0 - A_1, \\
\tilde{B}_0 &= B_0 + B_1 - C, \\
\tilde{C} &= 2\,B_1 - C,
\end{aligned} \tag{1.2.7}
$$

while A_1 and B_1 are not affected. In particular, the order of the pole at $\zeta = 0$ is the same as at $z = 0$ in (1.2.3).

All the properties of singularities of equation (1.2.2), (1.2.3) at infinity are transferred to the singularity at zero of equation (1.2.5), (1.2.6) by means of the transformation (1.2.4).

Turning to the *local solutions* at the finite singularities of the differential equation (1.2.2), (1.2.3) it was the great discovery of Lazarus Fuchs that an ansatz for the local solutions at the singularity $z = 0$ has the form (Fuchs, 1866)

$$y_0(z) = \sum_{n=0}^{\infty} a_n\, z^{n+\alpha_0} = z^{\alpha_0} \sum_{n=0}^{\infty} a_n\, z^n . \tag{1.2.8}$$

The term z^{α_0} is denoted the *asymptotic factor*. Inserting the ansatz into the differential equation and equating each power in z to zero yields

$$\begin{aligned}
&[\alpha_0\,(\alpha_0 - 1) + A_0\,\alpha_0 + B_0\,]\; a_0 = 0,\\
&[(A_0 + \alpha_0)\,(\alpha_0 + 1) + B_0]\; \alpha_1\\
&\qquad - [2\,\alpha_0\,(\alpha_0 - 1) + 2\,(A_0 + B_0) + A_1 - C]\; a_0 = 0,\\
&[1]\; a_{n+1} + [0]\; a_n + [-1]\; a_{n-1} = 0, \qquad n = 1, 2, 3, \ldots,
\end{aligned}$$

with

$$\begin{aligned}
[1] &= (n + 1 + \alpha_0)\; [(n + \alpha_0) + A_0] + B_0,\\
[0] &= -2\,(n + \alpha_0)\; [(n - 1 + \alpha_0) - (2\,A_0 + A_1)] + (C - 2\,B_0),\\
[-1] &= (n - 1 + \alpha_0)\; [(n - 2 + \alpha_0) + A_0 + A_1] + B_0 + B_1 - C.
\end{aligned}$$

This may also be written in the form

$$[\alpha_0\,(\alpha_0 - 1) + A_0\,\alpha_0 + B_0\,]\; a_0 = 0, \tag{1.2.9}$$

$$\begin{aligned}
&[(A_0 + \alpha_0)\,(\alpha_0 + 1) + B_0]\; a_1\\
&\qquad - [2\,\alpha_0\,(\alpha_0 - 1) + 2\,(A_0 + B_0) + A_1 - C]\; a_0 = 0,
\end{aligned} \tag{1.2.10}$$

$$\begin{aligned}
&\left(1 + \frac{\alpha_1}{n} + \frac{\beta_1}{n^2}\right) a_{n+1} - \left(2 + \frac{\bar{\alpha}_0}{n} + \frac{\beta_0}{n^2}\right) a_n\\
&\qquad + \left(1 + \frac{\alpha_{-1}}{n} + \frac{\beta_{-1}}{n^2}\right) a_{n-1} = 0, \quad n = 1, 2, 3, \ldots,
\end{aligned} \tag{1.2.11}$$

with

$$\begin{aligned}
\alpha_1 &= A_0 + 2\,\alpha_0 + 1, \quad \beta_1 = (A_0 + \alpha_0)\,(\alpha_0 + 1) + B_0,\\
\bar{\alpha}_0 &= 2\,A_0 + A_1 + 2\,(2\,\alpha_0 - 1),\\
\beta_0 &= (2\,A_0 + A_1)\,\alpha_0 + 2\,\alpha_0\,(\alpha_0 - 1) + 2\,B_0 - C,\\
\alpha_{-1} &= A_0 + A_1 + 2\,\alpha_0 - 3,\\
\beta_{-1} &= (A_0 + A_1)\,(\alpha_0 - 1) + \alpha_0\,(\alpha_0 - 3) + B_0 + B_1 - C + 2.
\end{aligned}$$

With $a_0 \neq 0$ ($a_0 = 0$ would mean getting the trivial solution $y(z) \equiv 0$), two values of α_0 result from (1.2.9), namely

$$\begin{aligned}
\alpha_{01} &= \frac{1 - A_0 + \sqrt{(1 - A_0)^2 - 4\,B_0}}{2},\\
\alpha_{02} &= \frac{1 - A_0 - \sqrt{(1 - A_0)^2 - 4\,B_0}}{2}.
\end{aligned} \tag{1.2.12}$$

Thus, fixing the value of the first coefficient a_0 of the series in (1.2.8) [that actually may be used for normalising the particular solution $y(z)$] allows us to calculate the value of the second coefficient a_1 from the initial condition (1.2.10). This, in turn, allows us to recursively calculate as many coefficients a_n from (1.2.11) as are needed.

As can be seen, there appear two particular solutions of the differential equation (1.2.2) from the ansatz (1.2.8). However, these are only two linearly independent solutions, i.e., a fundamental system of the differential equation, if the difference between the two characteristic exponents (1.2.12) is not an integer. Otherwise, there may appear a logarithmic term in one of the two particular solutions (see, e.g., Bieberbach, 1965, pp. 128, 136). These particular solutions are called *Frobenius solutions*. The quantity α_0 is called the *characteristic exponent*. $y_{01}(z)$ and $y_{02}(z)$ are linearly independent, thus, the pair $\{y_{01}(z), y_{02}(z)\}$ may also serve as a fundamental system of solutions of the differential equation (1.2.2), (1.2.3), i.e., the general solution $y^{(g)}(z)$ of (1.2.2), (1.2.3) may be written in the form

$$y^{(g)}(z) = C_1\, y_{01}(z) + C_2\, y_{02}(z),$$

where C_1 and C_2 are the totality of complex-valued constants in z. The series in (1.2.8) is convergent with radius of convergence r ranging to the neighbouring singularity of the differential equation (1.2.2), (1.2.3), thus $r = 1$. This is a consequence of the fact that for linear differential equations, all the particular solutions at ordinary points of the differential equations are holomorphic there (cf. Bieberbach, 1965, p. 5).

There are also two local solutions at the singularity $z = 1$, given by

$$y_{1i}(z) = \sum_{n=0}^{\infty} a_n\,(z-1)^{n+\alpha_{1i}} = (z-1)^{\alpha_{1i}} \sum_{n=0}^{\infty} a_n\,(z-1)^n, \quad i = 1, 2,$$

with

$$\alpha_{11} = \frac{1 - A_1 + \sqrt{(1-A_1)^2 - 4\,B_1}}{2},$$
$$\alpha_{12} = \frac{1 - A_1 - \sqrt{(1-A_1)^2 - 4\,B_1}}{2}.$$

$y_{11}(z)$ and $y_{12}(z)$ are linearly independent, thus the pair $\{y_{11}(z), y_{12}(z)\}$ may also serve as a fundamental system of solutions of the differential equation (1.2.2), (1.2.3), i.e., the general solution $y^{(g)}(z)$ of (1.2.2), (1.2.3) may be written in the form

$$y^{(g)}(z) = C_1\, y_{11}(z) + C_2\, y_{12}(z),$$

where C_1 and C_2 are not specific constants but are the totality of complex-valued constants in z.

Last but not least, the characteristic exponents of the Frobenius solutions at infinity are given by

$$\alpha_{\infty 1} = \frac{\tilde{A}_0 + \tilde{A}_1 - 1 + \sqrt{[\tilde{A}_0 + \tilde{A}_1 - 1]^2 - 4(\tilde{B}_0 + \tilde{B}_1 + \tilde{C}_1)}}{2},$$

$$\alpha_{\infty 2} = \frac{\tilde{A}_0 + \tilde{A}_1 - 1 - \sqrt{[\tilde{A}_0 + \tilde{A}_1 - 1]^2 - 4(\tilde{B}_0 + \tilde{B}_1 + \tilde{C}_1)}}{2},$$

with $\tilde{A}_0$ and $\tilde{B}_0$ from (1.2.7) and

$$\tilde{A}_1 = A_1, \tilde{B}_1 = B_1, \tilde{C}_1 = C.$$

Eventually it should be mentioned that the sum of all the characteristic exponents is given by

$$\alpha_{01} + \alpha_{02} + \alpha_{12} + \alpha_{12} + \alpha_{\infty 2} + \alpha_{\infty 2} = 1.$$

It may be seen from (1.2.12) that, if B_0 vanishes, thus $B_0 = 0$, one of the two characteristic exponents vanishes as well, namely

$$\alpha_{02} = 0.$$

This, in turn, means that one of the two Frobenius solutions is holomorphic at the singularity $z = 0$ of the differential equation (1.2.2), (1.2.3), and thus may be represented by a pure Taylor series. The analogous happens with respect to the singularity at $z = 1$ in the case when $B_1 = 0$. This is an important mathematical mechanism for dealing with the singular boundary eigenvalue problem in §1.2.4.

The Single Confluent Case of the Gauss Equation

The differential equation

$$\frac{d^2y}{dz^2} + \left[\frac{A_0}{z} + G_0\right]\frac{dy}{dz} + \left[\frac{B_0}{z^2} + \frac{C}{z} + D_0\right] y(z) = 0, \quad z \in \mathbb{C}, \tag{1.2.13}$$

is called the *single confluent case of the Gauss equation* since it may be derived from the Gauss differential equation (1.2.2), (1.2.3)

$$\frac{d^2y}{dz^2} + \left[\frac{A_0}{z} + \frac{A_1}{z-1}\right]\frac{dy}{dz} + \left[\frac{B_0}{z} + \frac{B_1}{(z-1)^2} + \frac{C}{z} - \frac{C}{z-1}\right] y(z) = 0, \quad z \in \mathbb{C},$$

in such a way that the singularity at $z = 1$ is considered to be located at an arbitrary point $z = z_0$ that is driven to infinity:

$$\frac{d^2y}{dz^2} + \left[\frac{A_0}{z} + \frac{A_1}{z-z_0}\right]\frac{dy}{dz} + \left[\frac{B_0}{z} + \frac{B_1}{(z-z_0)^2} + \frac{C}{z} - \frac{C}{z-z_0}\right] y(z) = 0, \quad z \in \mathbb{C},$$

with $z_0 \to \infty$, thus by carrying out a limiting process. However, the result of this limiting process depends on how it is carried out, for it has to be recognised that in order to get the most general result, the coefficients A_0, A_1, B_0, B_1, C have to be dependent on z_0. Taking the constants A_0, A_1, B_0, B_1, C as independent of z_0 does not lead to the most general result. To correctly carry out this limiting process, one has to dive deeper into the topic and give some more definitions. Therefore, the technical carrying out of this process is postponed to Chapter 2.

The single confluent case of the Gauss differential equation (1.2.13) has two singularities, one located at the origin $z = 0$ and the other positioned at infinity, as seen by the inversion $\zeta = \frac{1}{z}$:

$$\frac{\mathrm{d}^2 y}{\mathrm{d}\zeta^2} + \tilde{P}(\zeta)\,\frac{\mathrm{d}y}{\mathrm{d}\zeta} + \tilde{Q}(\zeta)\,y = 0, \quad \zeta \in \mathbb{C},$$

with

$$\tilde{P}(\zeta) = \left[\frac{2}{\zeta} - \frac{1}{\zeta^2}\,P(\zeta)\right]\frac{\mathrm{d}y}{\mathrm{d}\zeta} = -\frac{G_0}{\zeta^2} + \frac{2 - A_0}{\zeta},$$
$$\tilde{Q}(\zeta) = \frac{1}{\zeta^4}\,Q(\zeta) = \frac{D_0}{\zeta^4} + \frac{C}{\zeta^3} + \frac{B_0}{\zeta^2}.$$

As seen, the nature of this singularity is different in that the order of the pole of $\tilde{P}(\zeta)$ at the origin $\zeta = 0$ is two and the order of the pole of $\tilde{Q}(\zeta)$ at the origin $\zeta = 0$ is four. This makes a significant difference with respect to the local solution, as may be seen below.

Local solutions at the origin $z = 0$ are given by

$$y(z) = z^{\alpha_0} \sum_{n=0}^{\infty} a_n\, z^n. \tag{1.2.14}$$

The quantity α_0 is called the *characteristic exponent*. As above, it is given by inserting the ansatz (1.2.14) into (1.2.13), (1.2.1) and setting the equation of zeroth power to zero, resulting in the quadratic algebraic equation called the *indicial equation*:

$$\alpha_0^2 + (A_0 - 1)\,\alpha_0 + B_0 = 0, \tag{1.2.15}$$

generally yielding the two values

$$\alpha_{01,02} = \begin{cases} \dfrac{1 - A_0 + \sqrt{(1 - A_0)^2 - 4\,B_0}}{2}, \\[2ex] \dfrac{1 - A_0 - \sqrt{(1 - A_0)^2 - 4\,B_0}}{2}. \end{cases} \tag{1.2.16}$$

The coefficients a_n of the ansatz (1.2.14) are given by means of the above-mentioned insertion and equating higher powers of z to zero, resulting in the difference equation

$$\begin{aligned} &\alpha_0^2 + (A_0 - 1)\,\alpha_0 + B_0 = 0, \\ &G_0\,(1 + \alpha_0)\,a_1 + (B_0 + D_0)\,a_0 = 0, \\ &(n + \alpha_0)(n - 1 + \alpha_0)\,a_{n+1} + [(n + \alpha_0)\,A_0 + B_0 + D_0]\,a_n \\ &\qquad + C\,a_{n-1} = 0, \qquad n = 1, 2, 3, \ldots, \end{aligned} \tag{1.2.17}$$

which may be turned into a recurrence equation by fixing a_0.

Now, turning to the local solutions at infinity, the ansatz has to change such that – according to the German mathematician **Ludwig Wilhelm Thomé** (1841–1910) –

an exponential term occurs in the asymptotic factor. He wrote this down for the first time in 1883:

$$y(z) = \exp(\alpha_{1\infty}\, z)\; z^{\alpha_{0\infty}} \sum_{n=0}^{\infty} C_n\, z^{-n}. \tag{1.2.18}$$

It is plausible that the series is to be written in descending powers, since the singularity is located at infinity. It is to be remarked here that even in this case, the infinite series in the ansatz (1.2.18) are one-sided infinite, that is by no means to be taken for granted. The quantity α_1 – calculated by the same procedure as above – is called the *characteristic exponent of the second kind* and is determined by an algebraic second-order equation

$$\alpha_{1\infty j}^2 + G_0\, \alpha_{1\infty j} + D_0 = 0, \quad j = 1, 2, \tag{1.2.19}$$

generally yielding the two values

$$\begin{aligned} \alpha_{1\infty 1} &= \frac{-G_0 + \sqrt{G_0^2 - 4\, D_0}}{2}, \\ \alpha_{1\infty 2} &= \frac{-G_0 - \sqrt{G_0^2 - 4\, D_0}}{2}. \end{aligned} \tag{1.2.20}$$

The *characteristic exponent of the first kind* $\alpha_{0\infty j}$ obeys the first-order rational equation

$$\alpha_{0\infty j}\left(G_0 + 2\,\alpha_{1\infty j}\right) + A_0\, \alpha_{1\infty j} + C = 0, \quad j = 1, 2,$$

which depends on the characteristic exponents of the second kind $\alpha_{1\infty j}$, thus generally yielding two values:

$$\alpha_{0\infty j} = -\frac{A_0\, \alpha_{1\infty j} + C}{G_0 + 2\,\alpha_{1\infty j}}, \quad j = 1, 2. \tag{1.2.21}$$

The Biconfluent Case of the Gauss Equation

For the sake of completeness, it has to be remarked that there is still one further confluent case of the Gauss equation, namely by moving both of the finite singularities to infinity. The result is the *biconfluent case of the Gauss equation*. It has no finite singularity any more and looks like

$$\frac{d^2 y}{dz^2} + (G_0 + G_1\, z)\,\frac{dy}{dz} + \left(D_0 + D_1\, z + D_2\, z^2\right) y = 0, \quad z \in \mathbb{C}. \tag{1.2.22}$$

As may be seen from (1.2.5) and (1.2.6), at infinity there is a singularity of this differential equation: $P(\zeta)$ has a third-order pole while $Q(\zeta)$ has a sixth-order one.

The special choice

$$G_0 = G_1 = 0$$

of its parameters does not touch the fundamental character of the differential equation with respect to its local solutions at its singularity, but yields a biconfluent case of the Gauss equation, a pair of fundamental solutions of which are the so-called and well-known *parabolic cylinder functions* (cf. Slavyanov, 1996, §3).

The local solutions of (1.2.22) at the infinite singularity (which is actually a Thomé one, see above) are given by

$$y(z) = \exp\left(\frac{\alpha_2}{2}\, z^2 + \alpha_1\, z\right) z^{\alpha_0} \sum_{n=0}^{\infty} \frac{C_n}{z^n},$$

where the *characteristic exponents of second kind and second order* α_2 are given by

$$\begin{aligned} \alpha_{21} &= -\frac{1}{2}\left(G_1 + \sqrt{G_1^2 - 4\,D_2}\right), \\ \alpha_{22} &= -\frac{1}{2}\left(G_1 - \sqrt{G_1^2 - 4\,D_2}\right), \end{aligned} \tag{1.2.23}$$

and the *characteristic exponents of second kind and first order* α_1 are given by

$$\begin{aligned} \alpha_{11} &= -\frac{G_0\,\alpha_{21} + D_1}{2\,\alpha_{21} + G_1}, \\ \alpha_{12} &= -\frac{G_0\,\alpha_{22} + D_1}{2\,\alpha_{22} + G_1}. \end{aligned} \tag{1.2.24}$$

Eventually, the characteristic exponents of first kind α_0 are given by

$$\begin{aligned} \alpha_{01} &= -\frac{\alpha_{21} + \alpha_{11} + G_0 + D_0}{G_0 + 2\,\alpha_{21}}, \\ \alpha_{02} &= -\frac{\alpha_{22} + \alpha_{12} + G_0 + D_0}{G_0 + 2\,\alpha_{22}}. \end{aligned} \tag{1.2.25}$$

1.2.2 Linear Transformations

There is a linear transformation of the independent variable of equation (1.2.2) as well as of the dependent variable, the characteristic property of which is that it leaves the essential properties of the differential equation untouched. These linear transformations play an important role in the theory. In the former case, it is a specific Moebius transformation, which changes the position of the singularities of a differential equation in the complex number plane without touching the essentials of the equation itself, and the other changes in asymptotic factors of the local solutions, also without changing the essential properties of the equation. Because of their importance, they should be explicitly mentioned here.

Moebius Transformations

To have a transformation of the independent variable of the differential equation (1.2.2) that changes the locations of its singularities without touching the essential properties is of calculatory importance. In the actual context this means that the transformed differential equation has the same number of singularities, keeping the pole orders of the coefficients (1.2.3). The only change allowed is in the locations of the singularities of the differential equation.

It is to be expected that such a transformation is not of substantial significance, since it does not touch the structure or character of the differential equation, or the local solutions in the neighbourhood of its singularities, wherever these are located. However, as will be seen below, placing the singularities in an appropriate manner results in calculatory simplifications that often make the theory clearer and more comprehensible. Therefore, it is not surprising that the Moebius transformation is frequently applied in the following.

The access to Moebius transformations may be alleviated by means of purely geometric considerations. It is well known that the Eulerian complex plane of numbers may be compactified by means of a stereographic mapping, i.e., a mapping of the plane onto a sphere, the south pole of which is placed at the origin of the Eulerian plane (see Behnke and Sommer, 1976, pp. 13–19). The result is a sphere-shaped space of the complex numbers, with point of infinity the north pole. This is to be kept in mind in the following considerations. If the compactified plane of complex numbers is meant in the following, it is spoken of as the 'complex sphere'.

The transformation

$$x = \frac{a\,z + b}{c\,z + d}, \quad a\,d \neq b\,c, \tag{1.2.26}$$

of the independent variable z is called the *Moebius transformation* (cf. Gutzwiller, 1990, p. 345). This transformation is named after the German mathematician **August Ferdinand Moebius** (1790–1868). It is the only bijective conformal mapping of the closed complex sphere onto itself (sometimes also called *isomorphic transformation* or *homographic transformation*; cf. Behnke and Sommer, 1976, p. 324). It is just this mapping, interpreted as a transformation, that meets the condition formulated above. A Moebius transformation is nothing other than a rotation, as well as a stretching of the complex sphere, irrespective of what is done mathematically on this sphere.

The inverse mapping of (1.2.26) is of the same sort:

$$z = \frac{d\,x - b}{-c\,x + a}. \tag{1.2.27}$$

This mapping can put three singularities of equation (1.2.2) onto three other singularities without touching the substance of the equation, i.e., the number and s-ranks, and thus the characteristics, of any of its singularities. This allows us always to put three arbitrary points of equation (1.2.2) lying in the complex plane to the points 0, 1, ∞ (cf. Figure 1.1). This will become important in the following.

These are the locations where the differential equations have the most simple formulae, whatever is to be calculated.

With the help of the derivatives

$$\frac{\mathrm{d}x}{\mathrm{d}z} = \frac{a\,d - b\,c}{(c\,z + d)^2} = \frac{(c\,x - a)^2}{a\,d - b\,c},$$

$$\frac{\mathrm{d}^2x}{\mathrm{d}z^2} = -2\,c\,\frac{a\,d - b\,c}{(c\,z + d)^3} = 2\,c\,\frac{(c\,x - a)^3}{(a\,d - b\,c)^2},$$

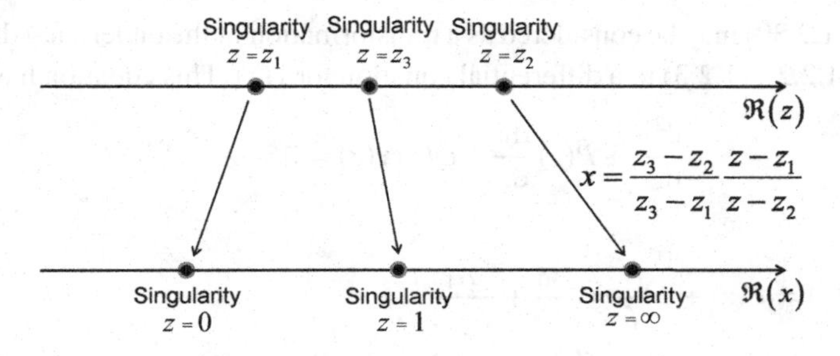

Figure 1.1 Moebius transformation placing three arbitrary points.

the coefficients $\tilde{P}(x)$ and $\tilde{Q}(x)$ of the transformed differential equation

$$\frac{\mathrm{d}^2 y}{\mathrm{d}x^2} + \tilde{P}(x)\,\frac{\mathrm{d}y}{\mathrm{d}x} + \tilde{Q}(x)\,y = 0$$

become

$$\tilde{P}(x) = \frac{P(x)}{\dfrac{\mathrm{d}x}{\mathrm{d}z}} + \frac{\dfrac{\mathrm{d}^2 x}{\mathrm{d}z^2}}{\left(\dfrac{\mathrm{d}x}{\mathrm{d}z}\right)^2} = P(x)\,\frac{a\,d - b\,c}{(c\,x - a)^2} + \frac{2\,c}{c\,x - a} \tag{1.2.28}$$

and

$$\tilde{Q}(x) = \frac{Q(x)}{\left(\dfrac{\mathrm{d}x}{\mathrm{d}z}\right)^2} = Q(x)\,\frac{(a\,d - b\,c)^2}{(c\,x - a)^4}. \tag{1.2.29}$$

s-Homotopic Transformations

The ansatz (1.2.8)

$$y(z) = z^{\alpha_0} \sum_{n=0}^{\infty} a_n\,z^n$$

for the solution of the differential equation (1.2.2), (1.2.3), may also be written in the form

$$y(z) = f(z)\,v(z) \tag{1.2.30}$$

with

$$\begin{aligned} f(z) &= z^{\alpha_0}, \\ v(z) &= \sum_{n=0}^{\infty} a_n\,z^n. \end{aligned} \tag{1.2.31}$$

Equation (1.2.30) may be considered as a transformation of the underlying differential equation (1.2.2), (1.2.3) to a differential equation for $v(z)$. This equation has the form

$$\frac{\mathrm{d}^2 v}{\mathrm{d}z^2} + \tilde{P}(z)\,\frac{\mathrm{d}v}{\mathrm{d}z} + \tilde{Q}(z)\,v(z) = 0, \quad z \in \mathbb{C},$$

with

$$\begin{aligned} \tilde{P}(z) &= P(z) + 2\,\frac{f'}{f} = \frac{\tilde{A}_0}{z} + \frac{\tilde{A}_1}{z-1}, \\ \tilde{Q}(z) &= Q(z) + P(z)\,\frac{f'}{f} + \frac{f''}{f} = \frac{\tilde{B}_0}{z^2} + \frac{\tilde{B}_1}{(z-1)^2} + \frac{\tilde{C}}{z} - \frac{\tilde{C}}{z-1}, \end{aligned} \tag{1.2.32}$$

whereupon

$$\begin{aligned} f' &= \frac{\mathrm{d}f}{\mathrm{d}z}, \quad f' = \frac{\mathrm{d}^2 f}{\mathrm{d}z^2}, \\ \tilde{A}_0 &= A_0 + 2\,\alpha_0, \\ \tilde{A}_1 &= A_1, \\ \tilde{B}_0 &= \alpha_0\,(\alpha_0 + A_0 - 1) + B_0, \\ \tilde{B}_1 &= B_1, \\ \tilde{C} &= C - \alpha_0\,A_1. \end{aligned} \tag{1.2.33}$$

The crucial point now is that for $\alpha_0 = \alpha_{01}$ from (1.2.12), as well as for $\alpha_0 = \alpha_{02}$, it holds that

$$\begin{aligned} \alpha_{01}\,(\alpha_{01} + A_0 - 1) + B_0 &= 0, \\ \alpha_{02}\,(\alpha_{02} + A_0 - 1) + B_0 &= 0. \end{aligned}$$

This means, because of (1.2.33), that for α_0 in (1.2.30), (1.2.31) being one of the characteristic exponents of the singularity of the equation (1.2.2), (1.2.3) at $z = 0$, the coefficent $\tilde{B}_0$ vanishes. Moreover, for $\alpha_0 = \alpha_{01}$ in (1.2.30), (1.2.31) it holds that

$$\begin{aligned} \tilde{\alpha}_{01} &= \frac{1 - \tilde{A}_0 + \sqrt{(1 - \tilde{A}_0)^2 - 4\,\tilde{B}_0}}{2} = \alpha_{01} - \alpha_{01} = 0, \\ \tilde{\alpha}_{02} &= \frac{1 - \tilde{A}_0 - \sqrt{(1 - \tilde{A}_0)^2 - 4\,\tilde{B}_0}}{2} = \alpha_{02} - \alpha_{01}, \end{aligned} \tag{1.2.34}$$

and for $\alpha_0 = \alpha_{02}$ in (1.2.30), (1.2.31), it holds that

$$\begin{aligned} \tilde{\alpha}_{01} &= \frac{1 - \tilde{A}_0 + \sqrt{(1 - \tilde{A}_0)^2 - 4\,\tilde{B}_0}}{2} = \alpha_{01} - \alpha_{02}, \\ \tilde{\alpha}_{02} &= \frac{1 - \tilde{A}_0 - \sqrt{(1 - \tilde{A}_0)^2 - 4\,\tilde{B}_0}}{2} = \alpha_{02} - \alpha_{02} = 0. \end{aligned} \tag{1.2.35}$$

Thus, in any case, one of the characteristic exponents vanishes under the transformation (1.2.30), (1.2.31).

As a result, it may be seen first that, as might have been expected, the coefficients $\tilde{A}_1$ and $\tilde{B}_1$ in (1.2.33) are not affected by the transformation (1.2.30), (1.2.31). Second, as may be seen from (1.2.34), (1.2.35), the transformed characteristic exponents $\tilde{\alpha}_{01}$ and $\tilde{\alpha}_{02}$ are the differences between the original α_{01}, α_{02}, respectively, and those taken in the transformation (1.2.30), (1.2.31). As may be seen from (1.2.34), (1.2.35), this means that in any case one of the characteristic exponents vanishes. This, in turn, means that one of the two Frobenius solutions becomes holomorphic at the origin, and thus has a pure Taylor series representation.

The transformation (1.2.30), (1.2.31) is called an s-*homotopic transformation*, a linear transformation of the dependent variable y with respect to one or more singularities that changes the characteristic exponents of the related singularity of a differential equation (1.2.2), (1.2.3) but leaves the location of the singularity unchanged.

Two subsequent s-homotopic transformations at the singularities $z = 0$ and $z = 1$ may be put together by applying

$$f(z) = z^{\alpha_0}\,(z-1)^{\alpha_1}.$$

In this case, the final result is that $\tilde{Q}(z)$ does not have a second-order pole any more, but only a first-order one. Or, in other words, it holds that $\tilde{B}_0 = 0$ and $\tilde{B}_1 = 0$ in (1.2.32):

$$\frac{d^2 v}{dz^2} + \left(\frac{\tilde{A}_0}{z} + \frac{\tilde{A}_1}{z-1}\right)\frac{dv}{dz} + \left(\frac{\tilde{C}}{z} - \frac{\tilde{C}}{z-1}\right) v(z) = 0, \quad z \in \mathbb{C}. \tag{1.2.36}$$

This means that there may be solutions of the differential equation that are holomorphic at $z = 0$ as well as at $z = 1$. Or, said another way, these particular solutions $y(z)$ are entire functions. Since the coefficient $P(z)$ in (1.2.3) at the singularity at infinity has a first-order pole, and the coefficient $Q(z)$ in (1.2.3) at the singularity at infinity has a second-order pole, as may be seen from (1.2.5), (1.2.6), this entire function must be a polynomial.

Traditionally, it is just this form that is called a *Gauss differential equation*, although – for didactical reasons – I use this notation for the slightly more general equation (1.2.2), (1.2.3).

It is clear that the differential equation (1.2.36) has a local solution at the origin $z = 0$ that may be written as in a Taylor series:

$$y(z) = \sum_{n=0}^{\infty} a_n\, z^n. \tag{1.2.37}$$

Inserting (1.2.37) into (1.2.36) yields a first-order linear difference equation

$$a_{n+1} = f(n)\, a_n, \quad n = 0, 1, 2, 3, \ldots, \tag{1.2.38}$$

with

$$f(n) = \frac{n^2 + (\tilde{A}_0 + \tilde{A}_1 - 1)\,n - \tilde{C}}{n^2 + (\tilde{A}_0 + 1)\,n + \tilde{A}_0} = 1 + \frac{\alpha}{n + \tilde{A}_0} + \frac{\beta}{n+1}, \tag{1.2.39}$$

whereby

$$\begin{aligned} \alpha &= \frac{\tilde{A}_0 (\tilde{A}_1 - 1) + \tilde{C}}{\tilde{A}_0 - 1}, \\ \beta &= -\frac{\tilde{A}_0 + \tilde{A}_1 + \tilde{C} - 2}{\tilde{A}_0 - 1}. \end{aligned} \tag{1.2.40}$$

Determining a_0 makes a linear two-term recurrence relation out of the first-order difference equation (1.2.38), whereas this determination means a normalisation of the solution (1.2.37) of (1.2.36). The resulting solution $y(z)$ via (1.2.37) is called a *hypergeometric function* (cf. Abramowitz and Stegun, 1970, Chapter 15).

Since (1.2.38) is a first-order equation, its general solution is given by the totality of values of a_0. All the particular solutions of (1.2.38) differ just by a factor:

$$a_n^{(g)} = L\, a_n, \quad n = 0, 1, 2, 3, \ldots,$$

where L is the totality of complex-valued contants in n, a_n is given by (1.2.38) and a_0 takes any fixed value, for example $a_0 = 1$.

Classical analysis tells us that the ratio of two consecutive terms a_{n+1} and a_n tends to the inverse of the radius of convergence of the Taylor series in (1.2.37), thus to unity. Here,

$$\lim_{n\to\infty} \left| \frac{a_{n+1}}{a_n} \right| = \frac{1}{r} = 1.$$

This is quite clear since the solution, represented by this ansatz, has the form

$$y_1(z) = (z-1)^{\alpha_0} \sum_{n=0}^{\infty} a_n (z-1)^n$$

at the singularity at $z = 1$. This, in general, is a function that is not holomorphic at $z = 1$. Generally, series in the ansatzes of local solutions of linear differential equations do converge up to the neighbouring singularity of the underlying differential equation.

The totality of particular solutions a_n of the difference equation (1.2.38) only comprises the non-trivial solutions. However, there are also trivial particular solutions, defined by a truncation of the sequence a_n:

$$a_{n+1} = f(n)\, a_n, \quad a_0 = 1, n = 0, 1, 2, 3, \ldots, m,$$

where

$$f(m) = 0, \quad m \in \mathbb{N}^0. \tag{1.2.41}$$

This means

$$m^2 + (\tilde{A}_0 + \tilde{A}_1 - 1)\, m - \tilde{C} = 0, \quad m = 0, 1, 2, \ldots. \tag{1.2.42}$$

In this case, $a_n = 0$ for $n > m$ holds. The corresponding series in (1.2.37) in this case become finite and the function $y(z)$ becomes a polynomial, which is actually called a *hypergeometric polynomial*.

Examples

- Consider the condition $f(1) = 0$, i.e., $m = 1$. Then

$$\tilde{C} = \tilde{A}_0 + \tilde{A}_1.$$

As a result, the particular solution of the difference equation (1.2.38) is a trivial one, namely

$$\begin{aligned} a_0 &= 1, \\ a_1 &= f(0)\, a_0 = -\frac{\tilde{C}}{\tilde{A}_0} = -\frac{\tilde{A}_0 + \tilde{A}_1}{\tilde{A}_0} = -1 - \frac{\tilde{A}_1}{\tilde{A}_0}, \\ a_2 &= f(1)\, a_1 = 0 \cdot a_1 = 0, \\ a_3 &= f(2)\, a_2 = 0, \\ &\dots \end{aligned}$$

The solution is the first-order hypergeometric polynomial

$$y(z) = 1 + a_1\, z = 1 - \left(1 + \frac{\tilde{A}_1}{\tilde{A}_0}\right) z.$$

- Consider the condition $f(2) = 0$, i.e., $m = 2$. Then

$$\tilde{C} = 2\,\left[\tilde{A}_0 + \tilde{A}_1 + 1\right].$$

As a result, the particular solution of the difference equation (1.2.38) is a trivial one, namely

$$\begin{aligned} a_0 &= 1, \\ a_1 &= f(0)\, a_0 = -\frac{\tilde{C}}{\tilde{A}_0} = -2\,\frac{\tilde{A}_0 + \tilde{A}_1 + 1}{\tilde{A}_0}, \\ a_2 &= f(1)\, a_1 = -\frac{\tilde{A}_0 + \tilde{A}_1 + 2}{2\,\tilde{A}_0 + 2}\, a_1 = 2\,\frac{\tilde{A}_0 + \tilde{A}_1 + 1}{\tilde{A}_0}\,\frac{\tilde{A}_0 + \tilde{A}_1 + 2}{2\,\tilde{A}_0 + 2}, \\ a_3 &= f(2)\, a_2 = 0, \\ a_4 &= f(3)\, a_3 = 0, \\ &\dots \end{aligned}$$

The solution is the second-order hypergeometric polynomial

$$y(z) = 1 + a_1\, z + a_2\, z^2.$$

1.2.3 Forms of Differential Equations

As may be seen from what is written above, it is crucial how many singularities there are of the differential equation (1.2.2), and of what sort these are. On the other hand, it is not essential for a differential equation where these singularities are located. This is nothing but a matter of appropriateness. Moreover, the concrete values of the characteristic exponents at the singularities of equation (1.2.2) are not substantially important, since these values might always be changed by an s-homotopic transformation. What is significant in this respect is the difference in the pair of characteristic exponents at a singularity of equation (1.2.2) (these are actually invariants of the s-homotopic transformations). Therefore, two differential equations (1.2.2) may look completely different, although they are essentially the same. Such a (trivial) distinction between differential equations (1.2.2) is called their *form*.

The *form of a differential equation* distinguishes equation (1.2.2) according to the locations of its singularities, as well as the concrete values of the pairs of characteristic exponents belonging to the singularities. As may be understood, this open up a rich variety of forms of equation (1.2.2). It is not reasonable to display them all here explicitly; the totality of forms of a differential equation (1.2.2) is generated by means of a Moebius as well as s-homotopic transformations. Both of these are linear transformations, the former of the independent variable and the latter of the dependent variable.

It is a matter of historical evolution that there are a few forms of equation (1.2.2) which have gained certain attention and thus are introduced below. The most important ones among them are determined by the following properties:

- The point at infinity is a singularity of the differential equation (1.2.2) with pole order the highest (or one of the highest) among all appearing singularities.
- At least one of the characteristic exponents at each of the finite singularities is zero.
- If the differential equation (1.2.2) has at least two singularities, then one of them is located at the origin $z = 0$.
- If the differential equation (1.2.2) has three or more singularities, then one of them is located at unity $z = 1$.

Such a differential equation is said to be in *standard form*. There are several other forms of equation (1.2.2) bearing a similar notation, the most important of which are mentioned in the following (cf. Slavyanov and Lay, 2000, pp. 4, 5, 13, 21):

- *General form*: No parameter of the differential equation is specified.
- *Natural form*: The singularity (or one of the singularities) having the highest pole order is located at infinity.
- *Normal form*: The coefficient in front of the first derivative vanishes.
- *Self-adjoint form*: The differential equation (1) is characterised by the condition

  $$\frac{\mathrm{d}P_0(z)}{\mathrm{d}z} = P_1(z)$$

  (see Courant and Hilbert, 1968, p. 238).
- *Canonical form*: All zeros of $P_0(z)$ in equation (1) are simple ones.

The totality of forms of a differential equation (1.2.2) is the totality of equations (1.2.2) that emerge from one another by means of Moebius, as well as s-homotopic, transformations. This totality is an uncountably infinite set.

1.2.4 Singular Boundary Eigenvalue Problems

In §1.2.2 we gave a criterion (cf. 1.2.41) for the parameters of the Gauss differential equation (1.2.36) that determines a particular solution that is holomorphic simultaneously at both of its finite singularities. Moreover, it was given an explicit method for its calculation. This is the most simple singular boundary eigenvalue problem.

Now we are going to treat the singular boundary eigenvalue problem of the single confluent case of the Gauss differential equation, a typical singular boundary eigenvalue problem, in order to display the aspect under which this mathematical problem is considered here, in this book, and, moreover, in order to give the method that is based on this aspect. The main point is that the method works not only for those differential equations the solutions of whose singular boundary eigenvalue problems yield classical special functions, but also for the full generality of the differential equation (1.2.2), (1.2.3):

$$\frac{\mathrm{d}^2 y}{\mathrm{d}z^2} + P(z)\,\frac{\mathrm{d}y}{\mathrm{d}z} + Q(z)\,y(z) = 0, \quad z \in \mathbb{C}, \tag{1.2.43}$$

with

$$P(z) = \frac{P_1(z)}{P_0(z)},\ Q(z) = \frac{P_2(z)}{P_0(z)} \tag{1.2.44}$$

and with $P_i(z)$, $i = 0, 1, 2$, being polynomials in z, from which result higher special functions by formulating singular boundary eigenvalue problems.

As already mentioned in the Introduction, the basic idea of this method for coping technically with this problem comes from George Cecil Jaffé, who, in connection with the calculation of the quantum mechanical energy levels of the hydrogen molecule ion, in 1933 applied this method for the first time (cf. Jaffé, 1933).

Jaffé did not do this with the single confluent case of the Gauss differential equation, as is done here for the sake of showing this method for a more simple equation, such that the principle comes out clearly without being bothered with complex calculations. The single confluent case of the Fuchsian differential equation with three singularities is the simplest non-trivial equation for which the method of singular boundary eigenvalue problems, presented in this book, applies. In the following this method is presented in detail for this differential equation that may be considered as exemplary.

Formulation

As was shown in §1.2.1, the single confluent case (1.2.13)

$$\frac{\mathrm{d}^2 y}{\mathrm{d}z^2} + \left[\frac{A_0}{z} + G_0\right]\frac{\mathrm{d}y}{\mathrm{d}z} + \left[\frac{B_0}{z^2} + \frac{C}{z} + D_0\right] y(z) = 0, \quad z \in \mathbb{C}, \tag{1.2.45}$$

of the Gauss differential equation has the characteristic exponents (1.2.16)

$$\begin{aligned} \alpha_{01} &= \frac{1 - A_0 + \sqrt{(1 - A_0)^2 - 4\,B_0}}{2}, \\ \alpha_{02} &= \frac{1 - A_0 - \sqrt{(1 - A_0)^2 - 4\,B_0}}{2}. \end{aligned} \tag{1.2.46}$$

at the singularity of the differential equation at the origin $z = 0$, from which result two different asymptotic behaviours of its particular solutions

$$\begin{aligned} y_{01}(z) &\sim z^{\alpha_{01}}, \\ y_{02}(z) &\sim z^{\alpha_{02}} \end{aligned} \tag{1.2.47}$$

on approaching the singularity, radially, on the positive real axis.

Moreover, as shown in §1.2.1, the single confluent case of the Gauss differential equation has the characteristic exponents

$$\begin{aligned} \alpha_{1\infty 1} &= \frac{-G_0 + \sqrt{G_0^2 - 4\,D_0}}{2}, \\ \alpha_{1\infty 2} &= \frac{-G_0 - \sqrt{G_0^2 - 4\,D_0}}{2} \end{aligned} \tag{1.2.48}$$

and

$$\begin{aligned} \alpha_{0\infty 1} &= -\frac{A_0\,\alpha_{1\infty 1} + C}{G_0 + 2\,\alpha_{1\infty 1}}, \\ \alpha_{0\infty 2} &= -\frac{A_0\,\alpha_{1\infty 2} + C}{G_0 + 2\,\alpha_{1\infty 2}} \end{aligned} \tag{1.2.49}$$

[cf. (1.2.20), (1.2.1)] at the infinite singularity of the differential equation, from which result two different asymptotic behaviours of its particular solutions, on approaching the singularity, radially, on the positive real axis:

$$\begin{aligned} y_{\infty 1}(z) &\sim \exp\left(\alpha_{1\infty 1}\, z\right) z^{\alpha_{0\infty 1}}, \\ y_{\infty 2}(z) &\sim \exp\left(\alpha_{1\infty 2}\, z\right) z^{\alpha_{0\infty 2}} \end{aligned} \tag{1.2.50}$$

as $z \to \infty$.

Suppose now that we are looking for a particular solution of the differential equation (1.2.45) that behaves like

$$y(z) = y_{02}(z) \tag{1.2.51}$$

and simultaneously like

$$y(z) = y_{\infty 2}(z) \tag{1.2.52}$$

on approaching the singularity at $z = 0$ and at infinity on the positive real axis, radially. Because of the linearity of the differential equation, it holds that

$$y_{02}(z) = c_1\, y_{\infty 1}(z) + c_2\, y_{\infty 2}(z). \tag{1.2.53}$$

Thus, the condition for this to happen is given by

$$c_1 = 0,$$

whereby this quantity c_1 is dependent on the parameters of the underlying differential equation but not dependent on the independent variable z.

This is the formulation of a boundary eigenvalue problem in searching for particular solutions that behave in a prescribed manner, simultaneously, on approaching the finite as well as the infinite singularity, radially, on the positive real axis.

The formulation of a criterion in dependence on the parameters of the differential equation is the solution of this problem. This criterion is called the *eigenvalue condition*. Such a condition is developed in the following.

One of the two Frobenius solutions at the finite singularity of the differential equation is to be connected with one of the two Thomé solutions at the infinite singularity of the differential equation. Hereby, it is assumed that the Thomé solution to be connected decreases exponentially as $z \longrightarrow \infty$ along the positive real axis, and thus is a recessive particular solution of the underlying differential equation.

Jaffé Ansatz I

The particular solution of the singular eigenvalue problem immediately above is a rather special one. The eigensolutions, viz. the solutions of the differential equation (1.2.13)

$$\frac{d^2y}{dz^2} + P(z)\,\frac{dy}{dz} + Q(z)\,y = 0 \tag{1.2.54}$$

with

$$\begin{aligned} P(z) &= \frac{A}{z} + G_0, \\ Q(z) &= \frac{B}{z^2} + \frac{C}{z} + D_0 \end{aligned} \tag{1.2.55}$$

that meet the boundary condition (i.e., the specific asymptotic behaviour as $z \longrightarrow \infty$ on the positive real axis), consist of an exponential factor, a power term, and an entire function $w(z)$:

$$y(z) = \exp\left(\alpha_{1\infty 2}\, z\right)\, z^{\alpha_{0\infty 2}}\, w(z) \tag{1.2.56}$$

with $\Re(\alpha_{1\infty 2}) < 0$. The differential equation for $w(z)$ is

$$\frac{d^2w}{dz^2} + \tilde{P}(z)\,\frac{dw}{dz} + \tilde{Q}(z)\,w = 0,$$

the coefficients $\tilde{P}(z)$ and $\tilde{Q}(z)$ of which are given by

$$\begin{aligned} \tilde{P}(z) &= \tilde{g}_0 + \frac{\tilde{g}_{-10}}{z}, \\ \tilde{Q}(z) &= \frac{\tilde{d}_{-10}}{z} \end{aligned} \tag{1.2.57}$$

with

$$\begin{aligned} \tilde{g}_0 &= G_0 + 2\,\alpha_{1\infty 2}, \\ \tilde{g}_{-10} &= A + 2\,\alpha_{0\infty 2}, \\ \tilde{d}_{-10} &= 2\,\alpha_{1\infty 2}\,\alpha_{0\infty 2} + A\,\alpha_{1\infty 2} + \alpha_{0\infty 2}\,G_0 + C. \end{aligned}$$

In order to get the eigenvalue condition, these particular solutions are to be connected to those, that meet the boundary condition at the origin. This is to be investigated in the following.

Jaffé Transformation

The singular boundary eigenvalue problem of the differential equation (1.2.45), and the differential equation (1.2.4), (1.2.57) under the ansatz (1.2.56), is considered. We apply the specific Moebius transformation

$$x = \frac{z}{z+1} \tag{1.2.58}$$

to the independent variable z of (1.2.4), (1.2.57), the crucial point of which is that the positive z-axis, i.e., the relevant interval of the singular boundary eigenvalue problem dealt with here, is compressed such that the points on the z-axis are shifted according to the following table:

$$\begin{array}{cccccc} z: & +\infty & +1 & 0 & -\frac{1}{2} & -1 \\ \downarrow & \downarrow & \downarrow & \downarrow & \downarrow & \downarrow \\ x: & +1 & +\frac{1}{2} & 0 & -1 & -\infty. \end{array}$$

As a result, the relevant interval of the boundary eigenvalue problem changes from

$$\{z \in \mathbb{R} | 0 \leq z < \infty\}$$

to

$$\{x \in \mathbb{R} | 0 \leq x \leq 1\}.$$

In order to carry out the transformation, it is necessary to write the coefficients $\tilde{P}(z)$ and $\tilde{Q}(z)$ in (1.2.57) in dependence on x, yielding, with the help of

$$\frac{1}{z} = -\frac{x-1}{x} = \frac{1}{x} - 1,$$

the coefficients

$$\tilde{P}(x) = \tilde{g}_0 - \tilde{g}_{-10} + \frac{\tilde{g}_{-10}}{x},$$

$$\tilde{Q}(x) = -\tilde{d}_{-10} + \frac{\tilde{d}_{-10}}{x}$$

and the differential equation for $w(x)$:

$$\frac{\mathrm{d}^2 w}{\mathrm{d}x^2} + \tilde{\tilde{P}}(x)\,\frac{\mathrm{d}w}{\mathrm{d}x} + \tilde{\tilde{Q}}(x)\,w = 0$$

with [cf. (1.2.28)]

$$\begin{aligned}\tilde{\tilde{P}}(x) &= \frac{\tilde{P}(x)}{(x-1)^2} + \frac{2}{x-1}\\ &= \frac{\tilde{g}_0 - \tilde{g}_{-10} + \frac{\tilde{g}_{-10}}{x}}{(x-1)^2} + \frac{2}{x-1}\\ &= \frac{\tilde{g}_0 - \tilde{g}_{-10}}{(x-1)^2} + \frac{\tilde{g}_{-10}}{x(x-1)^2} + \frac{2}{x-1}.\end{aligned}$$

By means of

$$\frac{1}{x(x-1)^2} = \frac{1}{(x-1)^2} - \frac{1}{x-1} + \frac{1}{x},$$

we can eventually write

$$\begin{aligned}\tilde{\tilde{P}}(x) &= \frac{\tilde{g}_0 - \tilde{g}_{-10}}{(x-1)^2} + \tilde{g}_{-10}\left(\frac{1}{(x-1)^2} - \frac{1}{x-1} + \frac{1}{x}\right) + \frac{2}{x-1}\\ &= \frac{\tilde{g}_0}{(x-1)^2} + \frac{2-\tilde{g}_{-10}}{x-1} + \frac{\tilde{g}_{-10}}{x}\\ &= \frac{\tilde{\tilde{g}}_{-2+1}}{(x-1)^2} + \frac{\tilde{\tilde{g}}_{-1+1}}{x-1} + \frac{\tilde{\tilde{g}}_{-10}}{x}\end{aligned} \tag{1.2.59}$$

with

$$\begin{aligned}\tilde{\tilde{g}}_{-2+1} &= \tilde{g}_0,\\ \tilde{\tilde{g}}_{-1+1} &= 2 - \tilde{g}_{-10},\\ \tilde{\tilde{g}}_{-10} &= \tilde{g}_{-10}.\end{aligned}$$

The coefficient $\tilde{\tilde{Q}}(x)$ [cf. (1.2.29)] is

$$\begin{aligned}\tilde{\tilde{Q}}(x) &= \frac{\tilde{Q}(x)}{(x-1)^4}\\ &= \frac{-\tilde{d}_{-10} + \frac{\tilde{d}_{-10}}{x}}{(x-1)^4}\\ &= -\frac{\tilde{d}_{-10}}{(x-1)^4} + \frac{\tilde{d}_{-10}}{x(x-1)^4}\end{aligned}$$

and with the help of

$$\frac{1}{x(x-1)^4} = \frac{1}{x} + \frac{1}{(x-1)^4} - \frac{1}{(x-1)^3} + \frac{1}{(x-1)^2} - \frac{1}{x-1}$$

it becomes eventually

$$\tilde{\tilde{Q}}(x) = \frac{\tilde{\tilde{d}}_{-10}}{x} + \frac{\tilde{\tilde{d}}_{-3+1}}{(x-1)^3} + \frac{\tilde{\tilde{d}}_{-2+1}}{(x-1)^2} + \frac{\tilde{\tilde{d}}_{-1+1}}{x-1}$$

with

$$\begin{aligned}\tilde{\tilde{d}}_{-3+1} &= -\tilde{d}_{-10},\\ \tilde{\tilde{d}}_{-2+1} &= \tilde{d}_{-10},\\ \tilde{\tilde{d}}_{-1+1} &= -\tilde{d}_{-10},\\ \tilde{\tilde{d}}_{-10} &= \tilde{d}_{-10}.\end{aligned}$$

Jaffé Ansatz II

In order to take into account the power term

$$z^{\alpha_{02}}$$

of the asymptotic factor of the Frobenius solution (1.2.8), we make a rather sophisticated ansatz, that I refer to as the *Jaffé ansatz*, since it was George Jaffé who seems to have applied it for the first time in Jaffé (1933, p. 539):

$$w(x) = f(x)\,v(x) \tag{1.2.60}$$

with

$$f(x) = (1-x)^{\beta} \tag{1.2.61}$$

and where $v(x)$ is to be holomorphic at the origin $x = 0$. This ansatz completes the solution method.

The resulting differential equation in $v(x)$ is

$$\frac{\mathrm{d}^2 v}{\mathrm{d}x^2} + \bar{P}(x)\,\frac{\mathrm{d}v}{\mathrm{d}x} + \bar{Q}(x)\,v = 0 \tag{1.2.62}$$

with [cf. (1.2.32)]

$$\begin{aligned}\bar{P}(x) &= \tilde{\tilde{P}}(x) + 2\,\frac{f'}{f},\\ \bar{Q}(x) &= \tilde{\tilde{Q}}(x) + \tilde{\tilde{P}}(x)\,\frac{f'}{f} + \frac{f''}{f},\end{aligned}$$

whereby

$$f' = \frac{\mathrm{d}f}{\mathrm{d}x},\ f'' = \frac{\mathrm{d}^2 f}{\mathrm{d}x^2}$$

and

$$\begin{aligned}f'(x) &= -\beta\,(1-x)^{\beta-1} = -\beta\,(1-x)^{-1}\,f = \frac{\beta}{x-1}\,f,\\ f''(x) &= \beta\,(\beta-1)\,(1-x)^{\beta-2} = \beta\,(\beta-1)\,(1-x)^{-2}\,f\\ &= \frac{\beta\,(\beta-1)}{(x-1)^2}\,f.\end{aligned}$$

The calculation of $\bar{P}(x)$ results in

$$\bar{P}(x) = \frac{\bar{g}_{-2+1}}{(x-1)^2} + \frac{\bar{g}_{-1+1}}{x-1} + \frac{\bar{g}_{-10}}{x}$$

and the calculation of $\bar{Q}(x)$ results in[1]

$$\begin{aligned}
\bar{Q}(x) &= \frac{\tilde{d}_{-10}}{x} - \frac{\tilde{d}_{-10}}{(x-1)^3} + \frac{\tilde{d}_{-10}}{(x-1)^2} - \frac{\tilde{d}_{-10}}{x-1} \\
&\quad + \left(\frac{\tilde{g}_0}{(x-1)^2} + \frac{2-\tilde{g}_{-10}}{x-1} + \frac{\tilde{g}_{-10}}{x}\right)\frac{\beta}{x-1} \\
&\quad + \frac{\beta(\beta-1)}{(x-1)^2} \\
&= \frac{\bar{d}_{-3+1}}{(x-1)^3} + \frac{\bar{d}_{-2+1}}{(x-1)^2} + \frac{\bar{d}_{-1-1}}{x+1} + \frac{\bar{d}_{-10}}{x}
\end{aligned}$$

with

$$\begin{aligned}
\bar{g}_{-2+1} &= \tilde{g}_0, \\
\bar{g}_{-1+1} &= 2 + 2\beta - \tilde{g}_{-10}, \\
\bar{g}_{-10} &= \tilde{g}_{-10}, \\
\bar{d}_{-3+1} &= \tilde{g}_0\,\beta - \tilde{d}_{-10}, \\
\bar{d}_{-2+1} &= \tilde{d}_{-10} + (2 - \tilde{g}_{-10})\,\beta + \beta\,(\beta - 1), \\
\bar{d}_{-1+1} &= \tilde{g}_{-10}\,\beta - \tilde{d}_{-10}, \\
\bar{d}_{-10} &= -(\tilde{g}_{-10}\,\beta - \tilde{d}_{-10}) = -\bar{d}_{-1+1}.
\end{aligned}$$

The reason for applying the Jaffé transformation (1.2.60), (1.2.61) is that it is possible now to put

$$\bar{d}_{-3+1} = 0$$

by choosing β accordingly:

$$\beta = \frac{\tilde{d}_{-10}}{\tilde{g}_0}. \tag{1.2.63}$$

The final forms of $\bar{P}(x)$ and of $\bar{Q}(x)$ are thus

$$\begin{aligned}
\bar{P}(x) &= \frac{\bar{g}_{-2+1}}{(x-1)^2} + \frac{\bar{g}_{-1+1}}{x-1} + \frac{\bar{g}_{-10}}{x}, \\
\bar{Q}(x) &= \frac{\bar{d}_{-2+1}}{(x-1)^2} + \frac{\bar{d}_{-1+1}}{x-1} - \frac{\bar{d}_{-1+1}}{x} = \frac{\bar{d}_{-2+1}}{(x-1)^2} + \frac{\bar{d}_{-1+1}}{x\,(x-1)}
\end{aligned}$$

and the differential equation for $v(x)$ is given by

$$x\,(x-1)^2\,\frac{\mathrm{d}^2 v}{\mathrm{d}x^2} + \sum_{i=0}^{2} \Gamma_i\,x^i\,\frac{\mathrm{d}v}{\mathrm{d}x} + \sum_{i=0}^{1} \Delta_i\,x^i\,v = 0 \tag{1.2.64}$$

[1] It should be recognised that $\frac{1}{x\,(x-1)} = \frac{1}{x-1} - \frac{1}{x}$.

with

$$\begin{aligned}\Gamma_2 &= \bar{g}_{-1+1} + \bar{g}_{-10},\\ \Gamma_1 &= -(\bar{g}_{-2+1} + \bar{g}_{-1+1} + 2\,\bar{g}_{-10}),\\ \Gamma_0 &= \bar{g}_{-10},\\ \Delta_1 &= \bar{d}_{-2+1} + \bar{d}_{-1+1},\\ \Delta_0 &= -\bar{d}_{-1+1}.\end{aligned}$$

The Meaning of β

It is quite clear that the exponent β in (1.2.61) reflects the connection of the two relevant local solutions of the differential equation (1.2.60), (1.2.61) at $z = 0$ and at infinity. This is to be substantiated in the following.

The power terms of the asymptotic factors of the two Jaffé ansatzes (1.2.56) and (1.2.60), (1.2.61) are

$$\begin{aligned} z^{\alpha_{02}}\left(\frac{2}{z+1}\right)^{\beta} &= z^{\alpha_{02}}\,(z+1)^{-\beta}\,2^{\beta}\\ &= z^{\alpha_{02}-\beta}\underbrace{\left(1+\frac{1}{z}\right)^{-\beta}}_{\to 1 \text{ as } z\to\infty} 2^{\beta}. \end{aligned} \tag{1.2.65}$$

According to (1.2.52), the power term of the asymptotic factor of the Thomé solution (1.2.18) is

$$z^{\alpha_{0\infty 2}} \tag{1.2.66}$$

as $z \to \infty$. Thus, it has to be

$$\beta = \alpha_{02} - \alpha_{0\infty 2} \tag{1.2.67}$$

if the power of z in (1.2.65) has to meet the power of z in (1.2.66). That this is the case may be seen from (1.2.63):

$$\begin{aligned}\beta &= \frac{\tilde{d}_{-10}}{\tilde{g}_0}\\ &= \frac{2\,\alpha_{1\infty 2}\,\alpha_{02} + A\,\alpha_{1\infty 2} + G_0\,\alpha_{02} + C}{G_0 + 2\,\alpha_{1\infty 2}}\\ &= \frac{A\,\alpha_{1\infty 2} + C}{G_0 + 2\,\alpha_{1\infty 2}} + \frac{G_0 + 2\,\alpha_{1\infty 2}}{G_0 + 2\,\alpha_{1\infty 2}}\,\alpha_{02}\\ &= -\alpha_{0\infty 2} + \alpha_{02},\end{aligned}$$

which is to be compared with (1.2.67):

$$\alpha_{0\infty 2} = -\frac{A\,\alpha_{1\infty 2} + C}{G_0 + 2\,\alpha_{1\infty 2}}$$

that is just the same and explains its meaning. The choice (1.2.63) of β guarantees that the two-step Jaffé ansatz (1.2.56) and (1.2.60), (1.2.61) admits the correct and complete asymptotic factor of the Thomé solution (1.2.18) when approaching the irregular singularity of the differential equation (1.2.22) (being located at infinity) along the positive real axis.

Difference Equation

Inserting the ansatz

$$v(x) = \sum_{n=0}^{\infty} a_n\, x^n \tag{1.2.68}$$

into the differential equation (1.2.64) results in the following calculation.

Taking the derivatives

$$\frac{\mathrm{d}v}{\mathrm{d}x} = \sum_{n=0}^{\infty} a_n\, n\, x^{n-1},$$

$$\frac{\mathrm{d}^2v}{\mathrm{d}x^2} = \sum_{n=0}^{\infty} a_n\, n\,(n-1)\, x^{n-2}$$

and inserting these as well as (1.2.68) into the differential equation

$$x\,(x-1)^2\,\frac{\mathrm{d}^2v}{\mathrm{d}x^2} + \sum_{i=0}^{2} \Gamma_i\, x^i\, \frac{\mathrm{d}v}{\mathrm{d}x} + \sum_{i=0}^{1} \Delta_i\, x^i\, v = 0,$$

results in

$$\begin{aligned}
&\sum_{n=0}^{\infty} a_n\, [n\,(n-1) + \Gamma_0\, n]\, x^{n-1}\\
&\quad + \sum_{n=0}^{\infty} a_n\, [-2\,n\,(n-1) + \Gamma_1\, n + \Delta_0]\, x^n\\
&\quad\quad + \sum_{n=0}^{\infty} a_n\, [n\,(n-1) + \Gamma_2\, n + \Delta_1]\, x^{n+1} = 0.
\end{aligned}$$

Equating coefficients of equal powers to zero yields the following second-order linear difference equations:

$$\begin{aligned}
&x^0\colon (1\cdot 0 + 1\,\Gamma_0)\,a_1 + (-2\cdot 0 + 0\cdot\Gamma_1 + \Delta_0)\,a_0 = 0,\\
&x^1\colon (2\cdot 1 + 2\,\Gamma_0)\,a_2 + (-2\cdot 1\cdot 0 + 1\cdot\Gamma_1 + \Delta_0)\,a_1 + \Delta_1\,a_0 = 0,\\
&x^2\colon (3\cdot 2 + 3\,\Gamma_0)\,a_3 + (-2\cdot 2\cdot 1 + 2\cdot\Gamma_1 + \Delta_0)\,a_2 + (1\cdot 0 + 1\cdot\Gamma_2 + \Delta_1)\,a_1 = 0,\\
&x^3\colon (4\cdot 3 + 4\,\Gamma_0)\,a_4 + (-2\cdot 3\cdot 2 + 3\cdot\Gamma_1 + \Delta_0)\,a_3 + (2\cdot 1 + 2\cdot\Gamma_2 + \Delta_1)\,a_2 = 0,\\
&x^4\colon (5\cdot 4 + 5\,\Gamma_0)\,a_5 + (-2\cdot 4\cdot 3 + 4\cdot\Gamma_1 + \Delta_0)\,a_4 + (3\cdot 2 + 3\cdot\Gamma_2 + \Delta_1)\,a_3 = 0,\\
&\ldots
\end{aligned}$$

Second-order difference equations of Poincaré–Perron type (cf. §1.3) for the coefficients

$$
\begin{aligned}
&\Gamma_0\, a_1 + \Delta_0\, a_0 = 0,\\
&(n+1)\,[n+\Gamma_0]\; a_{n+1}\\
&\quad + [-2\,(n-1)\,n + n\,\Gamma_1 + \Delta_0]\; a_n\\
&\qquad + [(n-2)\,(n-1) + (n-1)\,\Gamma_2 + \Delta_1]\; a_{n-1} = 0,\; n = 1,2,3,4,\ldots,
\end{aligned}
$$

may be written in the form of a three-term recurrence relation

$$
\begin{aligned}
&a_0 \text{ arbitrary},\\
&\Gamma_0\, a_1 + \Delta_0\, a_0 = 0,\\
&\left(1 + \frac{\alpha_1}{n} + \frac{\beta_1}{n^2}\right) a_{n+1} + \left(-2 + \frac{\alpha_0}{n} + \frac{\beta_0}{n^2}\right) a_n\\
&\qquad + \left(1 + \frac{\alpha_{-1}}{n} + \frac{\beta_{-1}}{n^2}\right) a_{n-1} = 0, \qquad n \geq 1,
\end{aligned}
\tag{1.2.69}
$$

where the coefficients of the difference equation are

$$
\begin{aligned}
\alpha_1 &= \Gamma_0 + 1, & \beta_1 &= \Gamma_0,\\
\alpha_0 &= \Gamma_1 + 2, & \beta_0 &= \Delta_0,\\
\alpha_{-1} &= \Gamma_2 - 3, & \beta_{-1} &= \Delta_1 - \Gamma_2 + 2.
\end{aligned}
\tag{1.2.70}
$$

The term ‘recurrence relation’ is justified because of the initial condition in (1.2.69): after having fixed the value of a_0, all the subsequent terms a_n, $n \geq 1$ may be calculated, recursively. Thus, the partial solutions of (1.2.69) may be calculated numerically, as precisely as needed.

In the following we discuss which specific particular solutions of the linear difference equation (1.2.69) are to be calculated in order to solve the singular boundary eigenvalue problem. Since the particular solutions of the difference equation (1.2.69), solving the singular boundary eigenvalue problem, are characterised by its index asymptotic behaviour as $n \to \infty$, equation (1.2.69) is investigated in this respect.

Summary

The solution path shown above is paradigmatic and applicable to any differential equation (1) on page ix. It is the central method (cf. §1.4.2 below) for the realisation of the mathematical principle (cf. §1.4.1) proposed in this book. Therefore, in the following, the method is summarised in a comprehensive form.

Differential equation:

$$
\frac{\mathrm{d}^2 y}{\mathrm{d}z^2} + P(z)\,\frac{\mathrm{d}y}{\mathrm{d}z} + Q(z)\,y(z) = 0, \quad z \in \mathbb{C}, \tag{1.2.71}
$$

with

$$P(z) = \frac{A_0}{z} + G_0,$$
$$Q(z) = \frac{B_0}{z^2} + \frac{C}{z} + D_0. \tag{1.2.72}$$

Characteristic exponents of the singularity at zero:

$$\alpha_{01} = \frac{1 - A_0 + \sqrt{(1 - A_0)^2 - 4\,B_0}}{2},$$
$$\alpha_{02} = \frac{1 - A_0 - \sqrt{(1 - A_0)^2 - 4\,B_0}}{2}. \tag{1.2.73}$$

Characteristic exponents of the singularity at infinity:

$$\alpha_{1\infty 1} = \frac{-G_0 + \sqrt{G_0^2 - 4\,D_0}}{2},$$
$$\alpha_{1\infty 2} = \frac{-G_0 - \sqrt{G_0^2 - 4\,D_0}}{2} \tag{1.2.74}$$

and

$$\alpha_{0\infty 1} = -\frac{A_0\,\alpha_{1\infty 1} + C}{G_0 + 2\,\alpha_{1\infty 1}},$$
$$\alpha_{0\infty 2} = -\frac{A_0\,\alpha_{1\infty 2} + C}{G_0 + 2\,\alpha_{1\infty 2}}. \tag{1.2.75}$$

Jaffé ansatz I:

$$y(z) = \exp\left(\alpha_{1\infty 2}\, z\right) z^{\alpha_{0\infty 2}}\, w(z) \tag{1.2.76}$$

with

$$\tilde{P}(z) = \tilde{g}_0 + \frac{\tilde{g}_{-10}}{z},$$
$$\tilde{Q}(z) = \frac{\tilde{d}_{-10}}{z}$$

and

$$\tilde{g}_0 = G_0 + 2\,\alpha_{1\infty 2},$$
$$\tilde{g}_{-10} = A + 2\,\alpha_{02},$$
$$\tilde{d}_{-10} = 2\,\alpha_{1\infty 2}\,\alpha_{02} + A\,\alpha_{1\infty 2} + \alpha_{02}\,G_0 + C.$$

Jaffé transformation: The relation (1.2.58)

$$x = \frac{z}{z+1}$$

results in

$$\tilde{P}(x) = \tilde{g}_0 - \tilde{g}_{-10} + \frac{\tilde{g}_{-10}}{x},$$
$$\tilde{Q}(x) = -\tilde{d}_{-10} + \frac{\tilde{d}_{-10}}{x}.$$

Thus we have

$$\tilde{\tilde{P}}(x) = \frac{\tilde{\tilde{g}}_{-2+1}}{(x-1)^2} + \frac{\tilde{\tilde{g}}_{-1+1}}{x-1} + \frac{\tilde{\tilde{g}}_{-10}}{x},$$
$$\tilde{\tilde{Q}}(x) = \frac{\tilde{\tilde{d}}_{-10}}{x} + \frac{\tilde{\tilde{d}}_{-3+1}}{(x-1)^3} + \frac{\tilde{\tilde{d}}_{-2+1}}{(x-1)^2} + \frac{\tilde{\tilde{d}}_{-1+1}}{x-1},$$

with

$$\tilde{\tilde{g}}_{-2+1} = \tilde{g}_0,$$
$$\tilde{\tilde{g}}_{-1+1} = 2 - \tilde{g}_{-10},$$
$$\tilde{\tilde{g}}_{-10} = \tilde{g}_{-10},$$
$$\tilde{\tilde{d}}_{-3+1} = -\tilde{d}_{-10},$$
$$\tilde{\tilde{d}}_{-2+1} = \tilde{d}_{-10},$$
$$\tilde{\tilde{d}}_{-1+1} = -\tilde{d}_{-10},$$
$$\tilde{\tilde{d}}_{-10} = \tilde{d}_{-10}.$$

Jaffé ansatz II: The relation

$$w(x) = f(x)\, v(x) \tag{1.2.77}$$

with

$$f(x) = (1-x)^{\beta} \tag{1.2.78}$$

results in

$$\bar{P}(x) = \frac{\bar{g}_{-2+1}}{(x-1)^2} + \frac{\bar{g}_{-1+1}}{x-1} + \frac{\bar{g}_{-10}}{x},$$
$$\bar{Q}(x) = \frac{\bar{d}_{-3+1}}{(x-1)^3} + \frac{\bar{d}_{-2+1}}{(x-1)^2} + \frac{\bar{d}_{-1-1}}{x+1} + \frac{\bar{d}_{-10}}{x}$$

with

$$
\begin{aligned}
\bar{g}_{-2+1} &= \tilde{g}_0, \\
\bar{g}_{-1+1} &= 2 + 2\beta - \tilde{g}_{-10}, \\
\bar{g}_{-10} &= \tilde{g}_{-10}, \\
\bar{d}_{-3+1} &= \tilde{g}_0\,\beta - \tilde{d}_{-10}, \\
\bar{d}_{-2+1} &= \tilde{d}_{-10} + (2 - \tilde{g}_{-10})\,\beta + \beta\,(\beta - 1), \\
\bar{d}_{-1+1} &= \tilde{g}_{-10}\,\beta - \tilde{d}_{-10}, \\
\bar{d}_{-10} &= -(\tilde{g}_{-10}\,\beta - \tilde{d}_{-10}) = -\bar{d}_{-1+1}.
\end{aligned}
$$

A Taylor expansion of the particular solution (that is convergent in the unit disc) results in a linear second-order difference equation, the coefficients of which are given by

$$
\begin{aligned}
\Gamma_2 &= \bar{g}_{-1+1} + \bar{g}_{-10}, \\
\Gamma_1 &= -(\bar{g}_{-2+1} + \bar{g}_{-1+1} + 2\,\bar{g}_{-10}), \\
\Gamma_0 &= \bar{g}_{-10}, \\
\Delta_1 &= \bar{d}_{-2+1} + \bar{d}_{-1+1}, \\
\Delta_0 &= -\bar{d}_{-1+1}.
\end{aligned} \tag{1.2.79}
$$

Index Asymptotics

As has been seen, the Jaffé transformation (1.2.58) and the Jaffé ansatz (1.2.60), (1.2.61) for getting the eigensolutions of the singular boundary eigenvalue problem, discussed above, lead to the irregular difference equation (1.2.69) of Poincaré–Perron type, that is symptomatic for this approach. With respect to the mathematical problem, these sort of difference equations have to be investigated in detail concerning their index-asymptotic behaviours, viz. as $n \to \infty$.

We start by dealing with second-order equations, although third-order as well as fourth-order equations occur as well, later in the text. Higher than fourth-order equations do not occur in this book, concretely; however, the general underlying principle becomes clear and is addressed in several places, particularly on a general level in §§1.4 and 3.5.

Given the irregular second-order difference equation (1.2.69) of Poincaré–Perron type

$$
\left(1 + \frac{\alpha_1}{n} + \frac{\beta_1}{n^2}\right) a_{n+1} + \left(-2 + \frac{\alpha_0}{n} + \frac{\beta_0}{n^2}\right) a_n + \left(1 + \frac{\alpha_{-1}}{n} + \frac{\beta_{-1}}{n^2}\right) a_{n-1} = 0 \tag{1.2.80}
$$

with $n = 1, 2, 3, \ldots$, dividing this equation by a_n yields

$$
\left(1 + \frac{\alpha_1}{n} + \frac{\beta_1}{n^2}\right) t_n + \left(-2 + \frac{\alpha_0}{n} + \frac{\beta_0}{n^2}\right) + \left(1 + \frac{\alpha_{-1}}{n} + \frac{\beta_{-1}}{n^2}\right) \frac{1}{t_{n-1}} = 0 \tag{1.2.81}
$$

with

$$
t_n = \frac{a_{n+1}}{a_n}.
$$

If the limit

$$t = \lim_{n \to \infty} t_n$$

exists, a carrying out of the limiting process yields a second-order algebraic equation for t:

$$t^2 - 2t + 1 = (t - 1)^2 = 0 \tag{1.2.82}$$

having a twofold solution

$$t_1 = t_2 = 1. \tag{1.2.83}$$

Equation (1.2.82) is called the *characteristic equation* of zeroth order of the difference equation (1.2.80) (cf. Adams, 1928, pp. 507–541).

If all solutions of the characteristic equation of zeroth order of a linear, homogeneous, ordinary difference equation are finite values, then the underlying difference equation is said to be of *Poinaré–Perron type*. If, furthermore, the characteristic equation of a difference equation of Poincaré–Perron type does have multiple roots, it is said to be *irregular*; otherwise, it is called *regular*. Thus, (1.2.80) is an irregular difference equation of Poincaré–Perron type.

The two foldedness of the solution (1.2.83) of the characteristic equation (1.2.82) means that the ratio

$$\frac{a_{n+1}}{a_n}$$

of two consecutive terms of each particular solution of the difference equation (1.2.80) tends to unity.

As may be expected at this level of zeroth-order asymptotics, it is not obvious from the characteristic equation (1.2.82) that the general solution of the difference equation (1.2.80) is a two-dimensional vector space since it is an irregular one of Poincaré–Perron type. The distinction between the two fundamental solutions of (1.2.80) becomes clear only when carrying out a leading-order asymptotics. This is done in the following.

Take one more order term in equation (1.2.82):

$$\left(1 + \frac{\alpha_1}{n}\right) t_n^2 + \left(-2 + \frac{\alpha_0}{n}\right) t_n + 1 + \frac{\alpha_{-1}}{n} = 0, \tag{1.2.84}$$

the asymptotic solutions of which are given by

$$t_{n1,2} = 1 \pm \sqrt{\frac{-\sum_{i=-1}^{i=+1} \alpha_i}{n}} + O\left(\frac{1}{n}\right) \quad \text{as } n \to \infty. \tag{1.2.85}$$

The O symbol may be seen in Slavyanov (1996, §I.1.1, pp. 1, 2). This symbol had already been coined in the 1930s (cf. Erdélyi, 1956, Chapter I).

Equation (1.2.84) is called the *characteristic equation* of the first order of the difference equation (1.2.80) (cf. Adams, 1928, p. 509). Here it is seen that there are two particular solutions of equation (1.2.80), since the ratio of two consecutive terms tends to unity:

$$\frac{a_{n+1}}{a_n} \longrightarrow 1$$

as $n \longrightarrow \infty$ for each of them, albeit, from two different, namely opposite, directions in the complex t-plane at a rate

$$O\left(\frac{1}{\sqrt{n}}\right)$$

as $n \longrightarrow \infty$. There are two possibilities. Either the discriminate in (1.2.46) is positive, then there is an exponentially increasing and an exponentially decreasing particular solution. This is shown in Figure 1.2. On the other hand, if the discriminant in (1.2.46) is a negative number, then both partial solutions are oscillating. In this case, the two arrows in Figure 1.2 are rotated by 90° at their arrowheads (cf. Figure 1.6). In the example given in §1.2.5 this case occurs and is considered in more detail there.

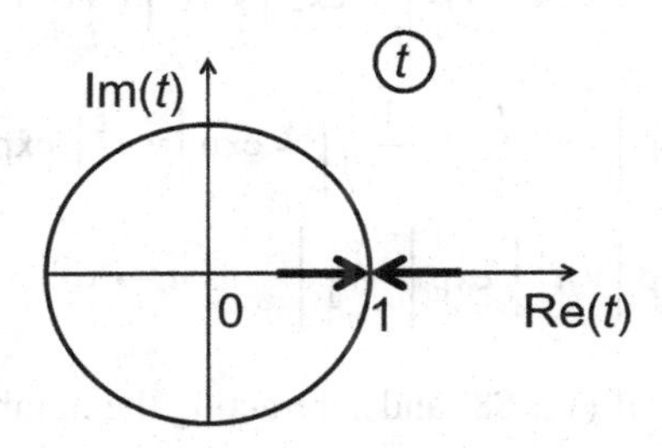

Figure 1.2 Solutions of the second-order characteristic equation.

Birkhoff Sets. Although the ratio of two consecutive terms

$$\frac{a_{n+1}}{a_n}$$

of all the particular solutions of the difference equation (1.2.80) tends to unity, there is a significant difference in their asymptotic behaviour as $n \longrightarrow \infty$. Either

$$\frac{a_{n+1,1}}{a_{n,1}} \sim 1 + \sqrt{\frac{-\sum_{i=-1}^{i=+1} \alpha_i}{n}} \tag{1.2.86}$$

or

$$\frac{a_{n+1,2}}{a_{n,2}} \sim 1 - \sqrt{\frac{-\sum_{i=-1}^{i=+1} \alpha_i}{n}} \tag{1.2.87}$$

as $n \longrightarrow \infty$. These ratios indicate the asymptotic behaviours of a pair of fundamental solutions of the difference equation (1.2.80) as $n \longrightarrow \infty$, that is considered in the following.

Proposition. *Suppose that two sequences of numbers $a_{n1,2}$, $n = 1,2,3,\ldots$, behave like*

$$a_{n1} \sim \exp\left(\gamma\, n^{\frac{1}{2}}\right), \tag{1.2.88}$$

$$a_{n2} \sim \exp\left(-\gamma\, n^{\frac{1}{2}}\right) \tag{1.2.89}$$

as $n \to \infty$, then both of them fulfil the difference equation (1.2.80):

$$\gamma = 2\sqrt{-\sum_{i=-1}^{i=+1} \alpha_i} \neq 0.$$

Proof Taking the first equation of (1.2.88) and adding the number 1 yields

$$\begin{aligned}
a_{n+1} &= \exp\left[\gamma\,(n+1)^{\frac{1}{2}}\right] = \exp\left[\gamma\, n^{\frac{1}{2}}\left(1+\frac{1}{n}\right)^{\frac{1}{2}}\right]\\
&\sim \exp\left[\gamma\, n^{\frac{1}{2}}\left(1+\frac{1}{2\,n}\right)\right] = \exp\left[\gamma\, n^{\frac{1}{2}}\right]\,\exp\left[\frac{\gamma\, n^{\frac{1}{2}}}{2\,n}\right]\\
&= \exp\left[\gamma\, n^{\frac{1}{2}}\right]\,\exp\left[\frac{\gamma}{2\,n^{\frac{1}{2}}}\right] \quad \text{as } n \to \infty.
\end{aligned}$$

Taking the first equation of (1.2.88) and subtracting the number 1 yields

$$\begin{aligned}
a_{n-1} &= \exp\left[\gamma\,(n-1)^{\frac{1}{2}}\right] = \exp\left[\gamma\, n^{\frac{1}{2}}\left(1-\frac{1}{n}\right)^{\frac{1}{2}}\right]\\
&\sim \exp\left[\gamma\, n^{\frac{1}{2}}\left(1-\frac{1}{2\,n}\right)\right] = \exp\left[\gamma\, n^{\frac{1}{2}}\right]\,\exp\left[-\frac{\gamma\, n^{\frac{1}{2}}}{2\,n}\right]\\
&= \exp\left[\gamma\, n^{\frac{1}{2}}\right]\,\exp\left[-\frac{\gamma}{2\,n^{\frac{1}{2}}}\right] \quad \text{as } n \to \infty.
\end{aligned}$$

Insertion of these expressions into the difference equation (1.2.80), which may be written in the form

$$\left(n^2 + \alpha_1\, n + \beta_1\right)\, a_{n+1} + \left(-2\, n^2 + \alpha_0\, n + \beta_0\right)\, a_n + \left(n^2 + \alpha_{-1}\, n + \beta_{-1}\right)\, a_{n-1} = 0,$$

leads to

$$\begin{aligned}
&\left(n^2 + \alpha_1\, n + \beta_1\right)\, \exp\left(\frac{\gamma}{2}\, n\right)\, \exp\left[\frac{\gamma}{2\, n^{\frac{1}{2}}}\right] + \left(-2\, n^2 + \alpha_0\, n + \beta_0\right)\, \exp\left(\frac{\gamma}{2}\, n\right)\\
&\quad + \left(n^2 + \alpha_{-1}\, n + \beta_{-1}\right)\, \exp\left(\frac{\gamma}{2}\, n\right)\, \exp\left[-\frac{\gamma}{2\, n^{\frac{1}{2}}}\right] = 0
\end{aligned}$$

or

$$
\begin{aligned}
& n^2\left(1+\frac{\gamma}{n^{\frac{1}{2}}}+\frac{\gamma^2}{2\,n^{\frac{2}{2}}}+\cdots\right)+\alpha_1\,n\left(1+\frac{\gamma}{n^{\frac{1}{2}}}+\frac{\gamma^2}{2\,n^{\frac{2}{2}}}+\cdots\right) \\
& +\beta_1\left(1+\frac{\gamma}{n^{\frac{1}{2}}}+\frac{\gamma^2}{2\,n^{\frac{2}{2}}}+\cdots\right)-2\,n^2\left(1+\frac{\gamma}{2\,n^{\frac{1}{2}}}+\frac{\gamma^2}{8\,n^{\frac{2}{2}}}+\cdots\right) \\
& +\alpha_0\,n\left(1+\frac{\gamma}{2\,n^{\frac{1}{2}}}+\frac{\gamma^2}{8\,n^{\frac{2}{2}}}+\cdots\right)+\beta_0\left(1+\frac{\gamma}{2\,n^{\frac{1}{2}}}+\frac{\gamma^2}{8\,n^{\frac{2}{2}}}+\cdots\right) \\
& +n^2+\alpha_{-1}\,n+\beta_{-1}=0.
\end{aligned}
\tag{1.2.90}
$$

Thus, a decomposition of the difference equation (1.2.90) into half-integer powers yields

$$
\begin{aligned}
& n^{\frac{4}{2}}+\gamma\,n^{\frac{3}{2}}+\frac{1}{2}\gamma^2\,n^{\frac{2}{2}}+\cdots \\
& +\alpha_1\,n^{\frac{2}{2}}+\alpha_1\,\gamma\,n^{\frac{1}{2}}+\frac{1}{2}\,\alpha_1\,\gamma^2\,n^{\frac{0}{2}}+\cdots \\
& +\beta_1\,n^{\frac{0}{2}}+\beta_1\,\gamma\,n^{-\frac{1}{2}}+\frac{1}{2}\,\beta_1\,\gamma^2\,n^{-\frac{2}{2}}+\cdots \\
& -2\,n^{\frac{4}{2}}-\gamma\,n^{\frac{3}{2}}-\frac{1}{4}\,\gamma^2\,n^{\frac{2}{2}}+\cdots \\
& +\alpha_0\,n^{\frac{2}{2}}+\frac{1}{2}\,\alpha_0\,\gamma\,n^{\frac{1}{2}}+\frac{1}{8}\,\alpha_0\,\gamma^2\,n^{\frac{0}{2}}+\cdots \\
& +\beta_0\,n^{\frac{0}{2}}+\frac{\beta_0\,\gamma}{2}\,n^{-\frac{1}{2}}+\frac{\beta_0\,\gamma^2}{8}\,n^{-\frac{2}{2}}+\cdots \\
& +n^{\frac{4}{2}}+\alpha_{-1}\,n^{\frac{2}{2}}+\beta_{-1}\,n^{\frac{0}{2}}=0.
\end{aligned}
\tag{1.2.91}
$$

In order for the left-hand side of this equation to be zero, it is necessary that all of the coefficients in front of the powers of n have to be zero. Therefore, every power of equation (1.2.91) is to be considered separately.

The highest power term is $n^{\frac{4}{2}}$:

$$
n^{\frac{4}{2}}:(1-2+1)\,n^{\frac{4}{2}}=0.
$$

This equation is fulfilled by the difference equation (1.2.80) itself! Thus, a condition cannot be made out of this term.

The next lower power term is $n^{\frac{3}{2}}$:

$$
n^{\frac{3}{2}}:(\gamma-\gamma)\,n^{\frac{3}{2}}=0.
$$

This equation is also fulfilled by the difference equation (1.2.80) itself!

The next lower power term is $n^{\frac{2}{2}}$:

$$
n^{\frac{2}{2}}:\left(\alpha_1+\alpha_0+\alpha_{-1}+\frac{1}{2}\gamma^2-\frac{1}{4}\gamma^2\right)n^{\frac{2}{2}}=0.
$$

From this equation result the two characteristic exponents $\gamma = \gamma_1$ and $\gamma = \gamma_2$:

$$\gamma_1 = +2\sqrt{-\sum_{i=-1}^{i=+1} \alpha_i}.$$

$$\gamma_2 = -2\sqrt{-\sum_{i=-1}^{i=+1} \alpha_i} = -\gamma_1.$$

Thus, both of the sequences in (1.2.88) fulfil the difference equation (1.2.80) as $n \longrightarrow \infty$ □

Thus, in dependence on the sign of the parameter γ, the sequence a_n, $n = 1, 2, 3, \ldots$, either tends to infinity or to zero as $n \longrightarrow \infty$. In both cases, the behaviour is of exponential type.

According to these lines, higher and higher orders may be incorporated into the calculations. The results are the coefficients of the so-called *Birkhoff set*, consisting of two formal expressions of two particular solutions $\{a_{n1}\}$ and $\{a_{n2}\}$ of the irregular difference equation (1.2.72) of Poincaré–Perron type and representing the full asymptotic behaviour of two linearly independent particular solutions of the difference equation as $n \longrightarrow \infty$ that may be considered exact. However, because of the considerable increase in calculatory expenditure, it is advisable to use an algebraic computer program. The Birkhoff set of the difference equation (1.2.72) is given by

$$\begin{aligned} a_{n1}(n) &= \exp\left(\gamma_1 n^{\frac{1}{2}}\right) n^{r_1} \left[1 + \frac{C_{11}}{n^{\frac{1}{2}}} + \frac{C_{12}}{n^{\frac{2}{2}}} + \cdots\right], \\ a_{n2}(n) &= \exp\left(\gamma_2 n^{\frac{1}{2}}\right) n^{r_2} \left[1 + \frac{C_{21}}{n^{\frac{1}{2}}} + \frac{C_{22}}{n^{\frac{2}{2}}} + \cdots\right] \end{aligned} \tag{1.2.92}$$

as $n \longrightarrow \infty$, with

$$\begin{aligned} \gamma_1 &= 2\sqrt{-\sum_{i=-1}^{i=+1} \alpha_i} \neq 0, \\ \gamma_2 &= -\gamma_1, \\ r_1 &= r_2 = -\frac{1}{4}\left(2\,\alpha_1 - 2\,\alpha_{-1} - 1\right). \end{aligned} \tag{1.2.93}$$

The coefficients C_{ij}, $i = 1, 2$, $j = 1, 2, 3, \ldots$, may be calculated but do not play any role in the following. The solutions $\{a_{n1}\}$, $\{a_{n2}\}$ of the Birkhoff set (1.2.92), (1.2.93) are called *Birkhoff series solutions*.

This sort of formal solution of linear ordinary difference equations is the analogue to the Thomé solution (cf. §1.2) for linear ordinary differential equations. These analogous ansatzes were shown, by **George David Birkhoff** and **Waldemar Juliet Trjitzinsky** (Birkhoff and Trjitzinsky, 1933), in the late 1920s and early 1930s, to be asymptotic solutions as $n \longrightarrow \infty$. Henri Poincaré showed this with respect to Thomé's formal series solutions a few decades earlier for differential equations.

Taking the difference equation (1.2.72) as a recurrence relation, the increase of the independent variable n is in a sense 'natural'. Thus, the index asymptotics here is the 'natural' behaviour of the independent variable n.

As a result of (1.2.93) it is now clear that if

$$\gamma = 2\sqrt{-\sum_{i=-1}^{i=+1} \alpha_i}$$

then (1.2.92) gives the possible asymptotic behaviours of the particular solutions a_n of the difference equation (1.2.80).

The question now appears which of the two particular solutions has to be chosen if the parameters adopt eigenvalues of the singular boundary eigenvalue problem, i.e., if the particular solution of the differential equation adopts the required asymptotic behaviour on approaching the singularity at $z = 0$ as well as at infinity along the positive real axis. This question is to be dealt with in the following.

A Proof

After having investigated the index-asymptotic behaviour of the particular solutions $\{a_n\}, n = 0, 1, 2, 3, \ldots$, of the irregular difference equation (1.2.69) of Poincaré–Perron type, the final step now is to be done in order to clarify the asymtotic behaviour of the function (1.2.68) as $z \longrightarrow \infty$ or $x \longrightarrow 1$.

General Remarks

It is quite clear from what is written above that the asymptotic behaviour of the function $v[x(z)]$ in (1.2.59) generally (viz. if the parameters of the differential equation $\underline{c}$ do not admit eigenvalues) is given by

$$v[x(z)] = \sum_{n=0}^{\infty} a_n \, x^n \sim \exp\left[(\alpha_{1\infty1} - \alpha_{1\infty2})\, z\right]$$

as $z \longrightarrow \infty$, since only then is the asymptotic behaviour of $y(z)$ in (1.2.56) in accordance with the dominant behaviour given by the Thomé solutions in formula (1.2.18) as $z \longrightarrow \infty$:

$$y(z) \sim \exp\left(\alpha_{1\infty2}\, z\right) \exp\left[(\alpha_{1\infty1} - \alpha_{1\infty2})\, z\right] = \exp\left(\alpha_{1\infty1}\, z\right).$$

On the other hand, it is clear that in the case when $\underline{c}$ admit eigenvalues, the asymptotic behaviour of the function $w(z)$ in (1.2.59) must be weaker than

$$\exp\left(\alpha\, z\right), \; |\alpha| = \alpha_{1\infty2}$$

as $z \longrightarrow \infty$, such that it is dominated by means of the asymptotic factor in (1.2.56):

$$\exp\left(\alpha_{1\infty2}\, z\right)$$

since only then is the asymptotic behaviour of $y(z)$ in (1.2.56) realised in formula (1.2.18) as $z \longrightarrow \infty$:

$$y(z) \sim \exp(\alpha_{1\infty 2}\, z)\, \exp[\alpha\, z^{\varepsilon}] \sim \exp(\alpha_{1\infty 2}\, z)$$

with $\{\varepsilon \in \mathbb{R}, -1 < \varepsilon < +1\}$ and α assumed to be $|\alpha| < \alpha_{1\infty 2}$. That this really happens is to be shown in detail in the following.

No-Eigenvalue Behaviour

In the following, these global aspects may be confirmed by a local consideration. We start with the non-eigenvalue case.

Proposition. *If the particular solution* $\{a_n\}$, $n = 0, 1, 2, 3, \ldots$, *of the difference* equation (1.2.72) *is given by*

$$a_n \sim \exp\left(\gamma_1\, n^{\frac{1}{2}}\right) \quad as\ n \longrightarrow \infty$$

then the function $v[x(z)]$ *in* (1.2.60) *behaves like*

$$v(x(z)) = \sum_{n=0}^{\infty} a_n\, x^n \sim \exp\left[(\alpha_{11} - \alpha_{12})\, z\right] \tag{1.2.94}$$

as $z \longrightarrow \infty$.

To prove this proposition, according to the Euler–Maclaurin formula (cf. also, e.g., deBruijn, 1961; Abramowitz and Stegun, 1970), the infinite sum in (1.2.94)

$$\sum_{n=0}^{\infty} a_n\, x^n = \sum_{n=0}^{\infty} a_n \left(\frac{z}{z+1}\right)^n$$

is replaced by the integral[2]

$$\int_{n=0}^{\infty} u_z(n)\, \mathrm{d}n$$

with

$$u_z(n) = \exp\left(\gamma_1\, n^{\frac{1}{2}} - \frac{n}{z}\right), \quad \gamma_1,\ z > 0, \text{real-valued}, n \in \mathbb{N}$$

as $n \longrightarrow \infty$. We hereby make use of Auxiliary I below.

Auxiliary I. *If*

$$x = \frac{z}{z+1} \tag{1.2.95}$$

then it holds that

$$x^n \sim \exp\left(-\frac{n}{z}\right) \tag{1.2.96}$$

[2] This is correct modulo a constant, which does not matter in the following.

as $z \longrightarrow \infty$. This means that the ratio

$$\frac{x^n}{\exp\left(-\frac{n}{z}\right)} \tag{1.2.97}$$

tends to unity as z tends to infinity for any natural value of n.

Proof of Auxiliary I We have[3]

$$\begin{aligned}
\ln\left(\frac{z}{z+1}\right) &= \ln\left(\frac{z}{z+\frac{z}{z}}\right) = \ln\left(\frac{1}{1+\frac{1}{z}}\right) \\
&= \ln 1 - \ln\left(1+\frac{1}{z}\right) = -\ln\left(1+\frac{1}{z}\right) \\
&\sim -\frac{1}{z} \text{ as } z \longrightarrow \infty
\end{aligned}$$

and thus

$$x^n = \exp\left[\ln\left(\frac{z}{z+1}\right)^n\right] = \exp\left[n\ln\left(\frac{z}{z+1}\right)\right] \sim \exp\left(-\frac{n}{z}\right)$$

as $z \longrightarrow \infty$. □

We formulate the following Auxiliary II.

Auxiliary II. *If the coefficients a_n in the Taylor series* (1.2.94)

$$v(x(z)) = \sum_{n=0}^{\infty} a_n \left(\frac{z}{z+1}\right)^n = \sum_{n=0}^{\infty} a_n \exp\left[n\ln\left(\frac{z}{z+1}\right)\right]$$

behave like

$$a_n \sim \exp\left(\gamma_1 n^{\frac{1}{2}}\right)$$

as $n \longrightarrow \infty$, then the function $v(x(z))$ behaves like

$$v[x(z)] \sim \exp\left[\left(\frac{\gamma_1}{2}\right)^2 z\right]$$

as $z \longrightarrow \infty$.

[3] See Bronstein et al. (2001, p. 1043): $\ln(1+x) \sim x$ as $x \rightarrow 0$, viz. $z \rightarrow \infty$.

Proof of Auxiliary II We consider that

$$\sum_{n=0}^{\infty} a_n\, x^n = \sum_{n=0}^{\infty} a_n \left(\frac{z}{z+1}\right)^n \sim \sum_{n=0}^{\infty} \exp\left(\gamma_1\, n^{\frac{1}{2}}\right) \left(\frac{z}{z+1}\right)^n$$

$$= \sum_{n=0}^{\infty} \exp\left(\gamma_1\, n^{\frac{1}{2}}\right) \exp\left[n \ln\left(\frac{z}{z+1}\right)\right]$$

$$= \sum_{n=0}^{\infty} \exp\left[\gamma_1\, n^{\frac{1}{2}} + \ln\left(\frac{z}{z+1}\right) n\right]$$

$$= \sum_{n=0}^{\infty} \exp\left[h(z,n)\right] = \sum_{n=0}^{\infty} u\,(z,n)$$

with

$$u\,(z,n) = \exp\left[h\,(z,n)\right] \overset{\text{Auxiliary I}}{\sim} \exp\left[\gamma_1\, n^{\frac{1}{2}} - \frac{n}{z}\right]$$

as $n,\ z \longrightarrow \infty$.

On the basis of the Euler–MacLaurin sum formula (see, e.g., deBruijn, 1961; Abramowitz and Stegun, 1970), we calculate

$$v[x\,(z)] = \int_{n=0}^{\infty} u(z,n)\,\mathrm{d}n = \int_{n=0}^{\infty} \exp\left[h(z,n)\right]\,\mathrm{d}n = \int_{n=0}^{\infty} \exp\left[\gamma_1\, n^{\frac{1}{2}} - \frac{n}{z}\right]\,\mathrm{d}n. \quad (1.2.98)$$

It is important to understand that for $\gamma_1 > 0$ or for $\Re(\gamma_1) > 0$, the function $u(z,n) = u_z(n)$ always has a hump that has two properties as $n \longrightarrow \infty,\ z = \text{const.}$:

- The apex of the hump tends to infinity as $n \longrightarrow \infty$.
- The hump becomes smaller and taller, viz. a peak, a delta function in the limit $n \longrightarrow \infty$.

These two properties are the conditions for applying the Lagrange integration method, which is done in the following.

According to this method, the function $u(z,n) = u_z(n)$ is approximated by a second-order polynomial in the limit $n \longrightarrow \infty,\ z = \text{const.}$ (cf., e.g., deBruijn, 1961):

$$u_z(n) \sim g_1(n) = a_1\, [n - n_0]^2 + b_1 \quad \text{as } n \longrightarrow \infty \quad (1.2.99)$$

with

$$\begin{aligned} a_1 &= a_1(n_0,\ \gamma_1,\ z), \\ b_1 &= b_1(n_0,\ \gamma_1,\ z), \\ n_0 &= n_0(\gamma_1,\ z), \end{aligned} \quad (1.2.100)$$

where n_0 is the location of the hump on the n-axis. In order to calculate an approximation of the integral function in (1.2.98), this function $g_1(n)$ may be integrated from the lower zero n_{11} of $g_1(n)$ to the upper one n_{12} yielding

$$\int_0^\infty g_1(z,n)\,\mathrm{d}n \sim \int_{n_{11}}^{n_{12}} g_1(z,n)\,\mathrm{d}n \sim v(z)$$

as $n \longrightarrow \infty$ and for fixed values of the independent variable z.

According to the Lagrange integration method, there are three determining conditions for the three parameters a_1, b_1, n_0:

$$g_1(n = n_0) = u_z(n = n_0),$$

$$\left.\frac{\mathrm{d}g_1(n)}{\mathrm{d}n}\right|_{n=n_0} = \left.\frac{\mathrm{d}u_z(n)}{\mathrm{d}n}\right|_{n=n_0} = 0,$$

$$\left.\frac{\mathrm{d}^2 g_1(n)}{\mathrm{d}n^2}\right|_{n=n_0} = \left.\frac{\mathrm{d}^2 u_z(n)}{\mathrm{d}n^2}\right|_{n=n_0}.$$

In the following, these three conditions are evaluated:

- The first condition yields

$$\exp\left[\gamma_1\, n_0^{\frac{1}{2}} - \frac{n_0}{z}\right] = a_1\,(n_0 - n_0)^2 + b_1 = b_1.$$

- The second condition yields

$$\frac{\mathrm{d}u}{\mathrm{d}n} = \frac{\gamma_1\, z - 2\sqrt{n}}{2\,z\,\sqrt{n}} \exp\left(\gamma_1\,\sqrt{n} - \frac{n}{z}\right),$$

thus

$$\left.\frac{\mathrm{d}u}{\mathrm{d}n}\right|_{n=n_0} = \frac{\gamma_1\, z - 2\sqrt{n_0}}{2\,z\,\sqrt{n_0}} \exp\left[\gamma_1\, n_0 - \frac{n_0}{z}\right] = 0,$$

resulting in

$$\gamma_1\, z - 2\sqrt{n_0} = 0$$

or

$$n_0(\gamma_1, z) = \left(\frac{\gamma_1\, z}{2}\right)^2.$$

On the other hand [cf. (1.2.99)]

$$\frac{\mathrm{d}g_1(n)}{\mathrm{d}n} = 2\,a_1\,(n - n_0),$$

from which follows

$$\left.\frac{\mathrm{d}g_1(n)}{\mathrm{d}n}\right|_{n=n_0} = 2\,a_1\,(n_0 - n_0) = 0,$$

as is shown to be correct.

- The third condition yields

$$\frac{\mathrm{d}^2 u}{\mathrm{d}n^2} = \frac{\gamma_1 z^2 (\gamma_1 \sqrt{n} - 1) - 4 \gamma_1 n z + 4 n^{\frac{3}{2}}}{4 z^2 n^{\frac{3}{2}}} \exp\left(\gamma_1 \sqrt{n} - \frac{n}{z}\right),$$

thus

$$\left.\frac{\mathrm{d}^2 u}{\mathrm{d}n^2}\right|_{n=n_0} = \frac{\gamma_1 z^2 (\gamma_1 \sqrt{n_0} - 1) - 4 \gamma_1 n_0 z + 4 n_0^{\frac{3}{2}}}{4 z^2 n_0^{\frac{3}{2}}} \exp\left(\gamma_1 \sqrt{n_0} - \frac{n_0}{z}\right)$$

and

$$\frac{\mathrm{d}^2 g_1}{\mathrm{d}n^2} = 2\, a_1,$$

from which follows

$$\begin{aligned} a_1 &= \frac{1}{2}\left(\frac{\gamma_1 z^2 (\gamma_1 \sqrt{n_0} - 1) - 4 \gamma_1 n_0 z + 4 n_0^{\frac{3}{2}}}{4 z^2 n_0^{\frac{3}{2}}} \exp\left(\gamma_1 \sqrt{n_0} - \frac{n_0}{z}\right)\right) \\ &= -\frac{\exp\left(\frac{\gamma_1^2 z}{4}\right)}{\gamma_1^2 z^3} \sim \exp\left(\frac{\gamma_1^2 z}{4}\right) \end{aligned}$$

as $n_0 \longrightarrow \infty$ or $z \longrightarrow \infty$.

Summarising the calculations, it is to be remarked that the function

$$u_z(n) = u(z, n) = \exp\left(\gamma_1 n^{\frac{1}{2}} - \frac{n}{z}\right)$$

in the limit $n \longrightarrow \infty$ is approximated by the quadratic parabola

$$g_1(n) = a_1 [n - n_0]^2 + b_1$$

with[4]

$$\begin{aligned} n_0(\gamma_1, z) &= \left(\frac{\gamma_1 z}{2}\right)^2, \\ a_1(n_0, \gamma_1, z) &= -\frac{b_1}{\gamma_1^2 z^3} = -\frac{\exp\left(\frac{\gamma_1^2 z}{4}\right)}{\gamma_1^2 z^3}, \\ b_1(n_0, \gamma_1, z) &= \exp\left(\gamma_1 n_0^{\frac{1}{2}} - \frac{n_0}{z}\right) = \exp\left(\frac{\gamma_1^2 z}{4}\right). \end{aligned}$$

For later use, we write

$$-\frac{b_1}{a_1} = \gamma_1^2 z^3.$$

The calculation of the zeros of $g_1(n)$ is done by

$$a_1 [n_1 - n_0]^2 + b_1 = 0$$

[4] Thus, for $z \longrightarrow \infty$, we have $a_1 \longrightarrow -b_1$.

or

$$a_1\, n_1^2 - 2\, a_1\, n_0\, n_1 + a_1\, n_0^2 + b_1 = 0.$$

Thus

$$\begin{aligned} n_{11} &= \frac{1}{2\, a_1} \left[2\, a_1\, n_0 - \sqrt{4 a_1^2\, n_0^2 - 4\, a_1 \left(a_1\, n_0^2 + b_1 \right)} \right] \\ &= n_0 - \sqrt{-\frac{b_1}{a_1}} = \left(\frac{\gamma_1\, z}{2} \right)^2 - \gamma_1\, \sqrt{z^3} = n_0 - \gamma_1\, \sqrt{z^3} \end{aligned}$$

and

$$\begin{aligned} n_{12} &= \frac{1}{2\, a_1} \left[2\, a_1\, n_0 + \sqrt{4 a_1^2\, n_0^2 - 4\, a_1 \left(a_1\, n_0^2 + b_1 \right)} \right] \\ &= n_0 + \sqrt{-\frac{b_1}{a_1}} = \left(\frac{\gamma_1\, z}{2} \right)^2 + \gamma_1\, \sqrt{z^3} = n_0 + \gamma_1\, \sqrt{z^3}. \end{aligned}$$

The final result is summarised by stating

$$\begin{aligned} v[x(z)] &= \sum_{n=0}^{\infty} a_n\, x^n \\ &\sim \sum_{n=0}^{\infty} \exp\left(\gamma_1\, n^{\frac{1}{2}} \right)\, x^n \text{ as } n \to \infty \\ &\sim \int_{n_{11}}^{n_{12}} \left[a_1\, (n - n_0)^2 + b_1 \right]\, \mathrm{d}n \\ &= \frac{a_1}{3}\, (n - n_0)^3 + b_1\, n \Big|_{n_{11}}^{n_{12}} \\ &= \frac{2}{3}\, b_1\, \sqrt{-\frac{b_1}{a_1}} \\ &= \frac{4}{3}\, \gamma_1\, \sqrt{z^3}\, \exp\left(\frac{\gamma_1^2\, z}{4} \right) \sim \exp\left(\frac{\gamma_1^2\, z}{4} \right) \end{aligned}$$

or

$$v[x(z)] \sim \exp\left[\left(\frac{\gamma_1}{2} \right)^2\, z \right] \tag{1.2.101}$$

as $z \to \infty$, which is to be shown. □

In order to prove the proposition, we need Auxiliary III below.

Auxiliary III. *For the coefficient*

$$\left(\frac{\gamma_1}{2} \right)^2 = \frac{\gamma_1^2}{4}$$

of the exponent of the dominant term of the series (1.2.101) *for* $v[x(z)]$ *of the single confluent case of the Gauss differential equation, it holds that*

$$\frac{\gamma_1^2}{4} = \alpha_{11} - \alpha_{12},$$

with α_{11}, α_{12} *being the characteristic exponents of second kind of the first order of the irregular singularity at infinity:*

$$\alpha_{11} = \frac{1}{2}\left(-G_0 + \sqrt{G_0^2 - 4\,D_0}\right),$$

$$\alpha_{12} = \frac{1}{2}\left(-G_0 - \sqrt{G_0^2 - 4\,D_0}\right).$$

Proof of Auxiliary III From the Birkhoff set (1.2.92), (1.2.93) it is given that

$$\frac{\gamma_1^2}{4} = -\sum_{i=-1}^{+1} \alpha_i.$$

The Jaffé expansion (1.2.68) yields the relations (1.2.70):

$$\begin{aligned} \alpha_1 &= \Gamma_0 + 1, \\ \alpha_0 &= \Gamma_1 + 2, \\ \alpha_{-1} &= \Gamma_2 - 3, \end{aligned}$$

resulting in

$$-\sum_{i=-1}^{+1} \alpha_i = -\sum_{i=0}^{2} \Gamma_i.$$

Furthermore

$$\begin{aligned} \Gamma_2 &= -4 + g_2 - g_1 + g_0 + 2\,g_{-1}, \\ \Gamma_1 &= 6 + g_1 - 2\,g_0 - 3\,g_{-1}, \\ \Gamma_0 &= -2 + g_0 + g_{-1}, \end{aligned}$$

yielding

$$-\sum_{i=0}^{2} \Gamma_i = \tilde{g}_{-2+1} = \tilde{g}_0 = G_0 + 2\,\alpha_{12} = \sqrt{G_0^2 - 4\,D_0}. \tag{1.2.102}$$

So

$$\frac{\gamma_1^2}{4} = -\sum_{i=-1}^{+1} \alpha_i = -\sum_{i=0}^{2} \Gamma_i = -g_2 = -(G_0 + 2\,\alpha_{11}) = \sqrt{G_0^2 - 4\,D_0}. \tag{1.2.103}$$

On the other hand [cf. (1.2.48)]

$$\alpha_{1\infty1} = \frac{1}{2}\left(-G_0 + \sqrt{G_0^2 - 4\,D_0}\right) = -\frac{1}{2}\,G_0 + \frac{1}{2}\,\sqrt{G_0^2 - 4\,D_0},$$

$$\alpha_{1\infty2} = \frac{1}{2}\left(-G_0 - \sqrt{G_0^2 - 4\,D_0}\right) = -\frac{1}{2}\,G_0 - \frac{1}{2}\,\sqrt{G_0^2 - 4\,D_0},$$

from which it is immediately seen that

$$\alpha_{1\infty1} - \alpha_{1\infty2} = \sqrt{G_0^2 - 4\,D_0},$$

which is the same as (1.2.102). Thus

$$\left(\frac{\gamma_1}{2}\right)^2 = \alpha_{1\infty1} - \alpha_{1\infty2},$$

as has to be shown. □

Proof of the Proposition Summarising the results of Auxiliaries I, II and III above, it becomes obvious that if the parameters $\underline{c}$ are supposed to be *no* eigenvalues, then

$$\begin{aligned}
y(z) &= \exp(\alpha_{12}\,z)\; z^{\alpha_{0r2}} \sum_{n=0}^{\infty} a_n \left(\frac{z}{z+1}\right)^n \\
&\overset{\text{as } n\to\infty}{\sim} \underbrace{\exp(\alpha_{12}\,z)}_{\text{asymptotic factor}} \sum_{n=0}^{\infty} \exp\left(\gamma_1\, n^{\frac{1}{2}}\right) \left(\frac{z}{z+1}\right)^n \\
&\overset{\text{Auxiliary I}}{\sim} \exp(\alpha_{12}\,z) \sum_{n=0}^{\infty} \exp\left(\gamma_1\, n^{\frac{1}{2}} - \frac{n}{z}\right) \\
&\overset{\text{Auxiliary II}}{\sim} \exp(\alpha_{12}\,z)\ \exp\left[\left(\frac{1}{2}\,\gamma_1\right)^2 z\right] \\
&\overset{\text{Auxiliary III}}{\sim} \exp(\alpha_{12}\,z)\ \exp\left[(\alpha_{11} - \alpha_{12})\; z\right] \\
&= \exp(\alpha_{12}\,z)\ \exp(\alpha_{11}\,z)\ \exp(-\alpha_{12}\,z) \\
&= \exp(\alpha_{11}\,z),
\end{aligned}$$

which proves the proposition. □

Boundary Eigenvalue Behaviour

In the following we admit that the parameters $\underline{c}$ are supposed to be eigenvalues, viz. we admit values such that the particular solution $y(z)$ of the differential equation (1.2.54), (1.2.55) meets the boundary conditions simultaneously at the regular singularity at the origin (already fulfilled by means of the ansatz (1.2.56), (1.2.58), (1.2.60), (1.2.61), (1.2.68)) as well as at the irregular singularity, located at infinity (to be fulfilled by means of a specific asymptotic behaviour of the particular solution $\{a_n, n = 0, 1, 2, 3, \ldots\}$ as $n \to \infty$ of the difference equation (1.2.69), possible only for special values $\underline{c}$ (eigenvalues) of the underlying differential equation (1.2.54), (1.2.55). In the eigenvalue case, asymptotics may be settled by means of an estimation.

Proposition 1.1. *Suppose that*

$$u_z(n) = \exp\left(-\alpha\, n^{\frac{1}{2}} - \frac{n}{z}\right),\ \alpha,\, z > 0, \textit{real-valued}$$

holds. Then the integral

$$\int_{n=0}^{\infty} u_z(n) = \int_{n=0}^{\infty} \exp\left(-\alpha\, n^{\frac{1}{2}} - \frac{n}{z}\right)\, \mathrm{d}n$$

is finite, defining a function

$$U(z)$$

that tends to a finite value as $z \to \infty$.[5]

Proof of the Proposition Estimating

$$u(z,n) < \exp\left(-\alpha\, n^{\frac{1}{2}} - \frac{n^{\frac{1}{2}}}{z}\right) \text{ for all } z > 0, \text{real-valued}$$

yields an integral

$$\int_1^{\infty} u(z,n)\, \mathrm{d}n < \int_1^{\infty} \exp\left[-n^{\frac{1}{2}}\left(\alpha + \frac{1}{z}\right)\right] \mathrm{d}n \text{ for all } z > 0, \text{real-valued}, \quad (1.2.104)$$

which may be solved exactly. By means of the substitution

$$N = \sqrt{n}$$

and

$$\frac{\mathrm{d}N}{\mathrm{d}n} = \frac{1}{2\sqrt{n}} = \frac{1}{2\,N},$$

the integral (1.2.104) becomes

$$\begin{aligned}\int_1^{\infty} \exp\left[-n^{\frac{1}{2}}\left(\alpha + \frac{1}{z}\right)\right] \mathrm{d}n &= 2\int_1^{\infty} N \exp\left[-N\left(\alpha + \frac{1}{z}\right)\right] \mathrm{d}N \\ &= 2\,\frac{\alpha + \frac{1}{z} + 1}{\left(\alpha + \frac{1}{z}\right)^2} \text{ for all } z > 0, \text{real-valued}.\end{aligned}$$

For $z \to \infty$, this yields

$$\lim_{z\to\infty} 2\,\frac{\alpha + \frac{1}{z} + 1}{(\alpha + \frac{1}{z})^2} = 2\,\frac{\alpha + 1}{\alpha^2} + O\left(\frac{1}{z}\right),$$

[5] It is possible to prove the proposition by means of the monotone convergence theorem; the result is $\int_{n=0}^{\infty} \exp\left(-\alpha\, n^{\frac{1}{2}} - \frac{n}{z}\right) \mathrm{d}n = 2\left(\frac{z}{\alpha}\right)^2$.

thus

$$v[x(z)] = \text{const.} + O\left(\frac{1}{z}\right)$$

as $z \to \infty$ if $\underline{c}$ adopts an eigenvalue, which has to be shown. □

As a summary of the foregoing considerations we state the following:

- If $\underline{c}$ is no eigenvalue, then the function $v(x)$ in (1.2.94) behaves like

$$v(x) = \sum_{n=0}^{\infty} a_n\, x^n \sim \exp\left[(\alpha_{11} - \alpha_{12})\, z\right]$$

as $z \to \infty$, whereby $x = \frac{z}{z+1}$ [cf. (1.2.58)].
- If $\underline{c}$ adopts an eigenvalue, then the function $v(x)$ in (1.2.94) behaves like

$$v(x) = \sum_{n=0}^{\infty} a_n\, x^n \sim \text{const.} + O\left(\frac{1}{z}\right)$$

as $z \to \infty$ or $x \to 1$. This means that the series in the recessive Thomé solution (1.2.18) at the infinite singularity of the underlying differential equation

$$y_{T2}(z) = \exp\left(\alpha_{1\infty 2}\, z\right)\, z^{\alpha_{0\infty 2}} \left(1 + \sum_{n=1}^{\infty} C_{n2}\, z^{-n}\right) \quad \text{as } z \to \infty,$$

which are generally asymptotic and divergent, become convergent series.

Besides the proof above, it is obvious that a series $\sum_{n=0}^{\infty} a_n$, where the a_n are exponentially decreasing, is finite.

Variable Asymptotics

Equation (1.2.54) is linear and of second order. Therefore, so is (1.2.69). Because of this linearity, the general solution of the difference equation (1.2.69) is given by

$$a_n^{(g)} = L_1\, a_{n1} + L_2\, a_{n2}, \quad n = 0, 1, 2, \ldots,$$

whereby a_{n1} and a_{n2} are any two linearly independent particular solutions of (1.2.69) for which it holds that

$$L_1\, a_{n1} + L_2\, a_{n2} = 0 \tag{1.2.105}$$

only for $L_1 = L_2 = 0$ and where $L_1 = L_1(\underline{c})$ and $L_2 = L_2(\underline{c})$ are any complex- or real-valued constants in n that may be dependent, however, on the parameters (denoted $\underline{c}$) of the differential equation (1.2.54), (1.2.55) (but not on n).

From the asymptotic behaviours (1.2.88) it is seen that $a_n^{(1)}$ and $a_n^{(2)}$ form a fundamental system of solutions of the difference equation (1.2.69). Then, as a result of the foregoing considerations, it has been shown that the eigenvalue condition of the singular boundary eigenvalue problem of the single confluent case (1.2.54), (1.2.55) of the Gauss differential equation (1.2.2), (1.2.3) is

$$L_1(\underline{c}) = 0 \qquad (1.2.106)$$

in (1.2.105) and the eigensolutions are given by (1.2.56), (1.2.58), (1.2.60), (1.2.61), (1.2.68), (1.2.69).

Thus, the singular boundary eigenvalue problem of the single confluent case (1.2.54), (1.2.55) of the Gauss differential equation (1.2.2), (1.2.3) is solved exactly. The Jaffé approach to the singular boundary eigenvalue problem, established above, seems a lot more complicated than the traditional approach, given before. However, the advantage of this approach is its generalisability. From a mathematical point of view, there is no limit on the order of the difference equation coming out of a Jaffé ansatz for the central two-point connection problem to be dealt with along the lines above.

Eigenvalue Condition

According to (1.2.106), the eigenvalue condition of the singular boundary eigenvalue problem is

$$L_1(\underline{c}) = 0.$$

The practical calculation is easy. For arbitrary $\underline{c}$ we need to calculate a value a_{N_0} for a sufficiently large index $n = N_0$. This is possible in a recursive way by means of the linear second-order difference equation (1.2.69), used as a three-term recurrence relation. Hereby, 'sufficiently large' is dependent on the required accuracy of the eigenvalues and in this context it means that the exponentially increasing fundamental solution a_{n1} for $n \to \infty$ is fully developed. Then, we equate $a_{N_0}(\underline{c})$ to zero (or, alternatively, a change of sign) in dependence on $\underline{c}$. This eventually determines the value of the eigenvalue parameter(s) that solve(s) the singular boundary eigenvalue problem.

1.2.5 Example: Laguerre Polynomials

It was pointed out at the beginning that there are special functions whose treatment has been known for a long time; these are called classical special functions and the method of their calculation is presented again here, for reasons of being able to test the new method displayed above. In this way it is possible to draw a comparison with problems whose results are known. One such classical (i.e., well-known) problem is the quantum mechanical calculation of the energy levels (i.e., eigenvalues) of spherically symmetric atoms. Such atoms are described by a so-called Coulomb potential of the stationary three-dimensional Schrödinger equation. This equation can be separated into its spatial dimensions, resulting in the single confluent case (1.2.13) of the Gauss differential equation (1.2.1) for the radial component.

First this classical solution will be calculated and then the new method, presented above, will be carried out and its results compared with those of the classical calculation. It will turn out that the classical solution is somewhat simpler than the new method; however, while the classical solution focuses on specific properties of the problem at hand and can, therefore, no longer be applied to problems that deviate even slightly from it, the new method is applicable to much more general problems, in principle to all singular boundary eigenvalue problems of differential equations (1.2.2), (1.2.1). It is precisely this limited applicability of the classical method, called the *Sommerfeld polynomial method* – based on focusing on the specific properties of each of the classical problems, that makes a separate treatment necessary, the characteristic property of which is taking care of the general aspects.

Classical Approach

The quantum mechanical treatment of the Schrödinger equation, the potential of which is the Coulomb potential (a spherical one, with first-order pole at its centre; cf. Figure 1.3), results in an algebraic boundary eigenvalue condition. This is shown in the following.

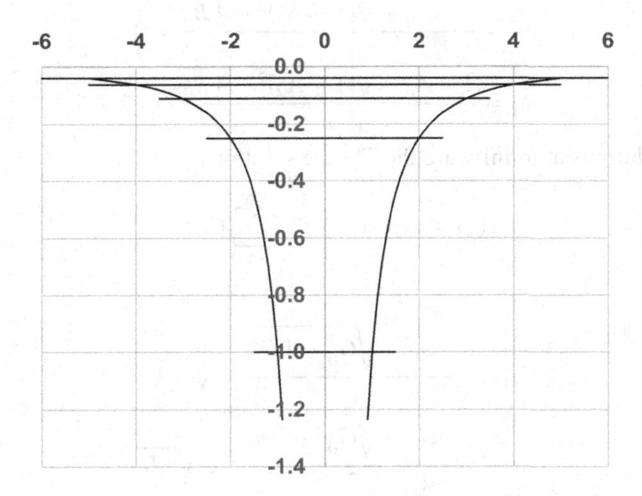

Figure 1.3 Quantum Coulomb problem.

The differential equation of the radial part is the single confluent case of the Gauss equation

$$\frac{d^2y}{dz^2} + P(z)\,\frac{dy}{dz} + Q(z)\,y = 0,\ z \in \mathbb{R}_0^+, \tag{1.2.107}$$

the coefficients of which are

$$P(z) = \frac{A_0}{z} + G_0 = \frac{2}{z},$$
$$Q(z) = \frac{B_0}{z^2} + \frac{C}{z} + D_0 = \frac{2}{z} + D_0, \tag{1.2.108}$$

whereby the coefficients are given by

$$G_0 = 0,$$
$$A_0 = 2,$$
$$B_0 = 0,$$
$$C = 2$$

and D_0 is the energy parameter. The local solutions at the two singularities are determined by the characteristic exponents. These are calculated in the following, starting with the singularity at the origin $z = 0$. The particular local solutions are Frobenius solutions

$$y(z) = z^{\alpha_0} \sum_{n=0}^{\infty} a_n\, z^n,$$

whereby the two characteristic exponents are given by

$$\alpha_{01} = \frac{1 - A_0 + \sqrt{(1 - A_0)^2 - 4\, B_0}}{2} = 0,$$
$$\alpha_{02} = \frac{1 - A_0 - \sqrt{(1 - A_0)^2 - 4\, B_0}}{2} = -1.$$

The local solutions at infinity are the Thomé solutions

$$y(z) = \exp(\alpha_1\, z)\, z^{\alpha_\infty} \sum_{n=0}^{\infty} C_n\, z^{-n}$$

with

$$\alpha_{11} = \frac{-G_0 + \sqrt{G_0^2 - 4\, D_0}}{2} = \sqrt{-D_0},$$
$$\alpha_{12} = \frac{-G_0 - \sqrt{G_0^2 - 4\, D_0}}{2} = -\sqrt{-D_0}$$

and

$$\alpha_{\infty 1} = -\frac{A_0\, \alpha_{11} + C}{G_0 + 2\, \alpha_{11}} = -1 - \frac{1}{\sqrt{-D_0}},$$
$$\alpha_{\infty 2} = -\frac{A_0\, \alpha_{11} + C}{G_0 + 2\, \alpha_{11}} = -1 + \frac{1}{\sqrt{-D_0}}.$$

The singular boundary eigenvalue problem of the physical problem is formulated as follows. Look for those values of D_0 that yield solutions of equation (1.2.107) that behave holomorphically at the origin $z = 0$ and, simultaneously, behave like $y(z) \sim \exp\left(-\sqrt{-D_0}\, z\right)$ as $z \to \infty$, radially, on the positive real axis.

The classical ansatz in order to solve this singular boundary eigenvalue problem is (cf. Schubert and Weber, 1980, p. 160)

$$y(z) = \exp(\alpha_{12}\, z)\, z^{\alpha_{01}} \sum_{n=0}^{\infty} a_n z^n = \exp\left(-\sqrt{-D_0}\, z\right) \sum_{n=0}^{\infty} a_n z^n, \tag{1.2.109}$$

resulting in a linear first-order difference equation

$$(n+1)(n+2)\, a_{n+1} - \left[n + 1 - \frac{1}{\sqrt{-D_0}}\right] a_n = 0, \quad n = 0, 1, 2, 3, \ldots, \tag{1.2.110}$$

that may be converted into a two-term recurrence relation by fixing $a_0 = 1$, obtaining

$$a_{n+1} = f(n, D_0)\, a_n = \frac{n + 1 - \frac{1}{\sqrt{-D_0}}}{(n+1)(n+2)}\, a_n, \quad n = 0, 1, 2, 3, \ldots.$$

The eigenvalue condition of the problem according to the Sommerfeld polynomial method is

$$f(n, D_0) = 0 \rightsquigarrow D_{0n} = -\frac{1}{(n+1)^2}, \quad n = 0, 1, 2, 3, \ldots,$$

which is actually just the condition for the difference equation (1.2.110) to admit trivial solutions. All the non-trivial particular solutions of the difference equation lead to an exponentially increasing behaviour of the solution $y(z)$ of the differential equation at infinity, as may be seen from the characteristic exponent α_{11}. In this case

$$\sum_{n=0}^{\infty} a_n z^n \sim \exp\left(2\sqrt{-D_0}\, z\right) \quad \text{as } z \to \infty$$

in (1.2.109).

The result are the eigenvalues

$$\begin{aligned} D_{00} &= -1, \\ D_{01} &= -\frac{1}{4}, \\ D_{02} &= -\frac{1}{9}, \\ &\ldots \\ D_{0n} &= -\frac{1}{n^2}, \\ &\ldots \end{aligned}$$

As may be seen, the quantum mechanical treatment of the Coulomb potential yields energy levels that cumulate at $D_0 = 0$ with

$$D_{0n} = -\frac{1}{n^2}, \quad n = 1, 2, 3, \ldots. \tag{1.2.111}$$

The corresponding eigensolutions of the underlying equation (1.2.107), (1.2.108) are given by

$$y_n(z) = \exp\left(-\sqrt{-D_0}\, z\right) L_n^{(1)}(z) = \exp\left(-\frac{z}{n+1}\right) L_n^{(1)}\left(\frac{2\,z}{n+1}\right), \quad n = 0, 1, 2, 3, \ldots,$$

where the $L_n^{(1)}(z)$ are polynomials. These polynomials are called *Laguerre polynomials* (see Figure 1.4), bearing the name of **Edmund Nicolas Laguerre** (1834–1886), a French mathematician of the nineteenth century, who discovered them in 1879. The explicit form of the Laguerre polynomials is

$$\begin{aligned}
L_0^{(1)}(z) &= 1, \\
L_1^{(1)}(z) &= 1 - z, \\
L_2^{(1)}(z) &= 1 - 2\,z + \frac{1}{2}\,z^2, \\
L_3^{(1)}(z) &= 1 - 3\,z + \frac{3}{2}\,z^2 - \frac{1}{6}\,z^3, \\
L_4^{(1)}(z) &= 1 - 4\,z + 3\,z^2 - \frac{2}{3}\,z^3 + \frac{1}{24}\,z^4, \\
&\ldots
\end{aligned}$$

or

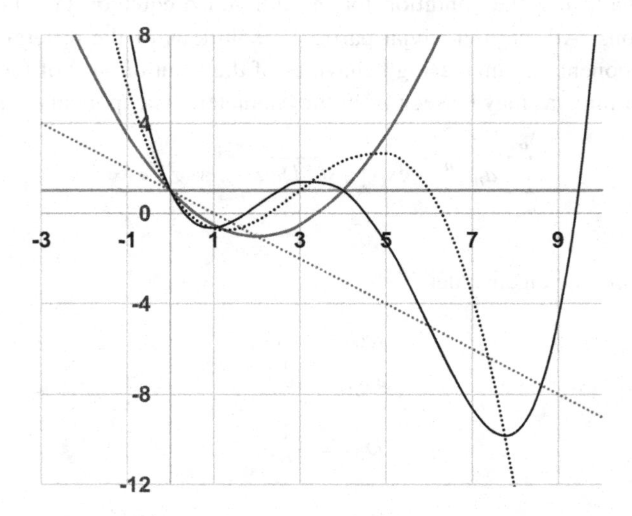

Figure 1.4 Laguerre polynomials.

$$L_n^{(1)}(z) = \sum_{k=0}^{n} \binom{n}{k} \frac{(-1)^k}{k!}\, z^k, \quad n = 0, 1, 2, 3, \ldots.$$

As a final result we give the eigensolutions belonging to the eigenvalues, the five lowest of which are given by (see Figure 1.5)

$$
\begin{aligned}
y_0(z) &= \exp\left(-\sqrt{-D_{00}}\, z\right) L_0^{(1)}(z) = \exp(-z),\\
y_1(z) &= \exp\left(-\sqrt{-D_{01}}\, z\right) L_1^{(1)}(z) = \exp\left(-\frac{1}{2}\, z\right)(1 - z),\\
y_2(z) &= \exp\left(-\sqrt{-D_{02}}\, z\right) L_2^{(1)}(z) = \exp\left(-\frac{1}{3}\, z\right)\left(1 - 2\, z + \frac{1}{2}\, z^2\right),\\
y_3(z) &= \exp\left(-\sqrt{-D_{03}}\, z\right) L_3^{(1)}(z) = \exp\left(-\frac{1}{4}\, z\right)\left(1 - 3\, z + \frac{3}{2}\, z^2 - \frac{1}{6}\, z^3\right),\\
y_4(z) &= \exp\left(-\sqrt{-D_{04}}\, z\right) L_4^{(1)}(z) = \exp\left(-\frac{1}{5}\, z\right)\left(1 - 4\, z + 3\, z^2 - \frac{2}{3}\, z^3 + \frac{1}{24}\, z^4\right),\\
&\ldots
\end{aligned}
$$

As one may see, the exponential term dominates the expression of the eigensolutions above.

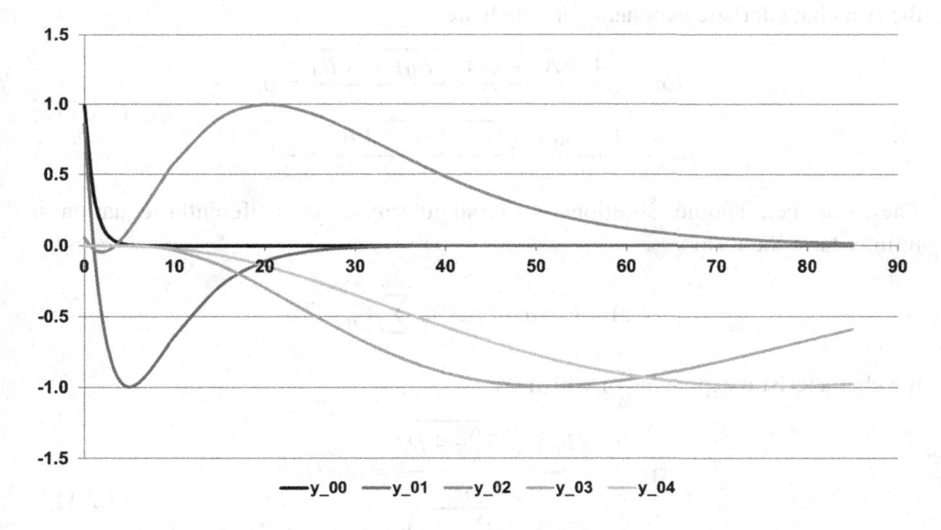

Figure 1.5 Radial eigensolutions of the quantum Coulomb problem.

Jaffé Method

The starting point is the single confluent case of the Gauss differential equation

$$
\frac{d^2 y}{dz^2} + P(z)\,\frac{dy}{dz} + Q(z)\, y = 0
$$

with

$$P(z) = \frac{A_0}{z} + G_0 = \frac{2}{z},$$
$$Q(z) = \frac{B_0}{z^2} + \frac{C}{z} + D_0 = \frac{2}{z} + D_0. \tag{1.2.112}$$

Thus

$$G_0 = 0,$$
$$A_0 = 2,$$
$$B_0 = 0,$$
$$C = 2.$$

The local (i.e., Frobenius) solutions at the singularity of the differential equation at the origin $z = 0$ have the form

$$y(z) = z^{\alpha_0} \sum_{n=0}^{\infty} a_n z^n,$$

the two characteristic exponents of which are

$$l\alpha_{01} = \frac{1 - A_0 + \sqrt{(1 - A_0)^2 - 4\,B_0}}{2} = 0,$$
$$\alpha_{02} = \frac{1 - A_0 - \sqrt{(1 - A_0)^2 - 4\,B_0}}{2} = -1. \tag{1.2.113}$$

The local (i.e., Thomé) solutions at the singularity of the differential equation at infinity have the form

$$y(z) = \exp(\alpha_1 z)\, z^{\alpha_{0\infty}} \sum_{n=0}^{\infty} C_n z^{-n},$$

the characteristic exponents of which are

$$\alpha_{11} = \frac{-G_0 + \sqrt{G_0^2 - 4\,D_0}}{2} = \sqrt{-D_0},$$
$$\alpha_{12} = \frac{-G_0 - \sqrt{G_0^2 - 4\,D_0}}{2} = -\sqrt{-D_0} \tag{1.2.114}$$

and

$$\alpha_{0\infty 1} = -\frac{A_0\,\alpha_{11} + C}{G_0 + 2\,\alpha_{11}} = -1 - \frac{1}{\sqrt{-D_0}},$$
$$\alpha_{0\infty 2} = -\frac{A_0\,\alpha_{12} + C}{G_0 + 2\,\alpha_{12}} = -1 + \frac{1}{\sqrt{-D_0}}. \tag{1.2.115}$$

For physical reasons, the boundary eigenvalue problem consists of connecting the Frobenius solution at the origin, related to α_{01}, to the Thomé solution, related to α_{12}.

This determines the Jaffé ansatz that consists of two parts, the first of which is given by (Jaffé ansatz I) [cf. (1.2.56)]

$$y(z) = \exp(\alpha_{12}\, z)\, z^{\alpha_{01}}\, w(z) = \exp\left(-\sqrt{-D_0}\, z\right)\, w(z), \tag{1.2.116}$$

resulting in a differential equation, having the coefficients

$$\tilde{P}(z) = \tilde{g}_0 + \frac{\tilde{g}_{-10}}{z},$$
$$\tilde{Q}(z) = \frac{\tilde{d}_{-10}}{z},$$

with

$$\tilde{g}_0 = G_0 + 2\,\alpha_{12} = -2\,\sqrt{-D_0},$$
$$\tilde{g}_{-10} = A_0 + 2\,\alpha_{01} = 2,$$
$$\tilde{d}_{-10} = 2\,\alpha_{12}\,\alpha_{01} + A_0\,\alpha_{12} + G_0\,\alpha_{01} + C = -2\,\sqrt{-D_0} + 2.$$

This determines the exponent β [cf. (1.2.63)] to be

$$\rightsquigarrow \beta = \frac{\tilde{d}_{-10}}{\tilde{g}_0} = 1 - \frac{1}{\sqrt{-D_0}}.$$

A Jaffé transformation (1.2.58)

$$x = \frac{z}{z+1} \tag{1.2.117}$$

changes the coefficients (1.2.112) of the differential equation (1.2.5) according to

$$\tilde{P}(x) = \tilde{g}_0 - \tilde{g}_{-10} + \frac{\tilde{g}_{-10}}{x},$$
$$\tilde{Q}(x) = -\tilde{d}_{-10} + \frac{\tilde{d}_{-10}}{x}.$$

The Jaffé transformation (1.2.117) results in a differential equation, the coefficients of which are given by

$$\tilde{\tilde{P}}(x) = \frac{\tilde{\tilde{g}}_{-2+1}}{(x-1)^2} + \frac{\tilde{\tilde{g}}_{-1+1}}{x-1} + \frac{\tilde{\tilde{g}}_{-10}}{x},$$
$$\tilde{\tilde{Q}}(x) = \frac{\tilde{\tilde{d}}_{-10}}{x} + \frac{\tilde{\tilde{d}}_{-3+1}}{(x-1)^3} + \frac{\tilde{\tilde{d}}_{-2+1}}{(x-1)^2} + \frac{\tilde{\tilde{d}}_{-1+1}}{x-1},$$

with

$$\begin{aligned}
\tilde{\tilde{g}}_{-2+1} &= \tilde{g}_0 = -2\sqrt{-D_0},\\
\tilde{\tilde{g}}_{-1+1} &= 2 - \tilde{g}_{-10} = 0,\\
\tilde{\tilde{g}}_{-10} &= \tilde{g}_{-10} = 2,\\
\tilde{\tilde{d}}_{-3+1} &= -\tilde{d}_{-10} = 2\sqrt{-D_0} - 2,\\
\tilde{\tilde{d}}_{-2+1} &= \tilde{d}_{-10} = -2\sqrt{-D_0} + 2,\\
\tilde{\tilde{d}}_{-1+1} &= -\tilde{d}_{-10} = 2\sqrt{-D_0} - 2,\\
\tilde{\tilde{d}}_{-10} &= \tilde{d}_{-10} = -2\sqrt{-D_0} + 2.
\end{aligned}$$

The second part of the Jaffé ansatz (Jaffé ansatz II; cf. (1.2.60), (1.2.61)) is

$$w[x(z)] = z^{\alpha_{0\infty 2}-\alpha_{01}}\, v(x) = z^{\alpha_{0\infty 2}}\, v(x) \tag{1.2.118}$$

and results in a differential equation (1.2.107), the coefficients of which are given by

$$\begin{aligned}
\bar{P}(x) &= \frac{\bar{g}_{-2+1}}{(x-1)^2} + \frac{\bar{g}_{-1+1}}{x-1} + \frac{\bar{g}_{-10}}{x},\\
\bar{Q}(x) &= \frac{\bar{d}_{-3+1}}{(x-1)^3} + \frac{\bar{d}_{-2+1}}{(x-1)^2} + \frac{\bar{d}_{-1-1}}{x+1} + \frac{\bar{d}_{-10}}{x},
\end{aligned}$$

with

$$\begin{aligned}
\bar{g}_{-2+1} &= \tilde{g}_0 = -2\sqrt{-D_0},\\
\bar{g}_{-1+1} &= 2 - \tilde{g}_{-10} = 0,\\
\bar{g}_{-10} &= \tilde{g}_{-10} = 2,\\
\bar{d}_{-3+1} &= \tilde{g}_0\,\beta - \tilde{d}_{-10} = 0,\\
\bar{d}_{-2+1} &= \tilde{d}_{-10} + (2 - \tilde{g}_{-10})\,\beta + \beta\,(\beta - 1) = 2 - 2\sqrt{-D_0} - \frac{1}{\sqrt{-D_0}} + \frac{1}{\sqrt{-D_0}^{\,2}},\\
\bar{d}_{-1+1} &= \tilde{g}_{-10}\,\beta - \tilde{d}_{-10} = 2\sqrt{-D_0} - \frac{2}{\sqrt{-D_0}},\\
\bar{d}_{-10} &= -(\tilde{g}_{-10}\,\beta - \tilde{d}_{-10}) = -\bar{d}_{-1+1} = \frac{2}{\sqrt{-D_0}} - 2\sqrt{-D_0}.
\end{aligned}$$

A Taylor expansion

$$v(x) = \sum_{n=0}^{\infty} a_n\, x^n \tag{1.2.119}$$

of the particular solution (that is convergent in the unit disc of the complex number plane) eventually results in the linear second-order difference equation for the coefficients a_n:

$$\begin{aligned}
&(n+1)\,[n+\Gamma_0]\,a_{n+1} + [-2\,(n-1)\,n + n\,\Gamma_1 + \Delta_0]\,a_n\\
&\quad + [(n-2)\,(n-1) + (n-1)\,\Gamma_2 + \Delta_1]\,a_{n-1} = 0, \quad n = 1,2,3,4,\ldots,
\end{aligned} \tag{1.2.120}$$

and an initial condition

$$\Gamma_0\, a_1 + \Delta_0\, a_0 = 0,$$

the coefficients of which are given by

$$\begin{aligned}
\Gamma_2 &= \bar{g}_{-1+1} + \bar{g}_{-10} = 2,\\
\Gamma_1 &= -(\bar{g}_{-2+1} + \bar{g}_{-1+1} + 2\,\bar{g}_{-10}) = 2\sqrt{-D_0} - 4,\\
\Gamma_0 &= \bar{g}_{-10} = 2,\\
\Delta_1 &= \bar{d}_{-2+1} + \bar{d}_{-1+1} = 2 + \frac{1}{\sqrt{-D_0}^{\,2}} - \frac{3}{\sqrt{-D_0}},\\
\Delta_0 &= -\bar{d}_{-1+1} = -2\sqrt{-D_0} + \frac{2}{\sqrt{-D_0}}.
\end{aligned} \tag{1.2.121}$$

The difference equation (1.2.120) may be put into a three-term recurrence relation

$$\begin{aligned}
&a_0 \text{ arbitrary},\\
&[3pt]\Gamma_0\, a_1 + \Delta_0\, a_0 = 0,\\
&\left(1 + \frac{\alpha_1}{n} + \frac{\beta_1}{n^2}\right) a_{n+1} + \left(-2 + \frac{\alpha_0}{n} + \frac{\beta_0}{n^2}\right) a_n\\
&\quad + \left(1 + \frac{\alpha_{-1}}{n} + \frac{\beta_{-1}}{n^2}\right) a_{n-1} = 0, \quad n \geq 1,
\end{aligned} \tag{1.2.122}$$

where the coefficients of the difference equation are

$$\begin{aligned}
\alpha_1 &= \Gamma_0 + 1, & \beta_1 &= \Gamma_0,\\
\alpha_0 &= \Gamma_1 + 2, & \beta_0 &= \Delta_0,\\
\alpha_{-1} &= \Gamma_2 - 3, & \beta_{-1} &= \Delta_1 - \Gamma_2 + 2,
\end{aligned} \tag{1.2.123}$$

which finalises the Jaffé method.

As can be seen from the Jaffé method, the classical approach, presented above, is only possible because $\alpha_{0\infty1}$ in (1.2.113) is an integer multiple of α_0 [with $D_0 = D_{0n}$ from (1.2.111)] in (1.2.114): $\alpha_{\infty1} = k\,\alpha_0$ with $k \in \mathbb{Z}$. The Jaffé method, and this is its strength, can be applied even if this condition is no longer fulfilled.

Crucial for the applicability of the classical method is that $\alpha_{0\infty2}$ in (1.2.116) [for the eigenvalues (1.2.111)] is integer. This is used implicitly in the classical method. In the Jaffé method, however, this fact becomes obvious. Therefore, it can also be understood as a condition, an eigenvalue condition:

$$\alpha_{0\infty2} = n, \quad n = 0, 1, 2, 3, \ldots .$$

This results in two conditions [cf. (1.2.115)]:

(1)

$$-1 + \frac{1}{\sqrt{-D_{0n}}} = n, \quad n = 0, 1, 2, 3, \ldots,$$

resulting in

$$D_{0n} = -\frac{1}{(n+1)^2}, \quad n = 0, 1, 2, 3, \ldots .$$

(2)

$$-1 - \frac{1}{\sqrt{-D_{0n}}} = n, \quad n = 0, 1, 2, 3, \ldots,$$

resulting in

$$D_{0n} = -\frac{1}{(n+1)^2}, \quad n = 0, 1, 2, 3, \ldots .$$

The crucial thing about the difference equation (1.2.121) is that all of its partial solutions $\{a_n\}$ oscillate when $n \longrightarrow \infty$. This will be considered in more detail below. The Birkhoff set (cf. §1.2.4) of the difference equation (1.2.120) is given by

$$\begin{aligned} a_{n1}(n) &= \exp\left(\gamma_1 n^{\frac{1}{2}}\right) n^{r_1} \left[1 + \frac{C_{11}}{n^{\frac{1}{2}}} + \frac{C_{12}}{n^{\frac{2}{2}}} + \cdots\right], \\ a_{n2}(n) &= \exp\left(\gamma_2 n^{\frac{1}{2}}\right) n^{r_2} \left[1 + \frac{C_{21}}{n^{\frac{1}{2}}} + \frac{C_{22}}{n^{\frac{2}{2}}} + \cdots\right] \end{aligned} \tag{1.2.124}$$

as $n \longrightarrow \infty$, with

$$\begin{aligned} \gamma_1 &= 2\sqrt{-\sum_{i=-1}^{i=+1} \alpha_i} \neq 0, \\ \gamma_2 &= -\gamma_1, \\ r_1 &= r_2 = \frac{1}{4}(2\,\alpha_1 - 2\,\alpha_{-1} - 1) = \frac{1}{4}(2\,\Gamma_0 + 2 - 2\,\Gamma_2 - 6 - 1) = -\frac{5}{4}. \end{aligned} \tag{1.2.125}$$

The relationship with the coefficients of the differential equation was given in the central proof of §1.2.4:

$$\begin{aligned} \frac{\gamma_1^2}{4} &= -\sum_{i=-1}^{+1} \alpha_i = -\sum_{i=0}^{2} \Gamma_i \\ &= -\bar{g}_{-2+1} = -(G_0 + 2\,\alpha_{11}) \\ &= -\sqrt{G_0^2 - 4\,D_0} = -4\sqrt{-D_0}, \end{aligned}$$

$$\gamma_1 = 4\sqrt{-\sqrt{-D_0}} = 4\sqrt[4]{|D_0|}\,\imath = \frac{4}{\sqrt[4]{n+1}}\,\imath, \quad n = 0, 1, 2, 3, \ldots .$$

This is seen in Figure 1.6.

However, this means that with

$$\gamma_1 = -\gamma_2 = \gamma$$

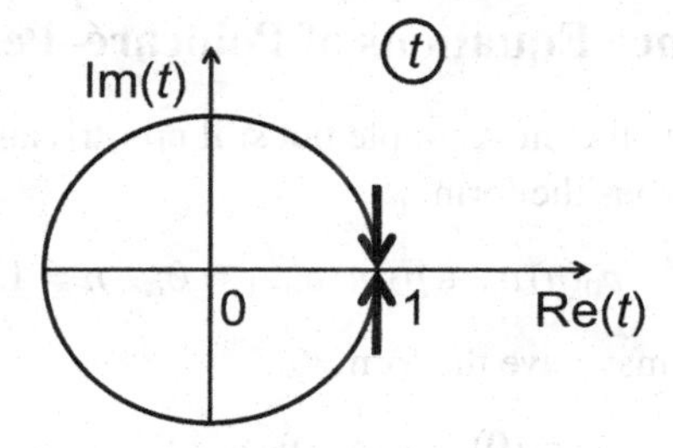

Figure 1.6 Solutions of the first-order characteristic equation.

and because of Euler's formula

$$\exp\left(\imath\gamma n^{\frac{1}{2}}\right) = \cos\left(\gamma n^{\frac{1}{2}}\right) + \imath \sin\left(\gamma n^{\frac{1}{2}}\right),$$
$$\exp\left(-\imath\gamma n^{\frac{1}{2}}\right) = \cos\left(\gamma n^{\frac{1}{2}}\right) - \imath \sin\left(\gamma n^{\frac{1}{2}}\right)$$

and thus

$$\exp\left(\imath\gamma n^{\frac{1}{2}}\right) + \exp\left(-\imath\gamma n^{\frac{1}{2}}\right)$$
$$= \cos\left(\gamma n^{\frac{1}{2}}\right) + \imath \sin\left(\gamma n^{\frac{1}{2}}\right) + \cos\left(\gamma n^{\frac{1}{2}}\right) - \imath \sin\left(\gamma n^{\frac{1}{2}}\right) = 2\cos\left(\gamma n^{\frac{1}{2}}\right),$$

an oscillating behaviour.

Since the difference equation has only real-valued coefficients, it also has real-valued solutions. Therefore, the imaginary parts in the partial solutions of the difference equation must cancel each other out. This, in turn, means that for the general solution

$$a_n^{(g)} = L_1\, a_n^{(1)} + L_2\, a_n^{(2)},$$

the two constants L_1 and L_2 must be equal: $L_1 = L_2 = \frac{1}{2}L$, meaning

$$a_n^{(g)} \sim L_1 \left[\exp\left(\imath\gamma n^{\frac{1}{2}}\right) + \exp\left(-\imath\gamma n^{\frac{1}{2}}\right)\right] n^r$$
$$= L_1 \left[\cos\left(\gamma n^{\frac{1}{2}}\right) + \imath \sin\left(\gamma n^{\frac{1}{2}}\right) + \cos\left(\gamma n^{\frac{1}{2}}\right) - \imath \sin\left(\gamma n^{\frac{1}{2}}\right)\right] n^r$$
$$= L \cos\left(\gamma n^{\frac{1}{2}}\right) n^r \text{ as } n \to \infty$$

and this is an oscillating behaviour. Because $r < 0$, this is asymptotically decreasing.

As a result, we keep in mind that in this example, in the series (1.2.119), the boundary eigenvalue condition is not hidden in the Taylor series of the solution set, but in the exponent $\alpha_{0\infty 2}$ in (1.2.115). Also, the series in (1.2.119) converges on the entire interval $0 \leq x \leq 1$ over which the singular boundary eigenvalue problem converges, because for $x = 1$ the following holds:

$$v(1) = \sum_{n=0}^{\infty} a_n \sim \sum_{n=0}^{\infty} \cos(\gamma n)\, n^r < \infty,$$

since $r < 0$, as it is an alternating series.

1.3 Difference Equations of Poincaré–Perron Type

For the sake of simplicity, the most simple but still non-trivial difference equation is of second order and thus has the form

$$p_1(n)\, a_{n+1} + p_0(n)\, a_n + p_2(n)\, a_{n-1} = 0, \quad n = 1, 2, 3, \ldots, \tag{1.3.1}$$

and the initial condition may have the form

$$p_1(0)\, a_1 + p_2(0)\, a_0 = 0. \tag{1.3.2}$$

If, in this case, a_0 is fixed,[6] then (1.3.2) yields the start of a recursive calculation of the a_n for as many terms as are needed by means of the three-term recurrence relation (1.3.1).

Since the treatment of linear difference equations by Henri Poincaré and Oskar Perron (Poincaré, 1885, 1886; Perron, 1909, 1910, 1911), it has been customary not to treat the solutions as sequences of numbers

$$a_0,\ a_1,\ a_2,\ a_3,\ \ldots,\ a_n,\ \ldots$$

but as the ratio of two consecutive terms

$$t_n = \frac{a_{n+1}}{a_n}. \tag{1.3.3}$$

For our purposes here, the asymptotic behaviour of this ratio for large values of the index n,

$$\lim_{n\to\infty} t_n,$$

is of particular interest, since this behaviour determines the solution of the differential equation in the neighbourhood of the singularity of the differential equation at z_0 [cf. the ansatz (1.2.119)], which generally represents the behaviour of the function at the still undetermined end of the relevant interval.

This asymptotic behaviour of two consecutive terms of the differential equation is determined by an algebraic equation whose order is as large as that of the underlying differential equation. This equation is called a *characteristic equation*; it is the most important criterion for its possible particular solutions.

If a linear, ordinary, homogeneous difference equation has only coefficients that converge to finite values for large indices n, then it is called a *difference equation of Poincaré–Perron type*. For the difference equation (1.3.1), this means

$$\lim_{n\to\infty} p_i(n) = \text{const.} < \infty,\ i = 0, 1, 2.$$

If the characteristic equation of a linear, ordinary, homogeneous difference equation has only simple roots, such a difference equation is called *regular*, otherwise *irregular*.

Linear ordinary difference equations of the first order

$$p_1(n)\, a_n + p_2(n)\, a_{n-1} = 0, \quad n = 1, 2, 3, \ldots, \tag{1.3.4}$$

[6] This determination gives a normalisation of the function (1.2.119).

are to be regarded as trivial in the sense of the above explanations. The simplest non-trivial example is the second-order difference equation in which a new term is the sum of the two foregoing ones:

$$a_{n+1} = a_n + a_{n-1}, \quad n = 1, 2, 3, \ldots, \tag{1.3.5}$$

which may be written in the form

$$a_{n+1} - a_n - a_{n-1} = 0, \quad n = 1, 2, 3, \ldots. \tag{1.3.6}$$

Here, all the coefficients $p_0(n)$, $p_1(n)$, $p_2(n)$ are independent of n, thus are constants: $p_0(n) = -p_1(n) = -p_2(n) = 1$. This linear, second-order difference equation is the most instructive one with respect to the principle that is to be shown in this book. Therefore, it is considered in some detail below.

1.3.1 Fibonacci Difference Equation

Fibonacci Sequence

It is easily seen that the initial condition of fixing two consecutive terms, a_0 and a_1, say, transforms the linear, homogeneous, second-order difference equation (1.3.5) or (1.3.6) into a linear, three-term recurrence relation, allowing the successive calculation of as many terms a_n as are needed.

Starting with $a_0 = 0$ and $a_1 = 1$ in (1.3.5) the recursively calculated sequence of numbers

$$a_0 = 0,\ a_1 = 1,\ a_2 = 1,\ a_3 = 2,\ a_4 = 3,\ a_5 =, 5,\ a_6 = 8,\ a_7 = 13,\ a_8 = 21,\ a_9 = 34, \ldots \tag{1.3.7}$$

is the well-known *Fibonacci sequence*. Therefore, I refer to (1.3.6) as the *Fibonacci difference equation.*

The characteristic equation of the difference equation (1.3.6) is given by

$$t^2 - t - 1 = 0, \tag{1.3.8}$$

the two solutions of which are

$$t_1 = \frac{1 + \sqrt{5}}{2} \approx 1.61803, \tag{1.3.9}$$

$$t_2 = \frac{1 - \sqrt{5}}{2} = -\frac{1}{t_1} = 1 - t_1 \approx -0.61803. \tag{1.3.10}$$

The fact that equation (1.3.6) is of second order generates two solutions t_1 and t_2 of its characteristic equation (1.3.8). This is of extraordinary importance, since it shows that there are two fundamental particular solutions of the difference equation (1.3.6) and the general solution $a_n^{(g)}$ of the linear difference equation (1.3.6) has – because of its linearity – the form

$$a_n^{(g)} = C_1\, a_n^{(1)} + C_2\, a_n^{(2)}, \tag{1.3.11}$$

where $a_n^{(1)}$ and $a_n^{(2)}$ are two linearly independent particular solutions of the difference equation (1.3.6), meaning that one solution is not a multiple of the other. Moreover, C_1 and C_2 do not depend on n, $a_n^{(1)}$ is related to (1.3.9) and $a_n^{(2)}$ is related to (1.3.10), meaning that the ratio of two subsequent terms of the particular solution $a_n^{(1)}$ of (1.3.6) tends to t_1 as $n \to \infty$ and that the ratio of two subequent terms of $a_n^{(2)}$ tends to t_2 as $n \to \infty$.

It has been known for more than 300 years that

$$a_n^{(1)} = t_1^n \tag{1.3.12}$$

and

$$a_n^{(2)} = t_2^n. \tag{1.3.13}$$

This fact is expressed in the famous *de Moivre–Binet* formula, which, moreover, gives the constants C_1 and C_2 of (1.3.11):

$$a_n = \frac{1}{\sqrt{5}}\, t_1^n - \frac{1}{\sqrt{5}}\, t_2^n. \tag{1.3.14}$$

The French mathematician **Jacques Philippe Marie Binet** (1786–1856) published this formula in 1843, although it was already well known to the French mathematician **Abraham de Moivre** (1667–1754) and to the Swiss mathematicians **Daniel Bernoulli** (1700–1782) and **Leonhard Euler** (1707–1783). The de Moivre–Binet formula gives the exact term a_n in dependence on n for the Fibonacci sequence, i.e., for the initial condition $a_0 = 0$ and $a_1 = 1$. The formula is remarkable since it displays each term of an infinite sequence of integer numbers by a sum of two irrational ones.

It is clearly seen from the de Moivre–Binet formula (1.3.14) that the Fibonacci sequence (1.3.7) consists of two series, one of them increasing exponentially as $n \to \infty$, the other decreasing exponentially in this limit since

$$|t_1| > 1 \text{ and } |t_2| < 1 :$$

$$\begin{aligned} a_n &= \frac{1}{\sqrt{5}}\left[\exp(n \ln t_1) - (-1)^n \exp(-n \ln t_1)\right] \\ &= \begin{cases} \frac{2}{\sqrt{5}} \sinh(n \ln t_1) & \text{for } n \text{ even } (n = 0, 2, 4, 6, \ldots), \\ \frac{2}{\sqrt{5}} \cosh(n \ln t_1) & \text{for } n \text{ odd } (n = 1, 3, 5, 7, \ldots). \end{cases} \end{aligned}$$

This composition of the Fibonacci sequence is shown in Figure 1.7.

Thus

$$C_1 = \frac{1}{\sqrt{5}}, \quad a_n^{(1)} = \exp(n \ln t_1), \tag{1.3.15}$$

$$C_2 = -\frac{(-1)^n}{\sqrt{5}}, \quad a_n^{(2)} = \exp(-n \ln t_1). \tag{1.3.16}$$

It is quite understandable that the partial solution $a_n^{(2)}$ in (1.3.16) is often not recognised since the terms become exponentially small on increasing index n:

$$a_n^{(2)} = \exp(-n \ln t_1). \tag{1.3.17}$$

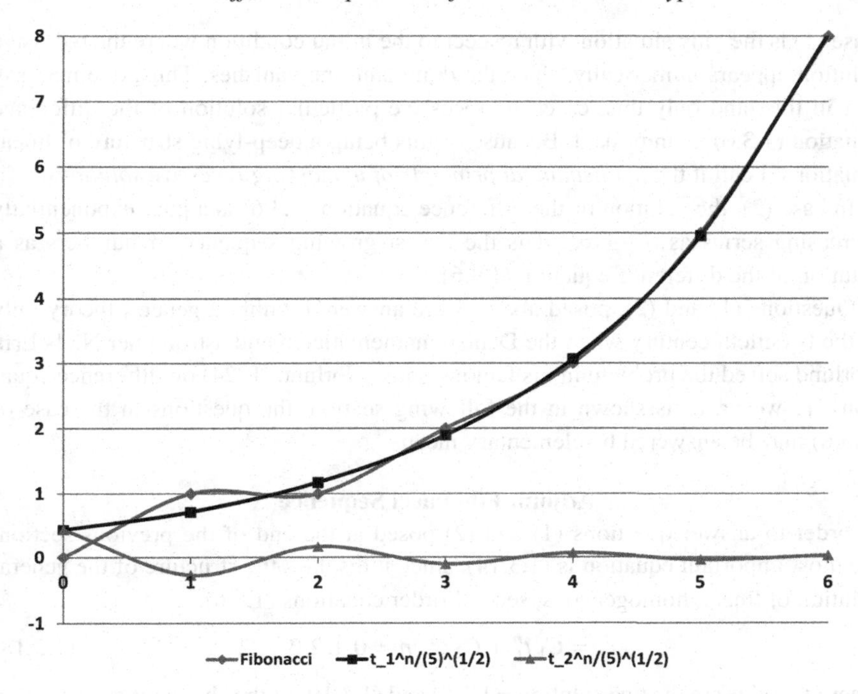

Figure 1.7 Composition of the Fibonacci sequence.

This is called the *recessive particular solution* of the difference equation (1.3.6), while the exponentially increasing particular solution

$$a_n^{(1)} = \exp(n \ln t_1)$$

is called the *dominant particular solution* of the difference equation (1.3.6). So, one can say that the recessive solution is masked by the dominant one. The masking of recessive particular solutions by dominant ones is a quite deep-lying phenomenon of linear equations with order higher than one.

The constants C_1 and C_2 of the general solution (1.3.11) of the difference equation (1.3.6) do not depend on n. These two quantities are exclusively dependent on the initial condition $\{a_0, a_1\}$ in a one-to-one relation, meaning that for each pair $\{a_0, a_1\}$ of initial conditions there is one and just one pair of constants $\{C_1, C_2\}$ and vice versa.

As the formula of de Moivre and Binet (1.3.14) shows, for $a_0 = 0$, $a_1 = 1$ the two constants C_1 and C_2 are equal to one another, having value $1/\sqrt{5}$:

$$a_0 = 0,\ a_1 = 1 \rightsquigarrow C_1 = -C_2 = \frac{1}{\sqrt{5}} \approx 0.44721.$$

There remain two interesting questions on this problem that are to be solved in the following:

(1) For which pair $\{a_0, a_1\}$ of initial conditions does it hold that $C_1 = 0$, $C_2 \neq 0$?
(2) For which pair $\{a_0, a_1\}$ of initial conditions does it hold that $C_1 \neq 0$, $C_2 = 0$?

Case (1) is the only situation with respect to the initial condition where the recessive solution appears numerically, since the dominant one vanishes. Thus, one may say that in this, and only this, case, the recessive particular solution of the difference equation (1.3.6) is unmasked. Because of this being a deep-lying structure of linear equations, I call it the *mathematical principle of unmasking recessive solutions*.

In case (2), the solution of the difference equation (1.3.6) is a pure exponentially increasing series as $n \to \infty$. It is the fastest growing sequence of numbers as a solution of the difference equation (1.3.6).

Questions (1) and (2), posed above, were answered within a general theory only in the twentieth century when the Danish mathematician and astronomer Niels Erik Nörlund solved the problem in his famous book (Nörlund, 1924) on difference equations. However, as is shown in the following section, the questions in the case of (1.3.6) may be answered by elementary means.

Adjoint Fibonacci Sequence

In order to answer questions (1) and (2) posed at the end of the previous section, the most important equation is (1.3.14), since it displays the structure of the general solution of linear, homogeneous, second-order equations (1.3.6):

$$a_n^{(g)} = C_1 t_1^n + C_2 t_2^n, \; n = 0, 1, 2, 3, \ldots, \tag{1.3.18}$$

where t_1 and t_2 are the two solutions (1.3.9) and (1.3.10) of the characteristic equation (1.3.8) of the difference equation (1.3.6). Writing explicitly the equations for a_0 and a_1 yields

$$n = 0: \quad a_0 = C_1 + C_2,$$
$$n = 1: \quad a_1 = C_1 t_1 + C_2 t_2.$$

Considering a_0 and a_1 as being given, this is a system of two linear equations of two unknown quantities C_1 and C_2 that may be solved by standard methods, yielding

$$C_1(a_0, a_1) = -\frac{a_1 - a_0 t_2}{t_2 - t_1} = -\frac{a_1}{t_2 - t_1} + \frac{a_0 t_2}{t_2 - t_1},$$
$$C_2(a_0, a_1) = \frac{a_1 - a_0 t_1}{t_2 - t_1} = \frac{a_1}{t_2 - t_1} - \frac{a_0 t_1}{t_2 - t_1},$$

thus linear equations in a_0 and a_1. For the sake of obtaining a one-dimensional problem, we put $a_1 = 1$ in the following (without losing generality), resulting in the solution of the posed problem:

$$(1)\; C_1 = 0 \overset{a_1=1}{\rightsquigarrow} a_0 = \frac{1}{t_2} = -t_1 \approx -1.61803$$
$$\rightsquigarrow C_2 = \frac{1}{t_2} = -t_1 \approx -1.61803, \tag{1.3.19}$$
$$(2)\; C_2 = 0 \overset{a_1=1}{\rightsquigarrow} a_0 = \frac{1}{t_1} = -t_2 \approx 0.61803$$
$$\rightsquigarrow C_1 = -\frac{1}{t_1} = t_2 \approx -0.61803. \tag{1.3.20}$$

In Figure 1.8 it is shown that the two constants C_1 and C_2 are dependent on a_0, while we take $a_1 = 1$. The most important pairs of values are given in the following table:

$a_0 = C_1 + C_2$	C_1	C_2
-2.00	$-\frac{1+2t_2}{t_2-t_1} \approx -0.1056$	$\frac{1+2t_1}{t_2-t_1} \approx -1.8944$
$-t_1 \approx -1.61803$	$0,00$	$-t_1$
-1.00	$-\frac{1+t_2}{t_2-t_1} \approx 0.1708$	$\frac{1+t_1}{t_2-t_1} \approx -1.1708$
$t_2 \approx -0.61803$	$-\frac{1-t_2\,t_1}{t_2-t_1} \approx 0.2764$	$\frac{1-t_2\,t_1}{t_2-t_1} \approx -0.8944$
0.00	$\frac{1}{\sqrt{5}} \approx 0.4472$	$= \frac{1}{\sqrt{5}} \approx -0.4472$
$-t_2 \approx 0.61803$	$-t_2$	0.00
1.00	$-\frac{1-t_2}{t_2-t_1} \approx 0.7236$	$\frac{1-t_1}{t_2-t_1} \approx 0.2764$
$t_1 \approx 1.61803$	$-\frac{1-t_1\,t_2}{t_2-t_1} \approx 0.8944$	$\frac{1-t_1^2}{t_2-t_1} \approx 0.7236$
2.00	1.00	1.00

Equation (1.3.19) answers the question as to the unmasking of the recessive solution (1.3.13) of (1.3.6). Taking the initial condition $a_0 = \frac{1}{t_2} \approx -1.61803$ and $a_1 = 1$ unmasks the recessive solution, the first few terms of which are given by

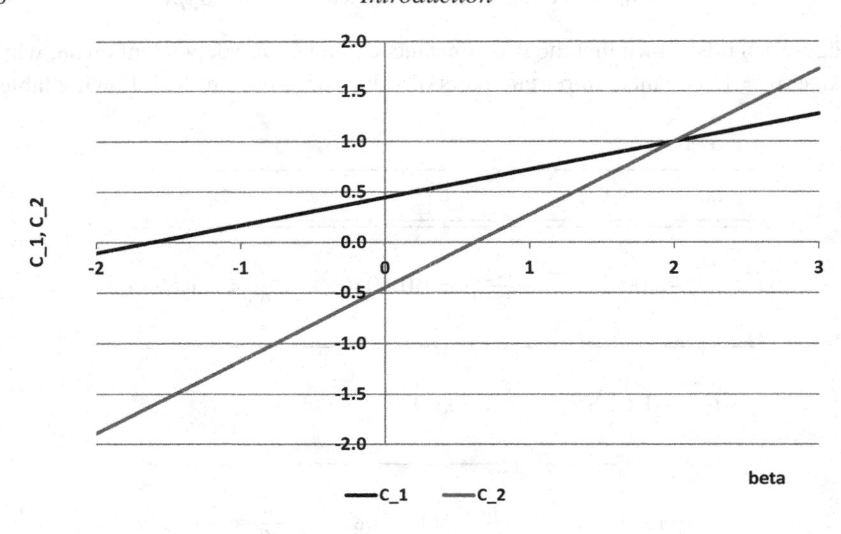

Figure 1.8 Coefficients of the general solution of the difference equation (1.3.6).

Number	Value
a_0	$\mathbf{-t_1 \approx -1.61803}$
a_1	$+\,\mathbf{1.00000}$
a_2	$t_2 \approx -0.61803$
a_3	$\approx +0.38197$
a_4	≈ -0.23607
a_5	$\approx +0.14590$
a_6	≈ -0.09017
a_7	$\approx +0.05573$
a_8	≈ -0.03444
a_9	$\approx +0.02129$
a_{10}	≈ -0.01316
a_{11}	$\approx +0.00813$
...	...

I refer to this sequence of numbers as the *adjoint Fibonacci sequence* for reasons which should be clear from the above execution. This behaviour is shown in Figure 1.9.

As can be seen, the adjoint Fibonacci sequence is an alternating one, as this is expected because of $t_2 < 0$. Its explicit calculation is given by

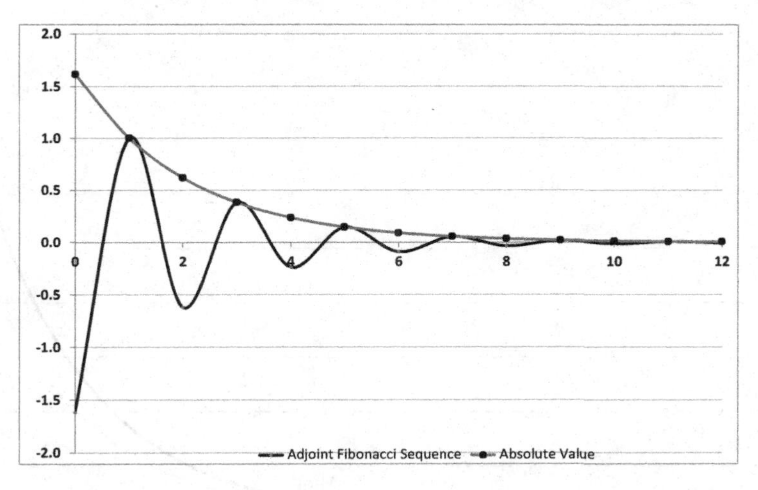

Figure 1.9 The adjoint Fibonacci sequence.

$$a_n^{(2)} = C_2\, t_2^n = -t_1\, t_2^n = t_2^{n-1},\ n = 0, 1, 2, 3, \ldots .$$

Equation (1.3.20) – in contrast to equation (1.3.19) – shows the purely exponentially increasing particular solution of the difference equation (1.3.6), where the exponentially decreasing particular solution is vanishing because $C_2 = 0$. This behaviour is shown in Figure 1.10. Its explicit calculation is given by

$$a_n^{(2)} = C_1\, t_1^n = -t_2\, t_1^n = t_1^{n-1},\ n = 0, 1, 2, 3, \ldots .$$

Final Remarks

It should be mentioned that the quite simple solution of problems (1) and (2) given above is possible only since the fundamental solutions (1.3.12) and (1.3.13) of the difference equation (1.3.1) are explicit, as shown by the formula of de Moivre and Binet. However, in the general case, i.e., if the linear second-order difference equation (1.3.1) has the form

$$a_{n+1} + p_n\, a_n + q_n\, a_{n-1} = 0, \quad n = 1, 2, 3, 4, \ldots, \tag{1.3.21}$$

there is an elaborate theory, based on the fact that this difference equation may be written formally as an infinite continued fraction; this is to be shown in the following. A linear second-order difference equation

$$a_{n+1} + p_n\, a_n + q_n\, a_{n-1} = 0, \quad n = 1, 2, 3, \ldots, \tag{1.3.22}$$

may be considered as a three-term recurrence relation if two subsequent numbers, a_0 and a_1, say, are fixed. This then determines a sequence a_n, $n = 1, 2, 3, \ldots$, of numbers. Dividing (1.3.22) by a_{n-1} yields

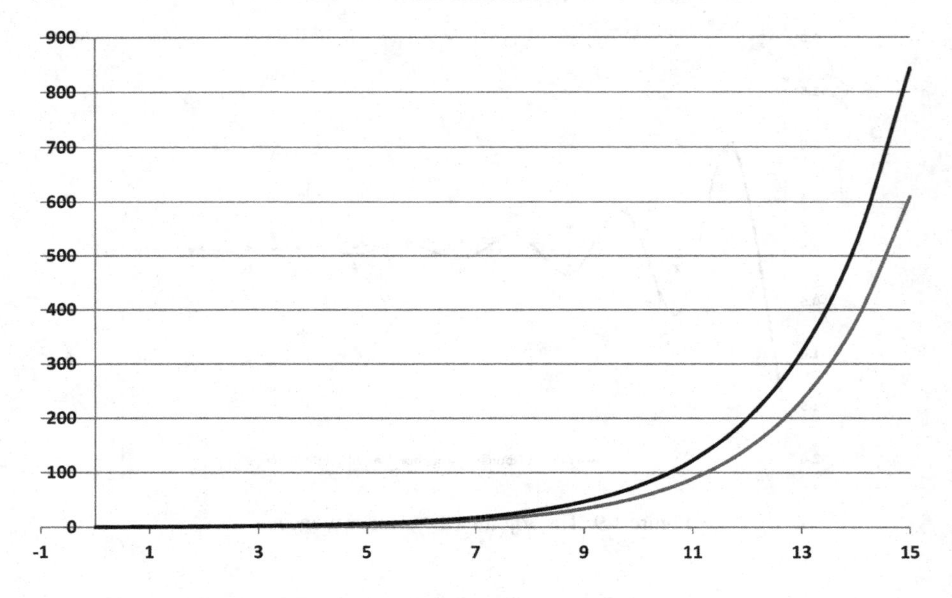

Figure 1.10 Purely exponentially increasing solution.

$$\frac{a_{n+1}}{a_{n-1}}\frac{a_n}{a_n} + p_n \frac{a_n}{a_{n-1}} + q_n = 0$$

or

$$\frac{a_n}{a_{n-1}}\left[\frac{a_{n+1}}{a_n} + p_n\right] + q_n = 0. \tag{1.3.23}$$

This is an equation for the ratio

$$\frac{a_n}{a_{n-1}}$$

of any two subsequent numbers a_{n+1} and a_n of this difference equation (1.3.21) that may be transformed into an infinite continued fraction. For this, equation (1.3.23) is written as

$$\begin{aligned}\frac{a_n}{a_{n-1}} &= -\frac{q_n}{p_n + \dfrac{a_{n+1}}{a_n}}\\ &= -\cfrac{q_n}{p_n - \cfrac{q_{n+1}}{p_{n+1} - \cfrac{q_{n+2}}{p_{n+2} - \cfrac{q_{n+3}}{\cdots}}}}\end{aligned}$$

an infinite continued fraction. For $n = 1$ this results in

$$\frac{a_1}{a_0} = -\cfrac{q_1}{p_1 - \cfrac{q_2}{p_2 - \cfrac{q_3}{p_3 - \cfrac{q_4}{\cdots}}}} \tag{1.3.24}$$

This theory was developed by Nörlund (1924); see in particular page 438.

The question of unmasking the recessive solutions is answered by the theory of Nörlund, such that equation (1.3.24) may be written as

$$\frac{a_1^{(2)}}{a_0^{(2)}} = -\cfrac{q_1}{p_1 - \cfrac{q_2}{p_2 - \cfrac{q_3}{p_3 - \cfrac{q_4}{\cdots}}}}.$$

The first theory of transcendental boundary eigenvalue conditions of singular boundary eigenvalue problems of linear, second-order differential equations was built on the basis of this (Gelfand and Schilow, 1964; Gelfand and Wilenkin, 1964).

The theory of Nörlund, in a sense, completed the mathematical questions posed by Fibonacci's presentation of the sequence that nowadays bears his name.

I do not want to finish this section without having pointed to the relevance of these results to what is called the *golden section*, sometimes called the golden ratio. Although already known in antiquity, it became the most important ratio in Renaissance times. It is the dividing of a length into two unequal parts such that the ratio of the whole distance to the larger part x is the same as the ratio of the larger part x to the smaller one $1 - x$:

$$\frac{1}{x} = \frac{x}{1-x}.$$

As is easily seen, this equation is similar to the characteristic equation (1.3.8) of the difference equation (1.3.5). The Fibonacci sequence is a particular solution of

$$x^2 + x - 1 = 0,$$

given by

$$x_1 = -\frac{1+\sqrt{5}}{2} = -t_1 \approx -1.61803,$$

$$x_2 = -\frac{1-\sqrt{5}}{2} = -t_2 = \frac{1}{t_1} = 1 - t_1 \approx 0.61803.$$

The example of the Fibonacci difference equation and its solutions has been dealt with in such detail since it shows clearly the mathematical principle of unmasking the recessive solution that is at the centre of this book. In particular, it may be seen by elementary means how we can carry out this unmasking, technically. Since the recessive solutions are normally eigensolutions of singular boundary eigenvalue problems, this is the mathematical mechanism underlying the solution method that is presented in this book. In the general case, the problem is a technical one. How can we get the coefficient of the dominant particular solution of the difference or differential equation to vanish in dependence on the parameters involved.

1.3.2 Regular Difference Equations

Consider the first-order linear difference equation

$$p_1(n+1)\,a_{n+1} - p_2(n)\,a_n = 0, \quad n = 0, 1, 2, 3, \ldots . \tag{1.3.25}$$

If one determines the value of a_0 here, then a linear two-term recurrence relation results from (1.3.25):

$$a_{n+1} - t_n\,a_n = 0, \quad n = 0, 1, 2, 3, \ldots, \tag{1.3.26}$$

with

$$t_n = \frac{p_2(n)}{p_1(n+1)}, \quad n = 0, 1, 2, 3, \ldots, \tag{1.3.27}$$

with the help of which one can calculate as many values of a_n in a recursive way as needed. If one writes (1.3.25) in the form

$$a_{n+1} = t_n\,a_n, \quad n = 0, 1, 2, 3, \ldots, \tag{1.3.28}$$

then it may be seen that the solution of (1.3.25) is given by

$$a_{n+1} = a_0\,\Pi_{i=1}^{n} t_i, \quad n = 0, 1, 2, 3, \ldots .$$

If the function t_n is a constant in n, i.e., $t_n = t$, then the result is

$$a_{n+1} = a_0\,t^n = a_0 \exp\left[n \ln(t)\right], \quad n = 0, 1, 2, 3, \ldots . \tag{1.3.29}$$

This was the case for the Fibonacci difference equation (1.3.6). For $t > 1$ this means an exponential growth of the solution a_n as $n \to \infty$ and for $t < 1$ an exponential decay of the solution a_n towards zero as $n \to \infty$.

For $t = 1$ the solution is given by a constant:

$$a_n = a_0, \quad n = 1, 2, 3, \ldots . \tag{1.3.30}$$

It is, in a certain sense, the transition from exponentially increasing to exponentially decreasing behaviour.

If t_n is not constant but tends to unity, $t \to 1$, then the solution a_n of the difference equation (1.3.28) is given by a power behaviour. For

$$t_n = 1 + \frac{r}{n}, \quad n = 1, 2, 3, \ldots, \tag{1.3.31}$$

we have

$$a_n = a_0\,\Pi_{i=1}^{n}\left(1 + \frac{r}{n}\right) = a_0\,n^r, \quad n = 1, 2, 3, \ldots . \tag{1.3.32}$$

This formula is also valid if $r < 0$ holds.

The characteristic behaviour displayed above remains the same for difference equations of higher order. However, one must now take into account that the function spaces of the solutions basically have vector-space structure, i.e., the general solution of the difference equation of mth order is the linear hull of its particular solutions:

$$a_n^{(g)} = \sum_{i=1}^{m} L_i \, a_n^{(i)} . \tag{1.3.33}$$

This will be focussed on in the following.

It was important above that t_n tends to a finite, non-vanishing number as $n \to \infty$. Therefore, we start with a second-order linear difference equation having the form

$$p_0(n) \, a_{n+1} + p_1(n) \, a_n + p_2(n) \, a_{n-1} = 0, \quad n = 1, 2, 3, \ldots, \tag{1.3.34}$$

with

$$\begin{aligned} \lim_{n\to\infty} p_0(n) &= p_0 = \text{const.}, \\ \lim_{n\to\infty} p_1(n) &= p_1 = \text{const.}, \\ \lim_{n\to\infty} p_2(n) &= p_2 = \text{const.} \end{aligned} \tag{1.3.35}$$

Dividing (1.3.34) by a_n, defining

$$t_n = \frac{a_{n+1}}{a_n}, \quad n = 0, 1, 2, 3, \ldots, \tag{1.3.36}$$

leads to

$$p_0(n) \, t_{n+1} + p_1(n) + \frac{p_2(n)}{t_{n-1}} = 0, \quad n = 1, 2, 3, \ldots . \tag{1.3.37}$$

Under the conditions (1.3.35), the limit $t = \lim_{n\to\infty} t_n$ exists, resulting in what is called the *characteristic equation*

$$p_0 \, t^2 + p_1 \, t + p_2 = 0, \quad n = 1, 2, 3, \ldots, \tag{1.3.38}$$

a second-order algebraic equation for t, indicating the possible asymptotic behaviours of the ratio of two consecutive term a_{n+1}, a_n of the particular solutions of the difference equation (1.3.34).

If the characteristic equation of a linear, ordinary, homogeneous difference equation has only simple roots, such a difference equation is called *regular*, otherwise it is called *irregular*.

1.3.3 Irregular Difference Equations

The characteristic equation of the Fibonacci difference equation (1.3.6) does have the two different values

$$t_1 = \frac{1 + \sqrt{5}}{2}$$

and

$$t_2 = \frac{1 - \sqrt{5}}{2} .$$

As can be seen, the quantity t on its own is decisive for the asymptotic behaviour of the partial solutions of the difference equation (1.3.6) as $n \to \infty$; and the behaviour for $n \to \infty$ is, after all, the natural behaviour of the difference equation, because it is used as a recursion and thus the index n grows in a natural way beyond all limits.

The characteristic of the particular solutions presented in §1.3.2 changes fundamentally, as soon as the characteristic equation of type (1.3.34) has multiple roots. This, of course, is possible only for difference equations whose order is larger than one. Let us therefore consider a second-order difference equation of the form

$$\left(1+\frac{\alpha_1}{n}\right)\, a_{n+1}+\left(-2+\frac{\alpha_0}{n}\right)\, a_n+\left(1+\frac{\alpha_{-1}}{n}\right)\, a_{n-1}=0, \quad n=1,2,3,\ldots.$$

Its characteristic equation

$$(t-1)^2=0 \tag{1.3.39}$$

has the twofold solution

$$t_1=t_2=1. \tag{1.3.40}$$

The considerations in §1.3.2 can no longer help here. What matters, and this is decisive, is *how* the twofold solution t_1 in (1.3.40) of the characteristic equation (1.3.39) is approximated by the two particular solutions of the difference equation. Information about this is therefore not provided by the characteristic equation (1.3.39), but by a characteristic equation extended by a term:

$$\left(1+\frac{\alpha_1}{n}\right)\, t^2(n)+\left(-2+\frac{\alpha_0}{n}\right)\, t(n)+\left(1+\frac{\alpha_{-1}}{n}\right)=0. \tag{1.3.41}$$

Its asymptotic solutions are

$$t_1=1+\sqrt{\frac{-\sum_{i=-1}^{+1}\alpha_i}{n}}+O\left(\frac{1}{n}\right),$$
$$t_2=1-\sqrt{\frac{-\sum_{i=-1}^{+1}\alpha_i}{n}}+O\left(\frac{1}{n}\right)$$

as $n \to \infty$. This shows that the one particular solution approaches the unit circle in the complex t-plane from inside and the other from outside.

A linear difference equation, the coefficients of which tend to finite values as the index tends to infinity and the characteristic equation of which has exclusively simple finite roots, is called a *regular difference equation of Poincaré–Perron type.*

A linear difference equation, the coefficients of which tend to finite values as the index tends to infinity and the characteristic equation of which has multiple finite roots, is called an *irregular difference equation of Poincaré–Perron type.*

A second-order characteristic equation (1.3.41) of an irregular difference equation is named after the American mathematician **Clarence Raymond Adams** (1898–1965), who applied it for the first time in Adams (1928).

1.4 Underlying Principle and Basic Method

The method for solving singular boundary eigenvalue problems of linear, ordinary, homogeneous differential equations of second order

$$P_0(z)\,\frac{d^2y}{dz^2} + P_1(z)\,\frac{dy}{dz} + P_2(z)\,y = 0, \quad z \in \mathbb{C}, \tag{1.4.1}$$

$P_i(z)$, $i = 0, 1, 2$, being polynomials in z, presented in this book relies essentially on approaches of generalised power series of the form

$$y(z) = f(z)\sum_{n=0}^{\infty} a_n\,(z - z_0)^n, \tag{1.4.2}$$

where $f(z)$ – being called an *asymptotic factor* – is either a power or an exponential function, at least an explicit function. These series (1.4.2) converge on dotted circular domains of the complex number plane around the point of expansion z_0. If one enters the differential equations with such power series, then the linear, ordinary, homogeneous differential equation of second order (1.4.1) is transformed into linear, ordinary, homogeneous difference equations for the coefficients a_n. These always have a start-up calculation in the form of an initial condition which turns the difference equation into a linear recursion, sometimes called a recurrence relation, with the help of which one can recursively calculate as many coefficients a_n of the power series as needed.

The special thing about this transformation is that the difference equation resulting from the differential equation is again linear, ordinary and homogeneous, but it does not necessarily have to be of second order. The order of the difference equation is not determined by the order of the differential equation behind it, but by the number, position and nature of its singularities. Therefore, it is useful to mention some properties of linear, ordinary, homogeneous difference equations, whereby their order should not be limited from the outset.

1.4.1 The Mathematical Principle of Unmasking Recessive Solutions

Consider the Frobenius solutions (1.2.8)

$$y_{F1}(z) = z^{\alpha_{01}} \sum_{n=0}^{\infty} a_{n1}\, z^n,$$
$$y_{F2}(z) = z^{\alpha_{02}} \sum_{n=0}^{\infty} a_{n2}\, z^n$$

about the singularity $z = 0$ of the differential equation (1.2.2), (1.2.1) as well as the Thomé solutions (1.2.18)

$$y_{T1}(z) = \exp(\alpha_{1\infty 1} z)\, z^{\alpha_{0\infty 1}} \left(1 + \sum_{n=1}^{\infty} C_{n1}\, z^{-n}\right) \quad \text{as } z \to \infty,$$

$$y_{T2}(z) = \exp(\alpha_{1\infty 2} z)\, z^{\alpha_{0\infty 2}} \left(1 + \sum_{n=1}^{\infty} C_{n2}\, z^{-n}\right) \quad \text{as } z \to \infty$$

at the irregular singularity at infinity. Both pairs of solutions may serve as a fundamental system in order to represent the general solution $y^{(g)}(z)$ of the differential equation (1.2.2), (1.2.1):

$$y^{(g)}(z) = c_1\, y_{F1}(z) + c_2\, y_{F2}(z)$$

and

$$y^{(g)}(z) = C_1\, y_{T1}(z) + C_2\, y_{T2}(z)$$

where c_1, c_2 as well as C_1, C_2 are arbitrary constants in z, viz. the totality of complex numbers. Thus, it is clear that the particular solution of the differential equation (1.2.2), (1.2.1) that is written by the Jaffé ansatz presented in §1.2.4 [cf. (1.2.56), (1.2.60), (1.2.61)]

$$y(z) = \exp(\alpha_{1\infty 2} z)\, z^{\alpha_{02}}\, (z+1)^{\alpha_{0\infty 2}-\alpha_{02}} \sum_{n=0}^{\infty} a_n \left(\frac{z}{z+1}\right)^n$$

is representable by means of either of these fundamental systems $y_{F1}(z)$, $y_{F2}(z)$ and $y_{T1}(z)$, $y_{T2}(z)$ by choosing specific, yet generally unknown, values $c_1 = c_{10}$, $c_2 = c_{20}$ or $C_1 = C_{10}$, $C_2 = C_{20}$:

$$y(z) = c_{10}\, y_{F1}(z) + c_{20}\, y_{F2}(z)$$

and

$$y(z) = C_{10}\, y_{T1}(z) + C_{20}\, y_{T2}(z).$$

Thus, it is always possible to write

$$\begin{aligned} y(z) &= \exp(\alpha_{1\infty 2} z)\, z^{\alpha_{02}}\, (z+1)^{\alpha_{0\infty 2}-\alpha_{02}} \sum_{n=0}^{\infty} a_n\, x^n \\ &= C_{10} \exp(\alpha_{1\infty 1} z)\, z^{\alpha_{01}} \left(1 + \sum_{n=0}^{\infty} C_{n1}\, z^{-n}\right) \\ &\quad + C_{20} \exp(\alpha_{1\infty 2} z)\, z^{\alpha_{02}} \left(1 + \sum_{n=0}^{\infty} C_{n2}\, z^{-n}\right) \end{aligned}$$

as $z \to \infty$, whereby the in general unknown constants C_{10}, C_{20} are dependent on the parameters of the differential equation (1.2.2), (1.2.1):

$$C_{10} = C_{10}(\underline{c}),\; C_{20} = C_{20}(\underline{c})$$

whereby $\underline{c}$ is the totality of parameters of the differential equation (1.2.2), (1.2.1). If the parameters of the differential equation $\underline{c}$ are arbitrary, then in general

$$C_{10} = C_{10}(\underline{c}) \neq 0.$$

Thus, because $\alpha_{1\infty 1} > 0$ and $\alpha_{1\infty 2} < 0$:

$$\exp(\alpha_{1\infty 2}\, z)\; z^{\alpha_{02}}\; (z+1)^{\alpha_{0\infty 2}-\alpha_{02}} \sum_{n=0}^{\infty} a_n\, x(z)^n$$
$$\sim C_1 \exp(\alpha_{1\infty 1}\, z)\; z^{\alpha_{01}} \left(1 + \sum_{n=0}^{\infty} C_{n1}\, z^{-n}\right)$$

as $z \to \infty$, from which results

$$\sum_{n=0}^{\infty} a_n\, x^n \sim C_1 \exp[(\alpha_{1\infty 1} - \alpha_{1\infty 2})\, z]\; z^{\alpha_{01}-\alpha_{02}}\; (z+1)^{-\alpha_{02}}$$
$$\times \left(1 + \sum_{n=0}^{\infty} C_{n1}\, z^{-n}\right)$$
$$\sim \exp[(\alpha_{1\infty 1} - \alpha_{1\infty 2})\, z]$$

as $z \to \infty$.

If the parameters of the differential equation $\underline{c}$ take on eigenvalues, then

$$C_{10} = C_{10}(\underline{c}) = 0.$$

Thus

$$\exp(\alpha_{1\infty 2}\, z)\; z^{\alpha_{02}}\; (z+1)^{\alpha_{0\infty 2}-\alpha_{02}} \sum_{n=0}^{\infty} a_n\, x^n$$
$$\sim C_2 \exp(\alpha_{1\infty 2}\, z)\; z^{\alpha_{0\infty 2}} \left(1 + \sum_{n=0}^{\infty} C_{n2}\, z^{-n}\right)$$

as $z \to \infty$, from which follows

$$\sum_{n=0}^{\infty} a_n\, x^n \sim C_2 \left(1 + \sum_{n=0}^{\infty} C_{n2}\, z^{-n}\right) \sim \text{const.}$$

as $z \to \infty$, whereby

$$(z+1)^{-\alpha_{0\infty 2}} = z^{-\alpha_{0\infty 2}} \left(1 + \frac{1}{z}\right)^{-\alpha_{0\infty 2}}$$

as $z \to \infty$.

The asymptotic considerations above [cf. (1.2.79)] tell us that there are two different asymptotic behaviours of the coefficients a_n as $n \longrightarrow \infty$:

$$a_{n1} = \exp\left(\gamma_1 \sqrt{n}\right) n^r \left(1 + \sum_{i=1}^{\infty} C_{1i}\, n^{-i/2}\right),$$

$$a_{n2} = \exp\left(\gamma_2 \sqrt{n}\right) n^r \left(1 + \sum_{i=1}^{\infty} C_{2i}\, n^{-i/2}\right).$$

Since the difference equation (1.2.69) is linear, the general solution of which is a two-dimensional vector space given by

$$a_n^{(g)} = L_1\, a_{n1} + L_2\, a_{n2} \tag{1.4.3}$$

where L_1 and L_2 in general is the totality of complex constants in n, but dependent on the parameters $\underline{c}$ of the difference equation (1.2.69) and thus of the differential equation (1.2.2), (1.2.1):

$$L_1 = L_1(\underline{c});\; L_2 = L_2(\underline{c}).$$

Thus, if

$$C_1(\underline{c}) \neq 0 \text{ then } L_1(\underline{c}) \neq 0 \text{ and } a_n = a_{n,1},$$

and if

$$C_2(\underline{c}) = 0 \text{ then } L_2(\underline{c}) = 0 \text{ and } a_n = a_{n,2}. \tag{1.4.4}$$

Hereby, the latter behaviour (1.4.4) solves the singular boundary eigenvalue problem of the single confluent case of the Gauss differential equation (1.2.2), (1.2.1), since the term on the right-hand side of (1.4.3) is dominated by the exponentially decreasing asymptotic behaviour of the Jaffé ansatz (1.2.56), (1.2.60), (1.2.61) as $z \longrightarrow \infty$. Therefore, the eigenvalue condition of the singular boundary eigenvalue problem

$$L_1(\underline{c}) = 0$$

in (1.4.3), which determines the eigenvalues.

The mechanism that makes the eigensolutions of the singular boundary eigenvalue problem – generally asymptotically dominated, viz. recessive – appear numerically as soon as the eigenvalue parameter adopts an eigenvalue, I call the mathematical principle of unmasking recessive solutions. It seems to be the crucial and rather general mechanism in solving boundary eigenvalue problems of linear differential equations.

1.4.2 The Method of Realising the Mathematical Principle

The mathematical principle of unmasking recessive solutions (of both difference and differential equations), as outlined in §1.4.1, is a very general one, which is, however, nothing more than a consequence of the structure of the function spaces that the particular solutions in question originate from: the linear structure of vector spaces.

This fundamental structure of abstract spaces is one thing; something quite different is its realisation in the concrete case. This requires a mathematical method. This method is now almost one hundred years old and comes from the mathematical physicists of quantum mechanics: Wolfgang Pauli, Friedrich Hund and above all George Cecil Jaffé. As explained in §1.2, it consists of several steps, aimed at finding the particular solution that solves the singular boundary eigenvalue problem of the differential equation in question, into a problem for a difference equation, which is first solved and, above all, whose solution can be interpreted in terms of the original boundary eigenvalue problem. The goal is a computationally solvable boundary eigenvalue condition.

Starting from the formulation of the problem by a differential equation, this solution proposed by Jaffé consists of four steps:

- First of all, an approach is needed that includes the asymptotic factors of the local solutions when radially approximating the two relevant singularities of the differential equation in question.
- If the relevant interval is not the unit interval, a Moebius transformation is performed that maps the relevant interval of the singular boundary eigenvalue problem to the unit interval.
- The connection must be established by introducing an additional power term.
- The differential equation resulting from these three steps has a particular solution that may be written in the form of a power series whose radius of convergence is the unit circle of the complex number plane. Its coefficients obey a linear ordinary difference equation of Poincaré–Perron type, which can be solved recursively. With this recursive solution, the boundary eigenvalue problem is formally solved, exactly.

The interpretation of the particular solution of the difference equation then succeeds on the basis of this approach in three further steps:

- Within the framework of an index asymptotics, it can be shown which behaviour the partial solutions exhibit for the difference equation mentioned above with the coefficients of the power series.
- Within the framework of a variable asymptotics, the index asymptotics of the coefficients of the series is related to the asymptotic behaviour of the particular solutions of the difference equation that solves the singular boundary eigenvalue problem.
- With this, a boundary eigenvalue condition can finally be formulated.

In this way, a solution of the problem is achieved, which also includes an understanding of the numerical procedure. The special feature – and this is the novelty – is the general validity of the method. It can be applied in principle to all singular boundary eigenvalue problems where linear, ordinary, homogeneous differential equations (1.2.2), (1.2.1) have polynomial-like coefficients, with the order of the polynomials not restricted, principally.

The analytical path from the differential equation to the difference equation is characterised by analytical and algebraic computations, which require a certain amount of calculatory effort. Already the new procedure for the classical case of the Coulomb potential of the Schrödinger equation requires – as explained – considerably more computational effort than the classical method. It is therefore advisable to use algebraic computer programs here. Otherwise, the generality of the method is limited by the manual computational effort, which can stand in the way here.

Higher special functions differ from the classical special functions in that the eigenvalue conditions are transcendental functions, whereas those of the classical special functions are of algebraic nature. It is immediately obvious that the variety of higher special functions is far greater than that of the classical ones. It is therefore a strategic advantage to specify a procedure for the extraction of the higher special functions and no longer to catalogue them explicitly, as was the case with the classical special functions. The main aim of this book is to indicate the procedure that will enable the reader to find solutions to his or her own problems. Thus, it is hoped that in this way new higher special functions will be found, which are hitherto unknown. Apart from algebraic and numerical computation possibilities, the procedure given in this book is a suitable means for this.

The principles and methods presented in this first chapter can be applied unchanged to more general differential equations. Only there is more computational effort necessary. This is done in Chapters 3 and 4 for the equations that are in a certain sense next to the Gaussian-type equations, namely, the differential equations of Heun type.

2
Singularities in Action

Singularities are the central quantities for the treatment of the boundary eigenvalue problems in this book, both the singularities of the differential equations and those of their solutions. Jules Henri Poincaré (1854–1912) was probably the first to recognise their importance and to treat them conceptually by introducing what he called the rank (cf. Poincaré, 1886, p. 305). This is adequately explained in Ince (1956). It has turned out, however, that the definition was not optimally chosen. Therefore, a slightly different definition is adopted in Slavyanov and Lay (2000), and it is that definition we adopt here. The quantity is named s-*rank* instead of 'rank', in order to distinguish it from the one defined by Poincaré. The 's' is meant to indicate 'singularity'. Within this definition, the non-elementary regular singularity is the standard one; it is defined to have s-rank one – $R = 1$. The definition of the s-rank of a singularity of a differential equation is aligned with the behaviour of the local solutions which are presented in §2.3.

In this concept, the singularities of differential equations (1.2.2), (1.2.1) thus always have half-integer s-ranks; one distinguishes between *regular singularities* and *irregular singularities*, depending on whether the s-rank is not larger or larger than one, respectively. It should be emphasised here that it is necessary to distinguish between the singularities of the underlying differential equation (discussed in §§2.1 and 2.2) and the singularities of the particular solutions of the underlying differential equation, discussed in §2.3. Accordingly, there are two types of regular singularities – those with s-rank 1 and those with s-rank $\frac{1}{2}$ – the latter being called *elementary singularities*. Among the irregular singularities there are those with integer s-rank and those with odd half-integer s-rank. Thus, $R = \frac{3}{2}$ is the smallest s-rank of an irregular singularity. So, the standard singularity is not – as with Poincaré – the elementary one, but the non-elementary regular singularity of the underlying differential equation; it has s-rank $R = 1$.

2.1 The Concept of s-Ranks

2.1.1 Linear Second-Order Differential Equations with Polynomial Coefficients

As already mentioned in the Introduction, this book deals with the *linear, homogeneous, second-order ordinary differential equation* with polynomial coefficients, defined on the complex plane:

$$P_0(z)\,\frac{d^2y(z)}{dz^2} + P_1(z)\,\frac{dy(z)}{dz} + P_2\,y(z) = 0, \quad z \in \mathbb{C}, \tag{2.1.1}$$

where $P_0(z)$, $P_1(z)$ and $P_2(z)$ are polynomials in z. Throughout this book it is assumed that no zeros of $P_0(z)$ coincide with zeros of $P_1(z)$ or $P_2(z)$. Otherwise, this would render particular investigations that are to be avoided.

At the zeros $z = z_i$, $i = 1, 2, 3, \ldots, n$, (where n is the order of $P_0(z)$) of the polynomial $P_0(z)$, the differential equation (2.1.1) degenerates such that the term of the second derivative vanishes. These points are called the *singularities of the differential equation* (2.1.1), or *singular points of the differential equation* (2.1.1).

That the points $z = z_i$ are singularities of the differential equation (2.1.1) may be seen better by dividing (2.1.1) by $P_0(z)$, to obtain

$$\boxed{\frac{d^2y}{dz^2} + P(z)\,\frac{dy}{dz} + Q(z)\,y(z) = 0, \quad z \in \mathbb{C},} \tag{2.1.2}$$

with

$$P(z) = \frac{P_1(z)}{P_0(z)}$$

and

$$Q(z) = \frac{P_1(z)}{P_0(z)}.$$

All the points of equation (2.1.2) in the range of its definition beyond singularities are called *ordinary points of the differential equation* (2.1.2). Singularities, being placed at a finite point of the complex plane, are called *finite singularities*, for short.

Here are some important statements on the singularities of equation (2.1.2):

- According to the definition above, the singularities of the coefficients $P(z)$ and $Q(z)$ of equation (2.1.2) become singularities of the differential equation (2.1.2).
- The singularities of equation (2.1.2) are isolated points of the complex plane.
- Equation (2.1.2) has a finite number of singularities.

There is a fundamental theorem of Paul Painlevé (cf. Bieberbach, 1965, pp. 10–12) that the solutions of (2.1.2) are uniquely determined by determining the functional value $y(z_0)$ as well as the derivative

$$\left.\frac{\mathrm{d}y}{\mathrm{d}z}\right|_{z=z_0}$$

at an ordinary point $z = z_0$ of the differential equation with $z \in \mathbb{C}$. As already defined above, the ordinary points of the differential equation (2.1.2) are all the points of the complex plane $z \in \mathbb{C}$ except the zeros $z = z_i$ of the polynomial $P_0(z)$.

The point at infinity requires particular attention. Thereby, the complex plane is thought to be compactified. This means that the plane is mapped onto a sphere by a *stereographic projection*. First it becomes imaginable that turning around a finite point clockwise means circling around the point at infinity anticlockwise (cf. Figure 2.1).

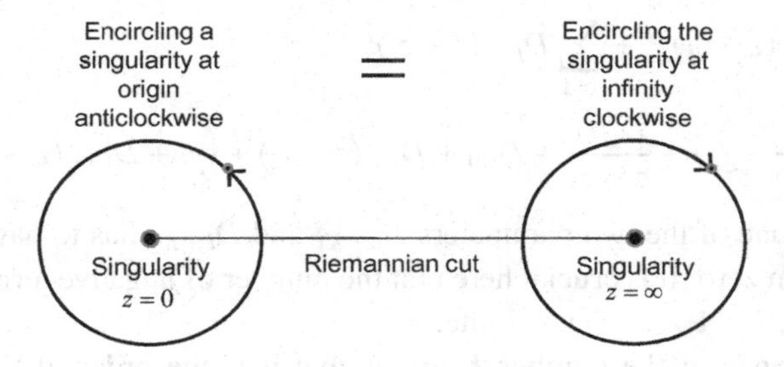

Figure 2.1 Encircling a point at infinity.

Since, strictly speaking, infinity is neither a point nor a number, it is dealt with by means of inverting the differential equation, viz. applying the transformation

$$\zeta = \frac{1}{z}, \tag{2.1.3}$$

thus getting

$$\frac{\mathrm{d}^2 y}{\mathrm{d}\zeta^2} + \left[\frac{2}{\zeta} - \frac{1}{\zeta^2} P(\zeta)\right] \frac{\mathrm{d}y}{\mathrm{d}\zeta} + \frac{1}{\zeta^4} Q(\zeta)\, y = 0, \quad \zeta \in \mathbb{C}, \tag{2.1.4}$$

with $P_0(\zeta) = P_0(1/z)$, $P_1(\zeta) = P_1(1/z)$ and $P_2(\zeta) = P_2(1/z)$ and then considering the point $\zeta = 0$ [cf. §1.2.1, equations (1.2.5), (1.2.6)]. All the properties of singularities of equation (2.1.2) at infinity are transferred to the singularity at zero of equation (2.1.4) by means of equation (2.1.3).

It is to be recognised that there is no differential equation (2.1.2) without singularities, as may be seen from (2.1.4).

2.1.2 The s-Rank of Singularities of Differential Equations

There are, by and large, two types of singularities of equation (2.1.2). This is to be outlined in the following. As is obvious, the coefficients of equation (2.1.2) are rational functions. Thus, $P(z)$ and $Q(z)$ may be written locally in a neighbourhood of the finite singularities z_i, $i = 1, 2, \ldots, m$, by means of Laurent series, generally having a finite number of negative power terms:

$$\begin{aligned} P(z; z_i) &\\ = \sum_{j=0}^{K_1} G_{i,-j}\,(z - z_i)^{-j} + \sum_{j=1}^{k_1} G_{i,j}\,(z - z_i)^j & \quad (2.1.5)\\ = \frac{G_{i,-K_1}}{(z - z_i)^{K_1}} + \cdots + \frac{G_{i,-1}}{z - z_i} + G_{i,0} + G_{i,1}\,(z - z_i) + \cdots + G_{i,k_1}\,(z - z_i)^{k_1}, & \end{aligned}$$

$$\begin{aligned} Q(z; z_i) &\\ = \sum_{j=0}^{K_2} D_{i,-j}\,(z - z_i)^{-j} + \sum_{j=1}^{k_2} D_{i,j}\,(z - z_i)^j & \quad (2.1.6)\\ = \frac{D_{i,-K_2}}{(z - z_i)^{K_2}} + \cdots + \frac{D_{i,-1}}{z - z_i} + D_{i,0} + D_{i,1}\,(z - z_i) + \cdots + D_{i,k_2}\,(z - z_i)^{k_2}, & \end{aligned}$$

where at least one of the two parameters $G_{i,-K_1}$ and $D_{i,-K_2}$ has to have a value that is different from zero. It is crucial here that the number of negative terms at all of the singularities z_i, $i = 1, 2, \ldots, m$, is finite.

The significance of the number $K_1(z_i)$ is that it is the order of the pole of the function $P_1(z)/P_0(z)$ at the point $z = z_i$. The significance of the number $K_2(z_i)$ is that it is the order of the pole of the function $P_2(z)/P_0(z)$ at the point $z = z_i$.

For an ordinary point $z = z_i$ of equation (2.1.2) it holds that $K_1 = K_2 = 0$ and the principal parts of the series (2.1.5) and (2.1.6) vanish. The Laurent series (2.1.5) and (2.1.6) show that the singularities of equation (2.1.2) may be characterised by means of the two numbers K_1 and K_2. In this sense, we define

$$R(z_i) = \max(1, K_1, K_2/2) \tag{2.1.7}$$

as the s-*rank of a singularity* of equation (2.1.2) at the point $z = z_i$. Thus, the s-rank of a singularity of equation (2.1.2) is either a half-integer (viz. odd half-integer) or an integer number, at least unity. The numbers k_1 in (2.1.5) and k_2 in (2.1.6) do not contribute to the definition of the s-rank of the singularity at z_i.

One might be tempted to omit the 1 under the maximum sign in (2.1.7) in order to define singularities, the s-rank of which is $R = \frac{1}{2}$. In fact, there are singularities having such an s-rank. However, for historical reasons (cf. Ince, 1956, p. 495), they are defined in a different way – considered a singularity of equation (2.1.2) at $z = z_i$, the s-rank of which is unity. Then, the coefficients $P(z; z_i)$ and $Q(z; z_i)$ have the form

$$\begin{aligned} P(z; z_i) &= \frac{A}{z - z_i} + \text{h.p.},\\ Q(z; z_i) &= \frac{B}{(z - z_i)^2} + \frac{C}{z - z_i} + \text{h.p.}, \end{aligned} \tag{2.1.8}$$

while 'h.p.' denotes the 'holomorphic part' of the function at $z = z_i$, thus for a function that is evolvable in a power series. According to Ince (1956, p. 495), such a singularity is called an *elementary singularity* of equation (2.1.2) if the coefficients A and B are related via

$$(A - 1)^2 - 4\,B = \frac{1}{4}. \tag{2.1.9}$$

Eventually, as already mentioned above, we distinguish an item that pervades the whole theory. If the s-rank of a singularity at $z = z_i$ of equation (2.1.2) is at most $R(z_i) = 1$ then it is called *regular*. Otherwise, the singularity of equation (2.1.2) is called *irregular*.

Examples. Consider a *non-elementary regular singularity* of equation (2.1.2) at $z = z_i$ having the coefficients

$$P(z; z_i) \sim \frac{G_{i,-1}}{z - z_i} \text{ as } z \to z_i, \tag{2.1.10}$$

$$Q(z; z_i) \sim \frac{D_{i,-2}}{(z - z_i)^2} \text{ as } z \to z_i, \tag{2.1.11}$$

the s-rank of which is thus $R(z_i) = 1$. This means that in general, $K_1(z_i) = 1$, $K_2(z_i) = 2$ [cf. equations (2.1.5), (2.1.6)].

In the case of an irregular singularity at $z = z_i$, the s-rank of which is $R(z_i) = 2$, the coefficients are

$$P(z; z_i) \sim \frac{G_{i,-2}}{(z - z_i)^2} \text{ as } z \to z_i,$$

$$Q(z; z_i) \sim \frac{D_{i,-4}}{(z - z_i)^4} \text{ as } z \to z_i.$$

The maximum negative powers in this case are $K_1(z_i) = 2$ and $K_2(z_i) = 4$; thus $R(z_i) = 2$.

Having an irregular singularity at $z = z_i$, the s-rank of which is $R(z_i) = 3/2$, the coefficients $P(z; z_i)$ and $Q(z; z_i)$ have the form

$$P(z; z_i) \sim \frac{G_{i,-1}}{z - z_i} \text{ as } z \to z_i,$$

$$Q(z; z_i) \sim \frac{D_{i,-3}}{(z - z_i)^3} \text{ as } z \to z_i.$$

Thus $K_1(z_i) = 1$ and $K_2(z_i) = 3$, yielding an odd half-integer s-rank of the singularity $R(z_i) = 3/2$.

2.1.3 Regular and Irregular Singularities of Differential Equations

There are regular as well as irregular singularities of equation (2.1.2). Regular singularities are to be distinguished as elementary and non-elementary regular singularities.

Irregular singularities are to be distinguished as having integer s-ranks and half-integer s-rank (viz. odd half-integer s-ranks).

The local behaviours of the coefficients $P(z; z_i)$ and $Q(z; z_i)$ of equation (2.1.2) at a non-elementary regular singularity at $z = z_i$ of equation (2.1.2) (the s-rank of which is $R = 1$) are given by (2.1.8). In the case when the singularity is elementary (regular), the behaviours of the coefficients $P(z; z_i)$ and $Q(z; z_i)$ of equation (2.1.2) at $z = z_i$ are the same. However, there is, in addition, a relation between A and B of (2.1.8) [cf. (2.1.9)]:

$$(A-1)^2 - 4\,B = \frac{1}{4}.$$

The coefficients $P(z; z_i)$ and $Q(z; z_i)$ of equation (2.1.2) at irregular singularities of equation (2.1.2) having s-rank two are given by

$$P(z, z_i) = \frac{G_{i,-2}}{(z-z_i)^2} + \frac{G_{i,-1}}{z-z_i} + \text{h.p.}, \tag{2.1.12}$$

$$Q(z; z_i) = \frac{D_{i,-4}}{(z-z_i)^4} + \frac{D_{i,-3}}{(z-z_i)^3} + \frac{D_{i,-2}}{(z-z_i)^2} + \frac{D_{i,-1}}{z-z_i} + \text{h.p.} \tag{2.1.13}$$

Eventually, this is considered an irregular singularity of equation (2.1.2) having a singularity at $z = z_i$, the s-rank of which is $R(z_i) = 3/2$:

$$P(z; z_i) = \frac{G_{i,-1}}{z-z_i} + \text{h.p.}, \tag{2.1.14}$$

$$Q(z; z_i) = \frac{D_{i,-3}}{(z-z_i)^3} + \frac{D_{i,-2}}{(z-z_i)^2} + \frac{D_{i,-1}}{z-z_i} + \text{h.p.} \tag{2.1.15}$$

2.2 Coalescence and Confluence Processes

This section is devoted to one of the most important processes of singularities of differential equations (2.1.2), namely the so-called *confluence process*, meaning that two (or even more) singularities coalesce under certain conditions in order to form one new singularity of the differential equation (2.1.2).

The crucial question in the context of confluence processes concerns the s-rank of this new singularity with respect to the s-ranks of the coalescing singularities, as well as the conditions that guarantee a reasonable confluence process at all.

2.2.1 Standard Confluence Process

The following considerations present an explicit calculation of a confluence process that turns out to be a standard one (cf. Ronveaux, 1995, p. 312).

Given a differential equation (2.1.2):

$$\frac{d^2y}{dz^2} + P(z;\epsilon)\frac{dy}{dz} + Q(z;\epsilon)\,y = 0, \qquad y = y(z),\ z \in \mathbb{C}, \tag{2.2.1}$$

having two non-elementary regular singularities, at $z = 0$ and at $z = \epsilon$, say. Thus, the s-ranks of these two singularities are both unity: $R_c(0) = 1$ and $R_c(\epsilon) = 1$ (the subscript c stands for 'coalescing'). Then, the coefficients $P(z;\epsilon)$ and $Q(z;\epsilon)$ of equation (2.2.1) may be written as

$$\begin{aligned} P(z;\epsilon) &= \hat{P}(z;\epsilon) + \tilde{P}(z) = \frac{A_0(\epsilon)}{z} + \frac{A_\epsilon(\epsilon)}{z-\epsilon} + \tilde{P}(z), \\ Q(z;\epsilon) &= \hat{Q}(z;\epsilon) + \tilde{Q}(z) = \frac{B_0(\epsilon)}{z^2} + \frac{B_\epsilon(\epsilon)}{(z-\epsilon)^2} + \frac{C_0(\epsilon)}{z} + \frac{C_\epsilon(\epsilon)}{z-\epsilon} + \tilde{Q}(z), \end{aligned} \tag{2.2.2}$$

where $\tilde{P}(z)$ and $\tilde{Q}(z)$ are the parts of $P(z;\epsilon)$ and of $Q(z;\epsilon)$ that are holomorphic at $z = 0$ and at $z = \epsilon$ and are assumed to not be dependent on ϵ.

Without loss of generality and for the sake of simplicity, ϵ is supposed to be positive real-valued and small. It quantifies the location of one of the two coalescing singularities and thus the distance between the two coalescing singularities.

It is crucial in the following that the quantities

$$A_0(\epsilon),\ A_\epsilon(\epsilon),\ B_0(\epsilon), \ldots, C_\epsilon(\epsilon) \tag{2.2.3}$$

in (2.2.2) are dependent on ϵ. If it were not for this dependence, the result of the confluence process would not be the general one (see below). This situation is the basis for carrying out what is called the *standard confluence process* $\epsilon \to 0$ in the following. It will be shown that the result of the confluence process is

$$\begin{aligned} \hat{P}(z) &= \lim_{\epsilon\to 0} \hat{P}(z;\epsilon) = \frac{G_2}{z^2} + \frac{G_1}{z}, \\ \hat{Q}(z) &= \lim_{\epsilon\to 0} \hat{Q}(z;\epsilon) = \frac{D_4}{z^4} + \frac{D_3}{z^3} + \frac{D_2}{z^2} + \frac{D_1}{z}, \end{aligned} \tag{2.2.4}$$

viz. the result is a differential equation (2.1.2) having a singularity at the origin, the s-rank of which is two: $R_r(0) = 2$ (the index r means 'resulting').

For substantiating this, first it can be shown that the most general form of the coefficients (2.2.3) is given by

$$
\begin{aligned}
A_0(\epsilon) &= \frac{A_{0,-1}}{\epsilon} + A_{0,0} + O(\epsilon),\\
A_\epsilon(\epsilon) &= \frac{A_{\epsilon,-1}}{\epsilon} + A_{\epsilon,0} + O(\epsilon),\\
B_0(\epsilon) &= \frac{B_{0,-2}}{\epsilon^2} + \frac{B_{0,-1}}{\epsilon} + B_{0,0} + O(\epsilon),\\
B_\epsilon(\epsilon) &= \frac{B_{\epsilon,-2}}{\epsilon^2} + \frac{B_{\epsilon,-1}}{\epsilon} + B_{\epsilon,0} + O(\epsilon),\\
C_0(\epsilon) &= \frac{C_{0,-3}}{\epsilon^3} + \frac{C_{0,-2}}{\epsilon^2} + \frac{C_{0,-1}}{\epsilon} + C_{0,0} + O(\epsilon),\\
C_\epsilon(\epsilon) &= \frac{C_{\epsilon,-3}}{\epsilon^3} + \frac{C_{\epsilon,-2}}{\epsilon^2} + \frac{C_{\epsilon,-1}}{\epsilon} + C_{\epsilon,0} + O(\epsilon)
\end{aligned}
\tag{2.2.5}
$$

as $\epsilon \longrightarrow 0$. The definition of the symbol O may be seen from Slavyanov (1996, §I.1.1, pp. 1, 2).

Second, we calculate the relations between the coefficients of these functions (2.2.5) and the coefficients

$$G_1,\ G_2,\ D_1,\ \ldots,\ D_4$$

in (2.2.4). The fundamental condition for this is to get a reasonable confluence process, viz. no quantity of $P(z;\epsilon)$ and of $Q(z;\epsilon)$ tends to infinity during the limiting process $\epsilon \longrightarrow 0$.

In order to carry out the standard confluence process $\epsilon \longrightarrow 0$, the coefficients $\hat{P}(z;\epsilon)$ and $\hat{Q}(z;\epsilon)$ are to be written in the form

$$\hat{P}(z;\epsilon) = \frac{A_0(\epsilon)\,(z-\epsilon) + A_\epsilon(\epsilon)\,z}{z\,(z-\epsilon)} = \frac{[A_0(\epsilon) + A_\epsilon(\epsilon)]\,z - \epsilon\,A_0(\epsilon)}{z\,(z-\epsilon)} \tag{2.2.6}$$

and

$$\hat{Q}(z;\epsilon) = \frac{F(z;\epsilon)}{z^2\,(z-\epsilon)^2} \tag{2.2.7}$$

with

$$
\begin{aligned}
F(z;\epsilon) = [C_0(\epsilon) + C_\epsilon(\epsilon)]\,z^3 &\\
+ [B_0(\epsilon) + B_\epsilon(\epsilon) - 2\,\epsilon\,C_0(\epsilon) - \epsilon\,C_\epsilon(\epsilon)]\,z^2 &\\
+ [-2\,\epsilon\,B_0(\epsilon) + \epsilon^2\,C_0(\epsilon)]\,z &\\
+ \epsilon^2\,B_0(\epsilon). &
\end{aligned}
$$

Carrying out the limiting process $\epsilon \longrightarrow 0$, the result of a rather lengthy calculation is

$$P(z) = \lim_{\epsilon\to 0} \hat{P}(z;\epsilon) + \tilde{P}(z) = \frac{-A_{0,1}}{z^2} + \frac{A_{0,0} + A_{\epsilon,0}}{z} + \tilde{P}(z)$$

and

$$
\begin{aligned}
Q(z) &= \lim_{\epsilon\to 0} \hat{Q}(z;\epsilon) + \tilde{Q}(z) \\
&= \frac{B_{0,-2}}{z^4} + \frac{-2\,B_{0,-1} + C_{\epsilon,-2}}{z^3} + \frac{B_{0,0} + B_{\epsilon,0} - 2\,C_{0,-1} - C_{\epsilon,-1}}{z^2} \\
&\quad + \frac{C_{0,0} + C_{\epsilon,0}}{z} + \tilde{Q}(z),
\end{aligned} \tag{2.2.8}
$$

yielding the final result

$$
\begin{aligned}
G_2 &= -A_{0,-1}, \\
G_1 &= A_{0,0} + A_{\epsilon,0}, \\
D_4 &= B_{0,-2}, \\
D_3 &= -2\,B_{0,-1} + C_{\epsilon,-2}, \\
D_2 &= B_{0,0} + B_{\epsilon,0} - 2\,C_{0,-1} - C_{\epsilon,-1}, \\
D_1 &= C_{0,0} + C_{\epsilon,0}.
\end{aligned} \tag{2.2.9}
$$

This result shows that the s-rank of the resulting singularity is at most $R_c(0) = 2$. It is just two if $G_2 \neq 0$ or $D_4 \neq 0$. Consequently, the s-rank of the singularity coming out of the standard confluence process generally is the sum of the s-ranks of the two coalescing singularities at $z = 0$ and at $z = \epsilon$.

Example. Suppose that the coefficients in (2.2.5) do not depend on ϵ. Then

$$
\begin{aligned}
A_0(\epsilon) &= A_{0,0}, \quad A_\epsilon(\epsilon) = A_{\epsilon,0}, \\
B_0(\epsilon) &= B_{0,0}, \quad B_\epsilon(\epsilon) = B_{\epsilon,0}, \\
C_0(\epsilon) &= C_{0,0}, \quad C_\epsilon(\epsilon) = C_{\epsilon,0}.
\end{aligned}
$$

The coefficients are then

$$
\begin{aligned}
P(z;\epsilon) &= \frac{A_0(\epsilon)}{z} + \frac{A_\epsilon(\epsilon)}{z-\epsilon} + \tilde{P}(z) \\
&= \frac{A_{0,0}}{z} + \frac{A_{\epsilon,0}}{z-\epsilon} + \tilde{P}(z)
\end{aligned}
$$

and

$$
\begin{aligned}
Q(z;\epsilon) &= \frac{B_0(\epsilon)}{z^2} + \frac{B_\epsilon(\epsilon)}{(z-\epsilon)^2} + \frac{C_0(\epsilon)}{z} + \frac{C_\epsilon(\epsilon)}{z-\epsilon} + \tilde{Q}(z) \\
&= \frac{B_{0,0}}{z^2} + \frac{B_{\epsilon,0}}{(z-\epsilon)^2} + \frac{C_{0,0}}{z} + \frac{C_{\epsilon,0}}{z-\epsilon} + \tilde{Q}(z),
\end{aligned}
$$

showing that the two coalescing singularities at $z = 0$ and at $z = \epsilon$ are each non-elementary regular, i.e., the s-rank of both is unity:

$$
R_c(0) = 1; \; R_c(\epsilon) = 1.
$$

The confluence process here is easy to carry out, but the result is by no means an irregular singularity the s-rank of which is two. Once again we have a non-elementary regular singularity:

$$P(z) = \frac{A_{0,0} + A_{\epsilon,0}}{z} + \tilde{P}(z), \tag{2.2.10}$$

$$Q(z) = \frac{B_{0,0} + B_{\epsilon,0}}{z^2} + \frac{C_{0,0} + C_{\epsilon,0}}{z} + \tilde{Q}(z), \tag{2.2.11}$$

thus

$$R_r(0) = 1.$$

This example bears a general aspect. If the coefficients $P(z;\epsilon)$ and $Q(z;\epsilon)$ in (2.2.1), viz. the functions in (2.2.5), do not depend on ϵ, then coalescing singularities result in a new singularity, the s-rank of which is the maximum of the s-ranks of the coalescing singularities. Thus, the s-ranks of coalescing singularities in a confluence process are generally subadditive (cf. Slavyanov and Lay, 2000, pp. 17–20).

If the confluence process of two (or even more) singularities results in a single singularity, the s-rank of which is the sum of the s-ranks of the coalescing singularities, it is called a *strong confluence process*. If the confluence process of two (or even more) singularities results in a single singularity, the s-rank of which is lower than the sum of the s-ranks of the coalescing singularities, it is called a *weak confluence process*.

Elementary Singularities

It is supposed in the following that the singularity of the differential equation (2.2.1) at $z = \epsilon$ is elementary, while the singularity at the origin $z = 0$ remains to be non-elementary regular. According to equation (2.1.9), this means that

$$[A_\epsilon(\epsilon) - 1]^2 - 4\, B_\epsilon(\epsilon) = \frac{1}{4} \tag{2.2.12}$$

holds. Inserting the explicit expressions from (2.2.5), this relation is given by

$$\frac{A_{0,-1}^2 - 4\, B_{0,-2}}{\epsilon^2} + \frac{2\, A_{0,-1}\left(A_{0,0} - 1\right) - 4\, B_{0,-1}}{\epsilon} + \left(A_{0,0} - 1\right)^2 - 4\, B_{0,0} = \tfrac{1}{4}.$$

Equation (2.2.12) may also be written in the form

$$B_\epsilon(\epsilon) = \frac{1}{4}\,[A_\epsilon(\epsilon) - 1]^2 - \frac{1}{16}. \tag{2.2.13}$$

Inserting the corresponding formulae from (2.2.5) yields

$$\begin{aligned}
B_\epsilon(\epsilon) &= \frac{1}{4}\left[A_\epsilon(\epsilon) - 1\right]^2 - \frac{1}{16} + O(\epsilon) \\
&= \frac{1}{4}\left[\frac{A_{\epsilon,-1}}{\epsilon} + A_{\epsilon,0} - 1\right]^2 - \frac{1}{16} + O(\epsilon) \\
&= \frac{1}{4}\left[\frac{A_{\epsilon,-1}^2}{\epsilon^2} + 2\,(A_{\epsilon,0} - 1)\frac{A_{\epsilon,-1}}{\epsilon} + (A_{\epsilon,0} - 1)^2\right] - \frac{1}{16} + O(\epsilon) \\
&= \frac{1}{4}\frac{A_{\epsilon,-1}^2}{\epsilon^2} + \frac{A_{\epsilon,0} - 1}{2}\frac{A_{\epsilon,-1}}{\epsilon} + \frac{(A_{\epsilon,0} - 1)^2}{4} - \frac{1}{16} + O(\epsilon)
\end{aligned}$$

as $\epsilon \longrightarrow 0$.

A comparison between this formula and the corresponding formula in (2.2.5)

$$B_\epsilon(\epsilon) = \frac{B_{\epsilon,-2}}{\epsilon^2} + \frac{B_{\epsilon,-1}}{\epsilon} + B_{\epsilon,0} + O(\epsilon) \text{ as } \epsilon \longrightarrow 0$$

leads to the following coefficients:

$$\begin{aligned}
B_{\epsilon,-2} &= \frac{1}{4}\,A_{\epsilon,-1}^2, \\
B_{\epsilon,-1} &= \frac{A_{\epsilon,-1}}{2}\,(A_{\epsilon,0} - 1), \\
B_{\epsilon,0} &= \frac{1}{4}\,(A_{\epsilon,0} - 1)^2 - \frac{1}{16}.
\end{aligned} \tag{2.2.14}$$

Thus, the relation (2.2.13) for $z = \epsilon$ to be an elementary singularity of the differential equation (2.2.1) is broken down into the coefficients of the functions $A_\epsilon(\epsilon)$ and $B_\epsilon(\epsilon)$ in (2.2.5).

2.2.2 Reducible Singularities

The s-homotopic transformation, defined for regular singularities in §1.2.2, may be generalised to irregular singularities (cf. Slavyanov and Lay, 2000, p. 10). So, it is reasonable to ask for the conditions that reduce the s-rank of an irregular singularity under the action of an s-homotopic transformation. This question is discussed in the following.

We consider the standard confluence process (cf. §2.2.1) that results in the coalescence of two non-elementary regular singularities of the differential equation

$$\frac{d^2y}{dz^2} + P(z;\epsilon)\,\frac{dy}{dz} + Q(z;\epsilon)\,y = 0, \qquad y = y(z), z \in \mathbb{C}. \tag{2.2.15}$$

It is assumed in the following that (2.2.15) has two non-elementary regular singularities, located at $z = 0$ and at $z = \epsilon$, whereby ϵ is taken to be real-valued and positive, without losing generality. Thus, the coefficients of equation (2.2.15) are given by

$$P(z;\epsilon) = \frac{A_0(\epsilon)}{z} + \frac{A_\epsilon(\epsilon)}{z-\epsilon} + p(z),$$

$$Q(z;\epsilon) = \frac{B_0(\epsilon)}{z^2} + \frac{B_\epsilon(\epsilon)}{(z-\epsilon)^2} + \frac{C_0(\epsilon)}{z} + \frac{C_\epsilon(\epsilon)}{z-\epsilon} + q(z),$$

where it is assumed that $p(z)$ and $q(z)$ describe functions that are holomorphic at the singularities at $z = 0$ as well as at $z = \epsilon$ and do not depend on ϵ. Thus, the s-rank of the coalescing singularities is $R_c(0) = 1$ and $R_c(\epsilon) = 1$.

These two singularities of the differential equation (2.2.15) are supposed to undergo the confluence process $\lim \epsilon \to 0$. In order for this limiting process to become a reasonable one, the functional terms in dependence on ϵ of the coefficients $P(z;\epsilon)$ and $Q(z;\epsilon)$ generally have to be (2.2.5). Moreover, the following conditions have to be met by these coefficients:

$$\begin{aligned} A_{0,-1} + A_{\epsilon,-1} &= 0, \\ C_{0,-3} + C_{\epsilon,-3} &= 0, \\ C_{0,-2} + C_{\epsilon,-2} &= 0, \\ C_{0,-1} + C_{\epsilon,-1} &= 0, \\ B_{0,-2} + B_{\epsilon,-2} - 2\,C_{0,-3} - C_{\epsilon,-3} &= 0, \\ B_{0,-1} + B_{\epsilon,-1} - 2\,C_{0,-2} - C_{\epsilon,-2} &= 0, \\ B_{0,-2} - \frac{1}{2}\,C_{0,-3} &= 0. \end{aligned} \tag{2.2.16}$$

Under these conditions the confluence process may be carried out, yielding one resulting singularity at $z = 0$ having s-rank two $[R_r(0) = 2]$:

$$P(z) = \frac{G_{-2}}{z^2} + \frac{G_{-1}}{z} + p(z),$$
$$Q(z) = \frac{D_{-4}}{z^4} + \frac{D_{-3}}{z^3} + \frac{D_{-2}}{z^2} + \frac{D_{-1}}{z} + q(z),$$

where it is assumed that $p(z)$ and $q(z)$ describe functions that are holomorphic at the resulting singularity at $z = 0$ and whereby the relations between the coefficients in $P(z)$ and in $Q(z)$ before and after the confluence process are given by

$$\begin{aligned} G_{-2} &= -A_{0,-1} = A_{\epsilon,-1}, & G_{-1} &= A_{0,0} + A_{\epsilon,0}, \\ D_{-4} &= B_{\epsilon,-2} = B_{0,-2}, & D_{-3} &= -2\,B_{0,-1} + C_{0,-2}, \\ D_{-2} &= B_{0,0} + B_{\epsilon,0} - 2\,C_{0,-1} - C_{\epsilon,-1}, & D_{-1} &= C_{0,0} + C_{\epsilon,0} \end{aligned} \tag{2.2.17}$$

such that the coefficients of equation (2.2.15) after the confluence process become

$$P(z) = \frac{A_{\epsilon,-1}}{z^2} + \frac{A_{0,0} + A_{\epsilon,0}}{z} + p(z) = -\frac{A_{0,-1}}{z^2} + \frac{A_{0,0} + A_{\epsilon,0}}{z} + p(z),$$

$$\begin{aligned} Q(z) &= \frac{B_{\epsilon,-2}}{z^4} + \frac{-2\,B_{0,-1} + C_{0,-2}}{z^3} + \frac{B_{0,0} + B_{\epsilon,0} - 2\,C_{0,-1} - C_{\epsilon,-1}}{z^2} \\ &\quad + \frac{C_{0,0} + C_{\epsilon,0}}{z} + q(z) \\ &= \frac{B_{0,-2}}{z^4} + \frac{-2\,B_{0,-1} + C_{0,-2}}{z^3} + \frac{B_{0,0} + B_{\epsilon,0} - 2\,C_{0,-1} - C_{\epsilon,-1}}{z^2} \\ &\quad + \frac{C_{0,0} + C_{\epsilon,0}}{z} + q(z). \end{aligned} \tag{2.2.18}$$

In a first step we now assume that the moving singularity of the differential equation (2.2.15) located at $z = \epsilon$ of two coalescing singularities under a confluence process $\epsilon \to 0$ is supposed to be elementary (regular), while the non-moving singularity located at $z = 0$ is still non-elementary regular. It was shown in §2.1.2 that in this case, the following relation holds [cf. eq. (2.2.14)]:

$$B_{\epsilon,-2} = \frac{1}{4}\,A_{\epsilon,-1}^2. \tag{2.2.19}$$

In a second step, we apply an s-homotopic transformation (cf.§1.2.1) that may be generalised to irregular singularities according to

$$y(z) = f(z)\,w(z) \tag{2.2.20}$$

with

$$f(z) = \exp\left(\frac{\gamma_1}{z}\right) z^{\gamma_0}. \tag{2.2.21}$$

This is applied to the differential equation (2.2.15), (2.2.18) with respect to the resulting singularity that is located at the origin, after having carried out the standard confluence process. The parameters γ_0 and γ_1 are to be determined in the following.

According to the third equation of (2.2.17), the differential equation for $w(x)$ retains the form (2.2.15), while the coefficients $P(z)$ and $Q(z)$ are transformed to $\tilde{P}(z)$ and $\tilde{Q}(z)$ according to

$$\begin{aligned} \tilde{P}(z) &= \frac{G_{-2} - 2\,\gamma_1}{z^2} + \frac{G_{-1} + 2\,\gamma_0}{z} + p(z), \\ \tilde{Q}(z) &= \frac{D_{-4} + \gamma_1^2 - G_{-2}\,\gamma_1}{z^4} + \frac{D_{-3} + \gamma_1\,(2 - G_{-1}) + \gamma_0\,(G_{-2} - 2\,\gamma_1)}{z^3} \\ &\quad + \frac{D_{-2} + \gamma_0\,(\gamma_0 - 1) - \gamma_1\,p_0 + G_{-1}\,\gamma_0}{z^2} \\ &\quad + \frac{D_{-1} + \gamma_0\,p_0 - \gamma_1\,p_1}{z} + \tilde{q}(z), \end{aligned} \tag{2.2.22}$$

whereby we have the relations (2.2.17).

As may be seen from the coefficients in (2.2.22), an irregular singularity of the differential equation (2.2.15) having s-rank $R(0) = 2$ is reduced to a singularity having s-rank $R(0) = 3/2$ if γ_1 is chosen such that

$$G_{-2} - 2\,\gamma_1 = 0$$

and

$$D_{-4} + \gamma_1^2 - G_{-2}\,\gamma_1 = 0.$$

The means that γ_1 has to be chosen as

$$\gamma_1 = \frac{1}{2}\,G_{-2}$$

and that we require

$$G_{-2}^2 = 4\,D_{-4}. \tag{2.2.23}$$

Because of the first and third equations in (2.2.17), this condition (2.2.23) requires the relation

$$B_{\epsilon,-2} = \frac{1}{4}\,A_{\epsilon,-1}^2. \tag{2.2.24}$$

As is eventually seen, the formula (2.2.24) is identical to the formula (2.2.19). Thus, the definition of the singularity being elementary is just the condition for determining the value of the parameter γ_1 of the s-homotopic transformation (2.2.20), (2.2.21).

This means that if the moving singularity of two coalescing singularities under a standard confluence process is elementary, then the resulting irregular *singularity is reducible*. This, in turn, means that there exists an s-homotopic transformation that reduces the s-rank of the resulting singularity from $R_r(0) = 2$ by one half, such that it becomes $R_r(0) = 3/2$.

Thus, it may be seen that with

$$\gamma_1 = \frac{1}{2}\,G_{-2}, \quad D_{-4} = \frac{1}{4}\,G_{-2}^2$$

we have

$$G_{-2} - 2\,\gamma_1 = 0,$$
$$D_{-4} + \gamma_1^2 - G_{-2}\,\gamma_1 = 0.$$

Moreover, we see that

$$\begin{aligned} \tilde{D}_{-3} &= D_{-3} + 2\,\gamma_1\,(1 - \gamma_0) + G_{-2}\,\gamma_0 - G_{-1}\,\gamma_1 \\ &= D_{-3} + G_{-2}\left(1 - \frac{1}{2}\,G_{-1}\right). \end{aligned} \tag{2.2.25}$$

Thus, the parameter γ_0 may not be chosen such that $\tilde{D}_{-3}$ vanishes, since it no longer occurs in this expression, meaning that the singularity cannot be reduced twice to become s-rank $R(0) = 1$. The resulting *singularity is irreducible*.

2.3 Local Solutions

2.3.1 Frobenius Solutions

Preliminary Remarks

Singularities of the differential equation (2.1.2)

$$\frac{\mathrm{d}^2y}{\mathrm{d}z^2} + P(z)\,\frac{\mathrm{d}y}{\mathrm{d}z} + Q(z)\,y(z) = 0, \quad z \in \mathbb{C}, \tag{2.3.1}$$

are to be strictly distinguished from singularities of the *solutions* $y = y(z)$ of the differential equation (2.3.1). This may clearly be seen by bearing in mind that there are no solutions of equation (2.3.1) having singularities at ordinary points of the differential equation (2.3.1): singularities of solutions of equation (2.3.1) may occur exclusively at singularities of equation (2.3.1) (cf. Bieberbach, 1965, p. 5). The locations of the singularities of the solutions of equation (2.3.1), in particular, do not depend on the initial conditions

$$y(z = z_0) = y_0, \qquad \left.\frac{\mathrm{d}y}{\mathrm{d}z}\right|_{z=z_0},$$

which themselves determine the particular solutions at ordinary points of equation (2.3.1). This is a consequence of the linearity of equation (2.3.1).

However, the reverse is by no means true: there may occur (particular) solutions $y = y(z)$ of equation (2.3.1) at singularities of equation (2.3.1) that do not have singularities there. It may even happen that the general solution of equation (2.3.1) does not have a singularity at a singularity of equation (2.3.1) but is holomorphic there. Such a singularity of equation (2.3.1) is denoted an *apparent singularity*.

The theory of local solutions of differential equations of type (2.3.1) was elaborated in the second half of the nineteenth century. A significant step came in the work of Lazarus Fuchs, and an important breakthrough was achieved in 1897 when Paul Painlevé proved a theorem that answers the question of analytic continuation of a local solution in the complex plane (cf. Bieberbach, 1965, p. 11).

In the following, we give the main facts and formulae of this brilliant theory. For a detailed study of the theory, the reader is referred to Ince (1956) and Bieberbach (1965), as well as Whittaker and Watson (1927).

Form of Solutions

In the following we consider local particular solutions about a non-elementary regular singularity at $z = z_0$ of equation (2.1.2) (i.e., the s-rank of which is unity):

$$\frac{\mathrm{d}^2y}{\mathrm{d}z^2} + P(z)\,\frac{\mathrm{d}y}{\mathrm{d}z} + Q(z)\,y(z) = 0, \quad z \in \mathbb{C}, \tag{2.3.2}$$

with

$$P(z) = \frac{A}{z - z_0} + p(z), \tag{2.3.3}$$

$$Q(z) = \frac{B}{(z - z_0)^2} + \frac{C}{z - z_0} + q(z), \tag{2.3.4}$$

where $p(z)$ and $q(z)$ are the parts of $P(z)$ and of $Q(z)$ that are holomorphic at $z = z_0$. Therefore, $p(z)$ as well as $q(z)$ may be written as convergent power series:

$$p(z) = \sum_{i=0}^{\infty} p_i \,(z - z_0)^i, \tag{2.3.5}$$

$$q(z) = \sum_{i=0}^{\infty} q_i \,(z - z_0)^i. \tag{2.3.6}$$

These series are convergent in the circular neighbourhood around the singularity z_0 extending to the neighbouring singularity.

It was an important step when Lazarus Fuchs discovered that a local solution of equation (2.3.2) at a regular singularity of equation (2.3.2) may be written as a generalised Laurent series about $z = z_0$ having a *finite* number of negative terms at most.[1] Thus, it may be written in the form

$$y(z) = (z - z_0)^{\alpha_{0r}} \, v(z) \tag{2.3.7}$$

where $v(z)$ is holomorphic at z_0. The quantity[2] α_{0r} is called the *characteristic exponent* of the singularity of equation (2.3.2) at $z = z_0$. $(z - z_0)^{\alpha_{0r}}$ is called the *asymptotic factor* of the solution since it determines the asymptotic behaviour of the solution on approaching radially the singularity at $z = z_0$ of the differential equation.

Entering the differential equation with this ansatz (2.3.7) and taking account of the derivatives

$$\frac{dy}{dz} = (z - z_0)^{\alpha_{0r}} \left[\frac{dv}{dz} + \frac{\alpha_{0r}}{z - z_0}\, v(z)\right]$$

and

$$\frac{d^2y}{dz^2} = (z - z_0)^{\alpha_{0r}} \left[\frac{d^2v}{dz^2} + \frac{2\,\alpha_{0r}}{z - z_0}\frac{dv}{dz} + \frac{\alpha_0\,(\alpha_{0r} - 1)}{(z - z_0)^2}\, v\right]$$

yields the following differential equation for $v(z)$:

$$\frac{d^2v}{dz^2} + \tilde{P}(z)\frac{dv}{dz} + \tilde{Q}(z)\, v = 0. \tag{2.3.8}$$

The coefficients $\tilde{P}(z)$ and $\tilde{Q}(z)$ of equation (2.3.8) are

$$\tilde{P}(z) = P(z) + \frac{2\,\alpha_{0r}}{z - z_0} = \frac{A + 2\,\alpha_{0r}}{z - z_0} + p(z)$$

[1] It should be remembered that a local solution of equation (2.3.2) at an ordinary point of equation (2.3.2) may be written as a pure power series, i.e., a Laurent series without negative terms.

[2] Here, ‘r’ means ‘regular’.

and

$$\tilde{Q}(z) = Q(z) + \frac{P(z)\,\alpha_{0r}}{z - z_0} + \frac{\alpha_{0r}\,(\alpha_{0r} - 1)}{(z - z_0)^2}$$
$$= \frac{\alpha_{0r}\,(\alpha_{0r} - 1) + B + A\,\alpha_{0r}}{(z - z_0)^2} + \frac{C + \alpha_{0r}\,p(z)}{z - z_0} + q(z).$$

Thus, the separation of the asymptotic factor in the ansatz (2.3.7) does not influence the structure of the original differential equation (2.3.2), so that it is possible to write (2.3.8) in the form

$$\frac{d^2v}{dz^2} + \left[\frac{\hat{p}_{-1}}{z - z_0} + \hat{p}_0 + \hat{p}_1\,(z - z_0) + \cdots\right]\frac{dv}{dz}$$
$$+ \left[\frac{\hat{q}_{-2}}{(z - z_0)^2} + \frac{\hat{q}_{-1}}{z - z_0} + \hat{q}_0 + \hat{q}_1\,(z - z_0) + \cdots\right] v = 0,$$

where

$$\hat{p}_{-1} = \hat{p}_{-1}(\alpha_{0r}) = A + 2\,\alpha_{0r},$$
$$\hat{p}_i = p_i,\ i = 0, 1, 2, \ldots,$$
$$\hat{q}_{-2} = \hat{q}_{-2}(\alpha_{0r}) = \alpha_{0r}\,(\alpha_{0r} - 1) + B + A\,\alpha_{0r},$$
$$\hat{q}_{-1} = \hat{q}_{-1}(\alpha_{0r}) = C + \alpha_{0r}\,p_0,$$
$$\hat{q}_i = \hat{q}_i(\alpha_{0r}) = q_i + \alpha_{0r}\,p_{i+1},\ i = 0, 1, 2, \ldots.$$

The crucial point now is that $v(z)$ may be written in the form of a power series. Although the structure of the differential equation (2.3.8) is the same as that of the original equation (2.3.2), this is possible because there is an additional parameter in equation (2.3.8) which may be disposed arbitrarily. Therefore we take the ansatz

$$v(z) = \sum_{n=0}^{\infty} a_n\,(z - z_0)^n, \tag{2.3.9}$$

resulting in the following difference equations of triangular shape:

$$\begin{aligned}
&\hat{q}_{-2}\,a_0 = 0,\\
&(1\cdot 0 + 1\,\hat{p}_{-1} + \hat{q}_{-2})\,a_1 + (0\,\hat{p}_0 + \hat{q}_{-1})\,a_0 = 0,\\
&(2\cdot 1 + 2\,\hat{p}_{-1} + \hat{q}_{-2})\,a_2 + (1\,\hat{p}_0 + \hat{q}_{-1})\,a_1 + \hat{q}_0\,a_0 = 0,\\
&(3\cdot 2 + 3\,\hat{p}_{-1} + \hat{q}_{-2})\,a_3 + (2\,\hat{p}_0 + \hat{q}_{-1})\,a_2\\
&\qquad + (\hat{p}_1 + \hat{q}_0)\,a_1 + \hat{q}_1\,a_0 = 0,\\
&(4\cdot 3 + 4\,\hat{p}_{-1} + \hat{q}_{-2})\,a_4 + (3\,\hat{p}_0 + \hat{q}_{-1})\,a_3\\
&\qquad + (2\,\hat{p}_1 + \hat{q}_0)\,a_2 + (\hat{p}_2 + \hat{q}_1)\,a_1 + \hat{q}_2\,a_0 = 0,\\
&(5\cdot 4 + 5\,\hat{p}_{-1} + \hat{q}_{-2})\,a_5 + (4\,\hat{p}_0 + \hat{q}_{-1})\,a_4 + (3\,\hat{p}_1 + \hat{q}_0)\,a_3\\
&\qquad + (2\,\hat{p}_2 + \hat{q}_1)\,a_2 + (\hat{p}_3 + \hat{q}_2)\,a_1 + \hat{q}_3\,a_0 = 0,\\
&\ldots
\end{aligned} \tag{2.3.10}$$

As is immediately seen from the first equation, this difference equation provides a non-trivial solution; viz. $a_n \neq 0$, $n = 0, 1, 2, 3, \ldots$, only if

$$\hat{q}_{-2} = 0. \tag{2.3.11}$$

Then the difference equation takes the form

$$\begin{aligned}
&(1 \cdot 0 + 1\,\hat{p}_{-1})\,a_1 + (0\,\hat{p}_0 + \hat{q}_{-1})\,a_0 = 0,\\
&(2 \cdot 1 + 2\,\hat{p}_{-1})\,a_2 + (1\,\hat{p}_0 + \hat{q}_{-1})\,a_1 + \hat{q}_0\,a_0 = 0,\\
&(3 \cdot 2 + 3\,\hat{p}_{-1})\,a_3 + (2\,\hat{p}_0 + \hat{q}_{-1})\,a_2 + (\hat{p}_1 + \hat{q}_0)\,a_1 + \hat{q}_1\,a_0 = 0,\\
&(4 \cdot 3 + 4\,\hat{p}_{-1})\,a_4 + (3\,\hat{p}_0 + \hat{q}_{-1})\,a_3\\
&\qquad + (2\,\hat{p}_1 + \hat{q}_0)\,a_2 + (\hat{p}_2 + \hat{q}_1)\,a_1 + \hat{q}_2\,a_0 = 0,\\
&(5 \cdot 4 + 5\,\hat{p}_{-1})\,a_5 + (4\,\hat{p}_0 + \hat{q}_{-1})\,a_4 + (3\,\hat{p}_1 + \hat{q}_0)\,a_3\\
&\qquad + (2\,\hat{p}_2 + \hat{q}_1)\,a_2 + (\hat{p}_3)\,a_1 + \hat{q}_3\,a_0 = 0
\end{aligned} \tag{2.3.12}$$

or

$$[n\,(n-1) + n\,\hat{p}_{-1}]\;a_n + \sum_{i=1}^{n} [(n-i)\,\hat{p}_{i-1} + \hat{q}_{i-2}]\;a_{n-i} = 0,\; n = 1, 2, 3, 4, \ldots.$$

Since it is assumed that $a_0 \neq 0$, the difference equation (2.3.12) becomes a one-sided, linear recurrence relation if and only if $\hat{q}_2 = 0$ holds such that all subsequent terms a_n, $n = 1, 2, 3, \ldots$, may be calculated recursively.

The condition (2.3.11)

$$\hat{q}_{-2} = 0$$

is the determining condition of the *characteristic exponent* $\alpha_{0\mathrm{r}}$, which is actually a quadratic equation:

$$\alpha_{0\mathrm{r}}^2 + (A - 1)\,\alpha_{0\mathrm{r}} + B = 0 \tag{2.3.13}$$

where A and B are from (2.3.3), (2.3.4).[3] Therefore, in general, there are two different values $\alpha_{0\mathrm{r}1}$ and $\alpha_{0\mathrm{r}2}$ of the characteristic equation (2.3.13) and two different sequences a_{nj}, $j = 1, 2$, $n = 0, 1, 2, 3, \ldots$, as linearly independent particular solutions $\{a_n^{(j)}\}$ of the difference equation (2.3.12) determining two local particular solutions $y_j(z)$ of the differential equation (2.3.2) that may be taken as a fundamental system (cf. Bieberbach, 1965, p. 136).

Equation (2.3.13) is called the *indicial equation* of the singularity, or *characteristic equation* or *fundamental equation* of the characteristic exponents $\alpha_{0\mathrm{r}1}$, $\alpha_{0\mathrm{r}2}$, the solutions of which are given by

$$\alpha_{0\mathrm{r}1,0\mathrm{r}2} = \frac{1 - A \pm \sqrt{(1-A)^2 - 4\,B}}{2}. \tag{2.3.14}$$

[3] It should be recognised that the coefficient C in (2.3.4) does not influence the characteristic exponents of the singularity of equation (2.3.13) at $z = z_0$.

In Figure 2.1 it is shown that the relations between the characteristic exponents form four regions in the $A - B$ plane with different signs, and where they are real- or complex-valued.

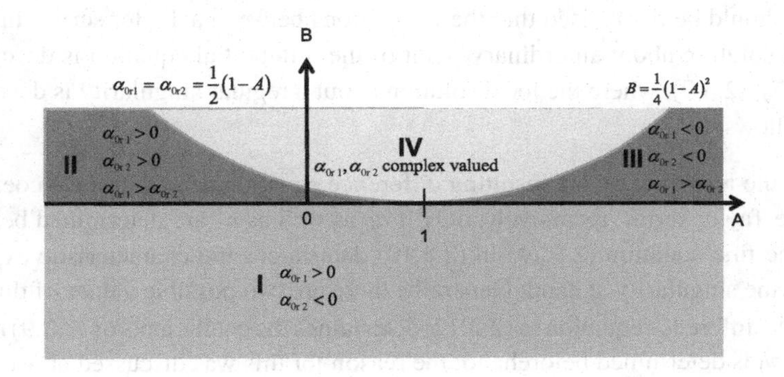

Figure 2.1 Relations between the characteristic exponents.

The series in (2.3.9) are generally convergent, with radius of convergence the distance between $z = z_0$ and the neighbouring singularity of equation (2.3.2) (cf. Bieberbach, 1965, p. 133).

It is remarkable that the two differential equations for $v_j(z)$, $j = 1, 2$,

$$\begin{aligned}\frac{\mathrm{d}^2 v_j}{\mathrm{d}z^2} &+ \left[\frac{\hat{p}_{-1}(\alpha_{0j})}{z - z_0} + \hat{p}_0 + \hat{p}_1(z - z_0) + \cdots\right]\frac{\mathrm{d}v_j}{\mathrm{d}z} \\ &+ \left[\frac{\hat{q}_{-1}(\alpha_{0j})}{z - z_0} + \hat{q}_0(\alpha_{0j}) + \hat{q}_1(\alpha_{0j})(z - z_0) + \cdots\right] v_j = 0\end{aligned} \tag{2.3.15}$$

still have a non-elementary regular singularity at $z = z_0$ if $\hat{p}_{-1} \neq 0$ and $\hat{q}_{-1} \neq 0$ as is supposed to be the case. However, at least one of its particular solutions about $z = z_0$ is holomorphic there.

It is plausible that (2.3.15) still has a singularity at $z = z_0$. A singularity of a differential equation (2.1.1) makes the second derivative vanishing there, locally; viz. locally, at the singularity, the differential equation becomes – roughly speaking – a first-order one. This means that a particular solution is determined by specifying the functional value $v(z_0) = v_0$ at $z = z_0$. In particular, it is not possible to specify the derivative

$$\frac{\mathrm{d}v}{\mathrm{d}z}$$

at $z = z_0$, since this is determined by the first equation of (2.3.15). However, this is just what happens when the determination of a_0 in the difference equation (2.3.12) is sufficient to determine its solution $\{a_n\}$, $n = 1, 2, 3, \ldots$, by starting the (one-sided) recurrence relation.

It is important for the following to understand that the difference

$$\alpha_{0r1} - \alpha_{0r2}$$

of the two characteristic exponents is given by

$$\alpha_{0r1} - \alpha_{0r2} = \sqrt{(1 - A)^2 - 4\,B}.$$

It should be recognised that the distinctions between a Taylor series, in which the local solution about an ordinary point of the differential equation is developed, and (2.3.7), (2.3.9), where the local solution about a regular singularity is developed, are as follows:

- In the former case, the resulting difference equation determines the coefficients of the Taylor series, recursively, only if a_0 as well as a_1 are determined beforehand.
- The first equation (2.3.11) in (2.3.10) determines the characteristic exponent α_{0r} of the singularity at hand. Generally, there are two possible values of this quantity.
- The difference equation in (2.3.12) determines the coefficients of (2.3.9) recursively if a_0 is determined beforehand; the reason for this was discussed above.

It is significant for the series in (2.3.9) that just one coefficient (namely a_0) is sufficient for determining the particular solution of equation (2.3.2), although it is a *second-order* differential equation. This is a consequence of the singularity and may be made plausible by remembering that just at the singularity, equation (2.1.1) becomes of first order.

The power series (2.3.9) converges in a disc about the singularity $z = z_0$, the radius of convergence of which goes as far as the neighbouring singularity. Thus, (2.3.7), (2.3.9) is a local solution of equation (2.3.2) about a regular singularity at $z = z_0$ that is called the *Frobenius solution* (Ince, 1956, p. 396).

A prominent feature of the Frobenius solution

$$y(z) = (z - z_0)^{\alpha_{0r}} \sum_{k=0}^{\infty} a_n\ (z - z_0)^n \tag{2.3.16}$$

is that it behaves multiplicatively on surrounding the singularity of equation (2.3.2) at $z = z_0$, viz. it is multiplied by

$$\exp\left(2\imath\pi\alpha_{0r}\right), \quad \imath = \sqrt{-1},$$

after having encircled the singularity once anticlockwise:

$$y(z\ \exp\left(2\imath\pi\right)) = \exp\left(2\imath\pi\alpha_0\right)\ y(z),$$

where z is taken to be a value in the neighbourhood of $z = z_0$. Therefore, the Frobenius solutions are called *multiplicative solutions* (cf. Bieberbach, 1965, p. 125). The situation is sketched in Figure 2.2.

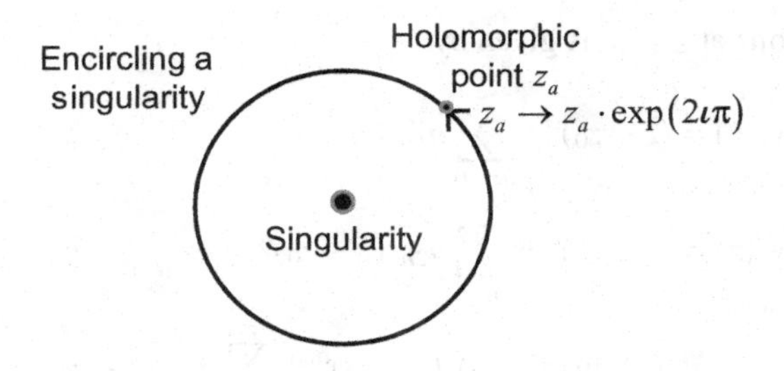

$$y(z_a) \to y(z_a \cdot \exp(2\iota\pi)) = \exp(2\iota\pi\alpha_0) \cdot y(z_a)$$

Figure 2.2 Multiplicative local solutions at regular singularities of differential equations.

In particular, if $\alpha_{0r} = 0$ then the solution is unchanged, and for

$$\alpha_{0r} = 1/2$$

the solution is multiplied by minus unity. These solutions have gained some recognition in the past, since they obey some symmetrical behaviour in regard to their radial behaviour with respect to the singularity. This, in particular, is the reason why the elementary singularities of equation (2.1.2) have been defined in a somewhat peculiar manner in (2.1.9).

The characteristic exponent α_{0r} may be restricted to the interval $[0, 1]$, since all other values may be incorporated into the numbering of the series (2.3.9).

It is to the merit of Lazarus Fuchs (1866) that he could show that (2.3.7), (2.3.9) is a convergent representation of a particular solution of equation (2.3.2) in the neigbourhood of its regular singularities $z = z_i$.

The procedure discussed above yields a first particular solution $y(z) = y_1(z)$ of equation (2.3.2) with a first characteristic exponent $\alpha_{0r} = \alpha_{0r1}$ and a first sequence a_{n1} of coefficients of the ansatz (2.3.8), (2.3.9).

If the difference

$$\alpha_{0r2} - \alpha_{0r1}$$

of the two solutions of the characteristic equation (2.3.13) is not an integer (including zero), then the procedure discussed above yields a further linearly independent particular solution of equation (2.3.2) at $z = z_0$ having the same form (2.3.7), (2.3.9) as the former one.

As a result, in this case $y_1(z)$ and $y_2(z)$ constitute a fundamental system of solutions of (2.3.2) at $z = z_0$, all linear combinations of which form the general solution of equation (2.3.2).

Eventually, it should be mentioned that in the case when the characteristic equation (2.3.13) has solutions the difference of which is an integer including zero, there may occur logarithms. Bieberbach (1965, p. 136) shows that in this case the fundamental system of solutions at $z = z_0$ is given by

$$y_1(z) = (z - z_0)^{\alpha_{0r1}} \sum_{n=0}^{\infty} a_{1n} (z - z_0)^n, \tag{2.3.17}$$

$$y_2(z) = (z - z_0)^{\alpha_{0r2}} \sum_{n=0}^{\infty} a_{2n} (z - z_0)^n$$

$$+ A \ln (z - z_0) (z - z_0)^{\alpha_{0r1}} \sum_{n=0}^{\infty} a_{1n} (z - z_0)^n. \tag{2.3.18}$$

The quantity A is either zero or unity. The reader should note here that the factor of the logarithm (besides A) is the multiplicative solution $y_1(z)$. Moreover, it is to be recognised that only if $A = 0$ holds will the second solution of (2.3.17) be multiplicative as well.

Finally, we make a remark about the coefficients a_n. These quantities obey a linear, homogeneous difference equation of triangular shape that is solvable – after having fixed the first coefficient a_0 – recursively, meaning that all the successive values a_n are determined by (2.3.12).

There remains the question about the form of the Frobenius solution in the case when

$$A = B = 0$$

in the coefficients (2.3.3) and (2.3.4). It seems to be plausible that in this case the s-rank of the singularity is $R = 1/2$; however, this is not the case. As may be seen from (2.1.9), the definition of the s-rank of elementary singularities of equation (2.3.2) is different. That this is reasonable may be seen by considering the form of the Frobenius solution.

Instead of (2.3.3), (2.3.4), we have

$$P(z) = p(z),$$
$$Q(z) = \frac{C}{z - z_0} + q(z).$$

From (2.3.13) it follows that the characteristic exponent α_{0r} takes on two possible values:

$$\alpha_{0r1} = 0,$$
$$\alpha_{0r2} = 1.$$

From this it follows that in both cases

$$\hat{q}_{-2} = 0,$$

and there are two values of p_{-1} depending on the value of α_{0r}:

$$\alpha_{0r1} = 0 : p_{-11} = 0, \quad q_{-11} = C,$$
$$\alpha_{0r2} = 1 : p_{-12} = 2, \quad q_{-12} = C + \hat{p}_0.$$

In the first case, a_0 has to be zero while a_1 may be chosen arbitrarily, and all the subsequent terms $a_2, a_3, \ldots$ may be calculated recursively from (2.3.12).

In the second case ($p_{-1} = p_{-12} = 2$), a_0 is arbitrary and all the subsequent terms $a_1, a_2, a_3, \ldots$ may be calculated recursively from (2.3.12).

In particular, it is seen that the characteristic exponent does not take the value $1/2$. Therefore, the singularity cannot consistently be defined as having s-rank $R = 1/2$.

The Point at Infinity

Singularities of differential equations (2.3.1) are first of all local quantities. Thus, up to now, we have exclusively been concerned with local properties with regard to the particular solutions. Now it is quite natural to ask what properties the particular solutions have at the point at infinity. In §1.2.1 this non-actual point was made an actual one by inverting the underlying differential equation and then proceeding with the point in the origin of the inverted differential equation as if this point were a finite one. First of all, there is nothing at all wrong with this. However, it must not be overlooked that in this kind of treatment the particular solutions are only local quantities with respect to the inverted differential equation, not with respect to the original one. The global implications of the non-actual point will be the focus of the following. For technical reasons, we consider differential equations with exclusively regular singularities. The question is posed under what conditions for the coefficients of the differential equation the non-actual point is a regular singularity or an ordinary point of the differential equation, i.e., is holomorphic there.

Thus, it is considered that the differential equation (2.3.1)

$$\frac{\mathrm{d}^2 y}{\mathrm{d}z^2} + P(z)\,\frac{\mathrm{d}y}{\mathrm{d}z} + Q(z)\,y(z) = 0, \quad z \in \mathbb{C}, \tag{2.3.19}$$

either has a regular singularity of the differential equation at infinity or an ordinary point. The traditional description of this derivation can be found in Bieberbach (1965, pp. 202–204).

If $z_1, z_2, z_3, \ldots, z_n$ are all the finite regular singularities of the differential equation (2.3.19), as a start, the decompositions of the coefficients $P(z)$ and $Q(z)$ into partial fractions have to admit the form

$$P(z) = \sum_{k=1}^{n} \frac{A_k}{z - z_k} + g_1(z), \quad Q(z) = \sum_{k=1}^{n} \left[\frac{B_k}{(z - z_k)^2} + \frac{C_k}{z - z_k}\right] + g_2(z).$$

Hereby, the coefficients A_k, B_k, C_k are (in general complex-valued) numbers, and $g_1(z)$ and $g_2(z)$ are entire functions. In addition to this, the condition has to be met that the point at infinity $z = \infty$ is either a regular singularity of the differential equation (2.3.19) or is holomorphic there. The condition for this is that the differential equation that occurs from the inversion $z = \frac{1}{\zeta}$ of (2.3.19)

$$\frac{\mathrm{d}^2 y}{\mathrm{d}\zeta^2} + \tilde{P}(\zeta)\,\frac{\mathrm{d}y}{\mathrm{d}\zeta} + \tilde{Q}(\zeta)\,y(\zeta) = 0, \quad \zeta \in \mathbb{C}, \tag{2.3.20}$$

with

$$\begin{aligned}\tilde{P}(\zeta) &= \frac{2}{\zeta} - \frac{1}{\zeta^2} P(\zeta),\\ \tilde{Q}(\zeta) &= \frac{1}{\zeta^4} Q(\zeta)\end{aligned} \tag{2.3.21}$$

either has a regular singularity at $\zeta = 0$ or is holomorphic there. This means that the function

$$\frac{2}{\zeta} - \frac{1}{\zeta^2} P(\zeta)$$

has at most a first-order pole at $\zeta = 0$ and the function

$$\frac{1}{\zeta^4} Q(\zeta)$$

has at most a second-order pole at $\zeta = 0$. For this to be so, it is necessary and sufficient that

$$\lim_{z\to\infty} P(z) = 0, \quad \lim_{z\to\infty} z\,Q(z) = 0$$

holds. For this to be so, it is necessary and sufficient, that

$$g_1(z) \equiv 0, \quad g_2(z) \equiv 0, \quad \sum_{k=1}^{n} C_k = 0$$

holds.

As a result, it may be stated that it is necessary and sufficient, for a differential equation (2.3.19) to be Fuchsian and to have at most finite regular singularities at $z_1, z_2, z_3, \ldots, z_k$ and at the point at infinity $z = \infty$, that this is the decomposition of its coefficients $P(z)$ and $Q(z)$ into partial fractions:

$$\begin{aligned}P(z) &= \sum_{k=1}^{n} \frac{A_k}{z - z_k},\\ Q(z) &= \sum_{k=1}^{n} \left[\frac{B_k}{(z - z_k)^2} + \frac{C_k}{z - z_k}\right],\\ &\sum_{k=1}^{n} C_k = 0.\end{aligned} \tag{2.3.22}$$

This version of the condition does not express that each of these points is actually a singularity of the differential equation (2.3.19). It may also be a holomorphic point, for the condition does not express which of the coefficients A_k, B_k, C_k vanishes.

If, in particular, the point $z = \infty$ is to be holomorphic, it is necessary and sufficient that

$$\frac{2}{\zeta} - \frac{1}{\zeta^2} P(\zeta) \text{ and } \frac{1}{\zeta^4} Q(\zeta)$$

are holomorphic at $\zeta = 0$. According to (2.3.22), this means that

$$\frac{2}{\zeta} - \frac{1}{\zeta}\sum_{k=1}^{n}\frac{A_k}{1 - z_k\,\zeta} \quad \text{and} \quad \frac{1}{\zeta^3}\sum_{k=1}^{n}\left[\frac{B_k}{(1 - z_k\,\zeta)^2} + \frac{C_k}{1 - z_k\,\zeta}\right]$$

are holomorphic at $\zeta = 0$. In turn, for this to be so, it is necessary and sufficient that

$$\begin{aligned} &\sum_{k=1}^{n} A_k = 2, \\ &\sum_{k=1}^{n} C_k = 0, \\ &\sum_{k=1}^{n} (B_k + C_k\, z_k) = 0, \\ &\sum_{k=1}^{n} \left(2\, B_k\, z_k + C_k\, z_k^2\right) = 0 \end{aligned} \tag{2.3.23}$$

hold.

It should be stated here that these considerations include the fact that there is no differential equation (2.3.19) without singularities: for (2.3.19) to have no finite singularity requires the coefficients $P(z)$ and $Q(z)$ to be entire functions, but (2.3.21) shows that in this case the point at infinity is a singularity of the differential equation (2.3.19).

At last we give the characteristic equation for the characteristic exponents of a differential equation having exclusively regular singularities. Suppose (2.3.19) has finite regular singularities at $z_1, z_2, \ldots, z_n$ and a regular singularity at infinity $z = \infty$, then the characteristic equation for the singularity at infinity is given by (Bieberbach, 1965, p. 128):

$$\alpha_{0\infty}\,(\alpha_{0\infty} - 1) + \left(2 - \sum_{k=1}^{n} A_k\right)\alpha_{0\infty} + \sum_{k=1}^{n} (B_k + C_k\, z_k) = 0 \quad \text{at } z = \infty, \tag{2.3.24}$$

resulting in

$$\alpha_{0\infty 1} = \frac{\sum_{k=1}^{n} A_k - 1 + \sqrt{\left(\sum_{k=1}^{n} A_k - 1\right)^2 - 4\,\sum_{k=1}^{n} (B_k + C_k\, z_k)}}{2},$$

$$\alpha_{0\infty 2} = \frac{\sum_{k=1}^{n} A_k - 1 - \sqrt{\left(\sum_{k=1}^{n} A_k - 1\right)^2 - 4\,\sum_{k=1}^{n} (B_k + C_k\, z_k)}}{2}.$$

The sum of all characteristic exponents of a Fuchsian differential equation is an invariant of linear transformations, Moebius as well as s-homotopic. In the case where infinity is an ordinary point of the differential equation [i.e., the coefficients $P(z)$ and $Q(z)$ are holomorphic there] this sum is $n - 2$, where n is the number of finite singularities of the differential equation; it is $n - 1$ in the case where infinity is a (regular) singularity of the differential equation, where n also is the number of finite singularities (cf. Bieberbach, 1965, pp. 204, 205).

2.3.2 Thomé Solutions

We now have the question of a convergent representation of particular solutions about irregular singularities of equation (2.1.2), for only in this case will an exact solution of the underlying singular boundary eigenvalue problems possibly await. There is indeed such an equivalent, however it is a rather complicated calculation involving two-sided infinite difference equations and their determinants (cf. Bieberbach, 1965, p. 139).

It is one of the fundamental statements in this book that convergent, two-sided ansatzes of particular solutions of differential equations (2.1.2) at irregular singularities for the purposes of singular boundary eigenvalue problems may be avoided: singular boundary eigenvalue problems of these differential equations may be solved exactly without taking into account two-sided infinite series. Therefore, it is not necessary to dive into these rather complex and voluminous algebraic and analytical calculations in order to get convergent representations of solutions of singular boundary eigenvalue problems involving irregular singularities, but it is sufficient to restrict to one-sided ansatzes. Although the local series ansatzes about the irregular singularities are not convergent, but just asymptotic in all cases, the singular boundary eigenvalue problems may be formulated on the basis of convergent series. The benefit, however, of this restriction to one-sided series is a significant reduction of calculative complexity. Therefore, in the following, the idea of one-sidedness of the series in the ansatzes for the particluar solutions at irregular singularities is the paramount agent for coping with irregular singularities in this book. We mention that an asymptotic representation of a function may be rather helpful, particularly in cases where functional behaviours are under discussion. A demonstration of this sort of usefulness is given in §3.3.1 concerning the *Stokes phenomenon* when the *Airy function* is discussed.

There were early efforts by the German mathematician **Ludwig Wilhelm Thomé** in the late nineteenth century to develop representations of local solutions about irregular singularities on the basis of one-sided infinite series. These ansatzes are – formally – quite similar to the Frobenius ones. They consist of an asymptotic factor and a one-sided infinite series of power type. The only formal difference is that the asymptotic factor contains an exponential function.

However, the fact that – as turns out to be the case – the series in general do not have finite radius of convergence confused Thomé and he spent a lot of effort trying to find criteria under which his series were convergent. The correct interpretation of Thomé's series as asymptotic representations of particular solutions of differential equations (2.1.2) at irregular singularities has, however, only been given by Jules Henri Poincaré. Here we present Thomé's series solutions with respect to the singular boundary eigenvalue problem.

Integer s-Ranks

In the following, Thomé's solutions of the form

$$y(z) = f(z)\left(1 + \sum_{n=1}^{\infty} \frac{C_n}{z^n}\right) \tag{2.3.25}$$

for irregular singularities with integer s-ranks of the differential equation (2.1.1) are given explicitly in tabular form.[4] The singularities of the differential equations always lie at infinity. The local representation of the coefficients $P(z)$ and $Q(z)$ in the vicinity of the singularity is followed by the specifications of the asymptotic factor $f(z)$ and then those of the characteristic coefficients contained therein.

s-Rank Two

Coefficients:

$$P(z) = p_0 + \sum_{i=1}^{\infty} \frac{p_{-i}}{z^i}, \tag{2.3.26}$$

$$Q(z) = q_0 + \sum_{i=1}^{\infty} \frac{q_{-i}}{z^i}. \tag{2.3.27}$$

Asymptotic factor:

$$f(z) = \exp(\alpha_1 z)\, z^{\alpha_0}. \tag{2.3.28}$$

Characteristic exponents:

$$\alpha_{1j} = \frac{1}{2}\left(-p_0 \pm \sqrt{p_0^2 - 4\, q_0}\right), \tag{2.3.29}$$

$$\alpha_{0j} = -\frac{p_{-1}\,\alpha_{1j} + q_{-1}}{p_0 + 2\,\alpha_{1j}}, \tag{2.3.30}$$

with $j = 1, 2$.

[4] There are Thomé solutions for irregular singularities having an odd half-integer s-rank in which occur half-integer powers of z. These have the form

$$y_m(z;\infty) = f(z;\infty) \sum_{n=0}^{\infty} a_{nm}\, z^{-n/2}, \quad m = 1, 2,$$

with

$$f(z;\infty) = \exp\left(\sum_{k=\frac{1}{2}}^{R(\infty)-1} \frac{\alpha_{km}(\infty)}{k}\, z^k\right) z^{\alpha_{0m}(\infty)},$$

but are not considered in detail here.

s-Rank Three

Coefficients:

$$P(z) = p_1 z + p_0 + \sum_{i=0}^{\infty} \frac{p_{-i}}{z^i},$$
$$Q(z) = q_2 z^2 + q_1 z + q_0 + \sum_{i=0}^{\infty} \frac{q_{-i}}{z^i}. \tag{2.3.31}$$

Asymptotic factor:

$$f(z) = \exp\left(\frac{\alpha_2}{2} z^2 + \alpha_1 z\right) z^{\alpha_0}. \tag{2.3.32}$$

Characteristic exponents:

$$\begin{aligned}
\alpha_{2j} &= \frac{1}{2}\left(-p_1 \pm \sqrt{p_1^2 - 4 q_2}\right),\\
\alpha_{1j} &= -\frac{p_0 \alpha_{2j} + q_1}{p_1 + 2 \alpha_{2j}},\\
\alpha_{0j} &= -\frac{\alpha_{2j} (p_{-1} + 1) + \alpha_{1j} (p_0 + \alpha_{1j}) + q_0}{p_1 + 2 \alpha_{2j}},
\end{aligned} \tag{2.3.33}$$

with $j = 1, 2$.

s-Rank Four

Coefficients:

$$P(z) = p_2 z^2 + p_1 z + p_0 + \sum_{i=1}^{\infty} \frac{p_{-i}}{z^i},$$
$$Q(z) = q_4 z^4 + q_3 z^3 + q_2 z^2 + q_1 z + q_0 + \sum_{i=1}^{\infty} \frac{q_{-i}}{z^i}.$$

Asymptotic factor:

$$f(z) = \exp\left(\frac{\alpha_3}{3} z^3 + \frac{\alpha_2}{2} z^2 + \alpha_1 z\right) z^{\alpha_0}. \tag{2.3.34}$$

Characteristic exponents:

$$\begin{aligned}
\alpha_{3j} &= \frac{1}{2}\left(-p_2 \pm \sqrt{p_2^2 - 4 q_4}\right),\\
\alpha_{2j} &= -\frac{p_1 \alpha_{3j} + q_3}{p_2 + 2 \alpha_{3j}},\\
\alpha_{1j} &= -\frac{\alpha_{3j} p_0 + \alpha_{2j} (p_1 + \alpha_{2j}) + q_2}{p_2 + 2 \alpha_{3j}},\\
\alpha_{0j} &= -\frac{\alpha_{3j} (p_{-1} + 2) + \alpha_2 p_0 + \alpha_{1j} (p_1 + 2 \alpha_{2j}) + q_1}{p_2 + 2 \alpha_{3j}},
\end{aligned} \tag{2.3.35}$$

with $j = 1, 2$.

From these concrete details, we can derive the scheme for the ansatz in the general case, in which the irregular singularity at infinity has s-rank $R(\infty) = s, \quad s = 2, 3, 4, \ldots$,:

$$y(z) = f(z)\left(1 + \sum_{n=0}^{\infty} \frac{C_n}{z^n}\right) \tag{2.3.36}$$

with

$$f(z) = \exp\left(\sum_{k=1}^{R(\infty)-1} \frac{\alpha_k}{k} z^k\right) z^{\alpha_0} \tag{2.3.37}$$

where the index k runs along integer numbers: $k = 1, 2, 3, \ldots$.

Thomé Solutions and the Stokes Phenomenon

Differential equations (2.3.1) having irregular singularities are called *confluent cases* of Fuchsian differential equations. This will be dealt with in detail in Chapter 3. Confluent cases of Fuchsian differential equations are no longer Fuchsian. By definition, they do have irregular singularities. There is no Frobenius solution about this sort of singularity any more. Irregular singularities of differential equations are points of indefiniteness. This means that convergent representations of local solutions comprise Laurent series solutions having an infinite number of positive as well as negative terms. This fact generates difference equations for the coefficients of these Laurent series that are two-sided infinite. This two-sidedness is a technical challenge for solving boundary eigenvalue problems according to the method presented in Chapters 1 and 3.

However, in order to solve the boundary eigenvalue problems it is not necessary to dive into this technically rather ambitious methods. In fact, it is sufficient to resort to Thomé solutions as introduced in §2.3.2 above. These ansatzes are sufficient in that they consist of two parts, an asymptotic factor and an (in general divergent) series of power type. As stated above, it was shown by Henri Poincaré that these ansatzes represent the local solutions about irregular singularities, asymptotically, approaching the singularity radially.

What is important for the method of solving the singular boundary eigenvalue problems is the asymptotic fractor of the ansatz, since this factor dominates the asymptotic behaviour of the solution along a straight line originating from the singularity at hand. Therefore, in the following, it is sufficient to keep in mind that there are local solutions of irregular singularities of differential equations (2.1.2), although these will not be necessary for treating the singular boundary eigenvalue problems according to the applied method, beyond its asymptotic factors.

However, there is a price to be paid for facilitating one-sidedness of infinite series in the case of irregular singularities. This price is the so-called *Stokes phenomenon*, to be discussed below.

There is some similarity between the Frobenius solutions (2.3.16)

$$y(z) = z^{\alpha_{0r}} \sum_{n=0}^{\infty} a_n z^n$$

and the Thomé solutions [cf. (2.3.25), (2.3.36)]

$$y(z) = \exp\left[g(z)\right]\, z^{\alpha_{0i}} \left(1 + \sum_{n=1}^{\infty} C_n\, z^{-n}\right),$$

since both of them consist of an asymptotic factor in front of an infinite series of power type.[5] In both of these asymptotic factors is contained a power α_0 of the independent variable z that is called the *characteristic exponent*. This similarity is an example of the reason for the warnings that are given to students of mathematical analysis that formal similarity does not mean similarity as regards content. The reason for this is to be discussed in the following.

Consider, as an example, the seemingly simple differential equation

$$\frac{d^2y}{dz^2} - z\,y = 0. \tag{2.3.38}$$

This equation has just one singularity, being placed at infinity. Thus, it does not have finite singularities at all. This singularity is irregular, having s-rank $R(\infty) = \frac{5}{2}$. Equation (2.3.38) is a famous differential equation, both from a mathematical and a physical point of view. It describes the intensity of light across a rainbow in the framework of the classical wave theory of light and emerges in what is nowadays known as the *Stokes phenomenon*, discovered in 1857 by **Sir George Gabriel Stokes**, one of the most prominent mathematical physicists ever. The differential equation (2.3.38) is called the *Airy equation*, since it was **George Bidell Airy** who seems to have written down for the first time one of its particular solutions with respect to the above-mentioned rainbow problem.

Because equation (2.3.38) does not have finite singularities, thus in particular the origin $z = 0$ is an ordinary point of the Airy equation (2.3.38), each particular solution of (2.3.38) may be written in the form of a convergent power series; the radius of convergence is infinite. Thus, the general solution of (2.3.38) may be written in the form

$$y_g(z) = c_1 \sum_{n=0}^{\infty} a_{n1}\, z^n + c_2 \sum_{n=0}^{\infty} a_{n2}\, z^n, \tag{2.3.39}$$

where the two constants c_1 and c_2 in z may be chosen arbitrarily. Both of these power series are convergent in the whole complex z-plane, and thus are entire functions. Hence, all particular solutions of the Airy equation are entire functions in z.

The two Thomé solutions at the singularity at infinity may generally be used to represent the general solution $y_g(z)$ of the Airy equation [cf. Slavyanov, 1996, pp. 10, 11, equations (2.6), (2.8)]:

[5] In the case of the Thomé solutions it is a matter of inverse power series, because the singularity of the underlying differential equation is supposed to lie at infinity.

$$y_g = C_1 \exp\left(\frac{2}{3} z^{\frac{3}{2}}\right) z^{-\frac{1}{4}} \left(1 + \sum_{n=0}^{\infty} c_{n1} z^{-\frac{n}{2}}\right)$$
$$+ C_2 \exp\left(-\frac{2}{3} z^{\frac{3}{2}}\right) z^{-\frac{1}{4}} \left(1 + \sum_{n=0}^{\infty} c_{n2} z^{-\frac{n}{2}}\right), \tag{2.3.40}$$

where the two constants C_1 and C_2 in z may be chosen arbitrarily. As long as these two Thomé solutions are linearly independent, this formula represents the general solution $y_g(z)$ as well.

Comparing the two representations (2.3.39) and (2.3.40) already indicates that there is something odd, since the power term $z^{-\frac{1}{4}}$ is by no means an entire function. This indeed lies at the heart of the Stokes phenomenon and is the origin of the problem, discussed in the following.

The power series of the representation (2.3.39) about the origin as a regular singularity of the differential equation (2.3.38) has a non-zero radius of convergence, actually reaching the neighbouring singularity of the underlying differential equation, and thus is unlimited.

The power series of the representation (2.3.40) of a Thomé solution at the irregular singularity of (2.3.38) does not necessarily have a non-zero radius of convergence. A consequence of this is that the constants c_1, c_2 are numbers for representations of fixed particular solutions of the Airy equation, valid in a complete neighbourhood (i.e., circular disc) of the origin (i.e., the whole complex plane $z \in \mathbb{C}$ here), while the values of the constants C_1, C_2 representing fixed particular solutions of the Airy equation are dependent on sector-shaped regions, the cusp of which is located at the origin (cf. Slavyanov and Veshev, 1997, §I.2.4). Thus, the representation (2.3.40) of the solutions of (2.3.38) by means of Thomé series solutions determines particular solutions of the Airy equation *only within these sectors*, while changing the sector means changing the values of the constants C_1, C_2 for a certain particular solution of the Airy equation, or, in other words, keeping the constants C_1, C_2 means changing the particular solution.

This is the reason why the Thomé series in (2.3.40) is generally divergent, since a power series has radius of convergence larger than zero only if it represents the function within a complete circular disc about the point of expansion.

A line separating two such sectors, discussed above, is called a *Stokes line* and the corresponding sector is called a **Stokes sector**. For the Airy equation, the Stokes lines are given by (cf. Slavyanov and Veshev, 1997, §I.2.4)

$$|\arg z| = \frac{2}{3}\pi. \tag{2.3.41}$$

This means that in crossing these lines surrounding the origin, the constants C_1 and C_2 change their values for one and the same particular solution of equation (2.3.38).

To illustrate with a concrete example, we consider the Airy function as a specific particular solution of the differential equation (2.3.38) on the positive and on the negative real axis, along with its asymptotic behaviours for $z \to +\infty$ and for $z \to -\infty$. Since the Airy equation (2.3.38) does not have any parameter, the values of

$$\left.\frac{\mathrm{d}y}{\mathrm{d}x}\right|_{z=0}$$

and of $y(z = 0)$ uniquely determine the particular solution, that is, as discussed above, an entire function. Consider the particular solution of (2.3.40) that decreases exponentially as $z \to \infty$ along the positive real axis, thus like

$$y = C_2 \exp\left(-\frac{2}{3} z^{\frac{3}{2}}\right) z^{-\frac{1}{4}} \left(1 + \sum_{n=0}^{\infty} c_{n2}\, z^{-\frac{n}{2}}\right), \tag{2.3.42}$$

viz. $C_1 = 0$ and $C_2 = \frac{1}{2\sqrt{\pi}} \neq 0$ (cf. Slavyanov, 1996, p. 13). In the form (2.3.39), this solution is determined by (cf. Slavyanov, 1996, p. 8)

$$c_1 = \frac{3^{-\frac{2}{3}}}{\Gamma\left(\frac{2}{3}\right)}, \qquad c_2 = \frac{3^{-\frac{1}{3}}}{\Gamma\left(\frac{1}{3}\right)}.$$

A question occurs over the change in constants C_1 and C_2 of this particular solution to C_1' and C_2' while changing from the positive to the negative real axis, viz. the asymptotics is taken along the negative real axis. The answer is (cf. Slavyanov, 1996, pp. 13, 14)

$$C_1' = \frac{\imath}{2\sqrt{\pi}}, \qquad C_2' = C_2 = \frac{1}{2\sqrt{\pi}}.$$

This may be shown by recognising that the Airy function as a particular solution of the Airy equation (being real-valued on the positive real axis) is an entire function and thus also real-valued as $z \to -\infty$ along the negative real axis.

This, in turn, may be seen by recognising that

$$\begin{aligned} &\frac{1}{z^{\frac{1}{4}}} \exp\left[-\frac{2}{3} z^{\frac{3}{2}}\right] + \imath \frac{1}{z^{\frac{1}{4}}} \exp\left[+\frac{2}{3} z^{\frac{3}{2}}\right] \\ &= \frac{2}{(-z)^{\frac{1}{4}}} \left\{\cos\left[\frac{2}{3}(-z)^{\frac{3}{2}}\right] \sin\left(\frac{\pi}{4}\right) - \sin\left[\frac{2}{3}(-z)^{\frac{3}{2}}\right] \sin\left(\frac{\pi}{4}\right)\right\} \end{aligned} \tag{2.3.43}$$

for real-valued negative z: $z < 0$. Formula (2.3.43) tells us that the Airy function is decaying with increasing positive real values of z and oscillating with increasing negative real values of z, while the amplitude of these oscillations is decaying according to $(-z)^{-\frac{1}{4}}$, $z < 0$. Thus, $(-z)^{-\frac{1}{4}}$, $z < 0$ is an envelope of the oscillations on the negative real axis, while it is dominated on the positive real axis by the exponentially increasing function. This is displayed in Figure 2.3.

Seen from a purely formal point of view, (2.3.43) is quite different from (2.3.40), which is by no means obvious or trivial. The Airy function as a specific particular solution of the Airy differential equation (2.3.38) is exponentially decreasing as $z \to$

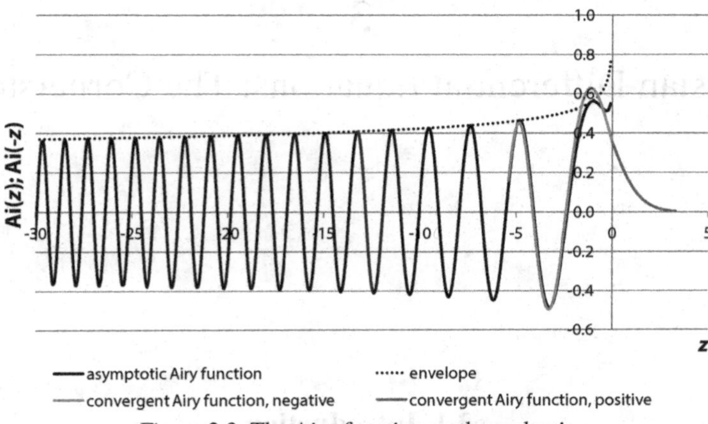

Figure 2.3 The Airy function on the real axis.

∞ along the positive real axis and is oscillating with slightly decreasing amplitudes along the negative real axis.

This example clearly shows the usefulness of asymptotic representations. It is quite plausible that it is hard to represent such behaviour by means of a convergent power series about the origin, resulting in even a modest rate of convergence.

As a result, it has to be stated that the Stokes phenomenon may occur with Thomé solutions as representations of particular solutions at irregular singularities of differential equations (2.1.2). Or, in other words, the Stokes phenomenon is the price for ansatzes of particular solutions at irregular singularities of differential equation (2.1.2), the series of which are one-sided infinite instead of two-sided infinite.

For calculational reasons, one-sidedness of infinite series as solutions of differential equations (2.1.2) at irregular singularities is a dogma in this book. It is the crux of the method presented for applying the mathematical principle of unmasking recessive particular solutions of differential equations (2.1.2) to solve singular boundary eigenvalue problems that the drawback occurring as a result of the Stokes phenomenon can be avoided. Although it is necessary to keep in mind that this phenomenon does exist, concrete calculations on the basis of the method are not at all touched by the phenomenon.

3

Fuchsian Differential Equations: The Cornerstones

3.1 Introduction

While Chapter 2 is devoted mainly to *local* properties of singularities of differential equations (2.1.2) and to the local solutions of differential equations (2.1.2) about their singularities, in the following, *global* properties come into play.

The necessity to distinguish between Frobenius and Thomé solutions entails the distinction between regular and irregular singularities of differential equations. On the level of differential equations this, in turn, entails the distinction between *Fuchsian differential equations* and *confluent cases of Fuchsian differential equations*.

The definition of a Fuchsian differential equation is that its singularities are exclusively regular. This is a rather useful definition, since Fuchsian differential equations are thus completely characterisable by just one number, the number of its singularities. So, a differential equation (2.1.2)

$$\frac{d^2y}{dz^2} + P(z)\,\frac{dy}{dz} + Q(z)\,y = 0, \quad z \in \mathbb{C}, \tag{3.1.1}$$

is defined as being a *Fuchsian differential equation* if all its singularities are regular ones. For a differential equation to be Fuchsian is rather a strong condition, so that they may be characterised in general, as we have already done in §2.3.1 (cf. Fuchs, 1866; Bieberbach, 1965, p. 203), by the following theorem of Fuchs.[1]

Theorem 3.1. *A differential* equation (3.1.1) *is Fuchsian with singularities precisely at* $z_1, z_2, \ldots, z_n, \infty$ *if and only if its coefficients* $P(z)$ *and* $Q(z)$ *have the following expansions into partial fractions – cf.* (2.3.22)*:*

$$\begin{aligned} P(z) &= \sum_{k=1}^{n} \frac{A_k}{z - z_k}, \\ Q(z) &= \sum_{k=1}^{n} \left[\frac{B_k}{(z - z_k)^2} + \frac{C_k}{z - z_k} \right] \end{aligned} \tag{3.1.2}$$

[1] Lazarus Fuchs is considered the founder of the theory of linear second-order ordinary differential equation. This theorem, which he proved in 1866, is regarded as a cornerstone of the theory.

and

$$\sum_{k=1}^{n} C_k = 0. \tag{3.1.3}$$

If the point at infinity is an ordinary point of the Fuchsian differential equation (3.1.1), (3.1.2), *then the parameters of the coefficients $P(z)$ and $Q(z)$ must, additionally to* (3.1.3), *meet three further conditions (cf. Bieberbach, 1965, p. 204 and §2.3.1):*

$$\boxed{\begin{aligned} \textstyle\sum_{k=1}^{n} A_k &= 2, \\ \textstyle\sum_{k=1}^{n} (B_k + C_k\, z_k) &= 0, \\ \textstyle\sum_{k=1}^{n} (2B_k\, z_k + C_k\, z_k^2) &= 0. \end{aligned}} \tag{3.1.4}$$

From a mathematical point of view, it is of no relevance whether the point at infinity is an ordinary point of the differential equation (3.1.1) (i.e., whether the coefficients $P(z)$ and $Q(z)$ are holomorphic at infinity) or a singular one; moreover, it is of no fundamental relevance where the finite singularities of a differential equation (3.1.1), (3.1.2) are located. However, it is, in a sense, 'natural' for a differential equation (3.1.1), (3.1.2) to have a singularity at the point at infinity; and even more, that this singularity at infinity has the highest s-rank among all the other singularities of the differential equation. Therefore, a differential equation (3.1.1), (3.1.2) is said to be in its *natural form* if it has a singularity at infinity, the s-rank of which is its highest one. If, as is the case for Fuchsian differential equations, more than one singularity has the highest s-rank, then any one of them is taken to lie at infinity.

Fuchsian differential equations in natural form having less than five singularities bear names:

- A Fuchsian differential equation having one singularity is called a *Laplace equation*.
- A Fuchsian differential equation having two singularities is called an *Euler equation*.
- A Fuchsian differential equation having three singularities is called a *Gauss equation*.
- A Fuchsian differential equation having four singularities is called a *Heun equation*.

Strictly speaking, the literature only calls these equations so if one has passed from the natural form of the Fuchsian differential equations through Moebius and through s-homotopic transformations to those forms that have the following properties:

- Three of its singularities are located at infinity, at the origin and at unity (in this sequence if the differential equation only has one or two singularities).
- At least one of the characteristic exponents of each pair at all of its finite singularities is zero.

However, I already call equations with the above listed designations the general differential equations of Fuchs class, while I call those which possess the latter properties *standard forms*. While singular boundary eigenvalue problems of the Gauss differential equation[2] and their confluent cases produce *classical special functions*, singular boundary eigenvalue problems of the Heun differential equation and its confluent cases produce *higher special functions*. Therefore, in this chapter, mainly Heun's differential equation and its confluent cases will be considered. §§3.2 and 3.3 consider the Fuchs equations, while §3.4 focuses on the confluent cases.

Confluent cases of Fuchsian differential equations are characterised by having at least one irregular singularity. Confluent cases of Fuchsian differential equations may be thought of as being created by means of confluence processes applied to Fuchsian differential equations.

In the literature, almost no Fuchsian differential equation with more than four singularities has been considered so far, although it became clear very early in the twentieth century that such equations also have some significance. These efforts go back mainly to **Edward Lindsay Ince** (1891–1941) and to **Felix Medland Arscott** (1922–1996). I therefore refer to the Fuchsian differential equation with five singularities as the *Arscott differential equation* and the differential equation with six singularities as the *Ince differential equation*.

Similarly, I speak of the *set of Arscott differential equations* and the *set of Ince differential equations* as the totality of the Arscott differential equations or the Ince differential equations, respectively, and their forms.

Moreover, I also refer to the *class of Arscott differential equations* and the *class of Ince differential equations* as the totality of the Arscott differential equations or the Ince differential equations, respectively, and their confluent and reduced cases.

Finally, I refer to the *differential equations of Arscott type* and the *differential equations of Ince type* as the totality of the class and the set of Arscott differential equations or the class and the set of Ince differential equations.

These two equations will be dealt with in more detail in §3.5.2, thereby opening the door to the land beyond Heun.

3.2 Heun Differential Equation

The Fuchsian differential equation having four singularities, all lying in the finite z-plane, is given by

$$\begin{aligned} \frac{d^2y}{dz^2} &+ \left(\frac{A_{z1}}{z-z_1} + \frac{A_{z2}}{z-z_2} + \frac{A_{z3}}{z-z_3} + \frac{A_{z4}}{z-z_4}\right)\frac{dy}{dz} \\ &+ \left[\frac{B_{z1}}{(z-z_1)^2} + \frac{B_{z2}}{(z-z_2)^2} + \frac{B_{z3}}{(z-z_3)^2} + \frac{B_{z4}}{(z-z_4)^2}\right. \\ &\left. + \frac{C_{z1}}{z-z_1} + \frac{C_{z2}}{z-z_2} + \frac{C_{z3}}{z-z_3} + \frac{C_{z4}}{z-z_4}\right] y = 0. \end{aligned} \tag{3.2.5}$$

[2] Singular boundary eigenvalue problems of Laplace and Euler equations produce elementary functions.

In order to get the conditions of the parameters for the point at infinity to be an ordinary point of the differential equation (3.2.5), one has to limit the formulae (3.1.2) and (3.1.4) to $n = 4$:

$$\begin{aligned}
&\sum_{k=1}^{4} A_{z_k} = 2,\\
&\sum_{k=1}^{4} C_{z_k} = 0,\\
&\sum_{k=1}^{4} (B_{z_k} + C_{z_i}\, z_k) = 0,\\
&\sum_{k=1}^{4} (2\, B_{z_k}\, z_k + C_{z_k}\, z_k^2) = 0.
\end{aligned} \tag{3.2.6}$$

Equation (3.2.5), (3.2.6) is the *Heun differential equation* in its *general form.* It is named after Karl Heun, who investigated the differential equation first and published this treatment in Heun (1889).

The Heun differential equation by definition has four finite singularities, the s-rank of which is unity. In the form (3.2.5), (3.2.6) all of these four singularities are finite ones. Because the differential equation is of second order, there is a pair of characteristic exponents at each of these four singularities which obeys a second-order algebraic, i.e., quadratic, equation (cf. Bieberbach, 1965, p. 204):

$$\alpha_{z_i}\,\left(\alpha_{z_i} - 1\right) + A_{z_i}\,\alpha_{z_i} + B_{z_i} = 0, \quad i = 1,\ldots,4, \tag{3.2.7}$$

from which result four pairs of characteristic exponents:

$$\alpha_{z_i 1},\, \alpha_{z_i 2}, \quad i = 1,\ldots,4. \tag{3.2.8}$$

If one of the singularities in (3.2.5) is located at infinity, then (3.2.5) is written in its *natural form*

$$\begin{aligned}
&\frac{d^2y}{dz^2} + \left(\frac{A_{z_1}}{z - z_1} + \frac{A_{z_2}}{z - z_2} + \frac{A_{z_3}}{z - z_3}\right)\frac{dy}{dz}\\
&\quad + \left[\frac{B_{z_1}}{(z - z_1)^2} + \frac{B_{z_2}}{(z - z_2)^2} + \frac{B_{z_3}}{(z - z_3)^2}\right.\\
&\qquad \left. + \frac{C_{z_1}}{z - z_1} + \frac{C_{z_2}}{z - z_2} + \frac{C_{z_3}}{z - z_3}\right] y = 0.
\end{aligned} \tag{3.2.9}$$

In order for the point at infinity to be a regular singularity of the differential equation, the parameters C_{z_i}, $i = 1, 2, 3$, have to satisfy the following condition:

$$\sum_{i=1}^{3} C_{z_i} = 0. \tag{3.2.10}$$

The Heun differential equation by definition has four finite singularities, the s-rank of which is unity. In the form (3.2.5), (3.2.6) all of these four singularities are finite ones. Because the differential equation is of second order, there is a pair of characteristic exponents at each of these four singularities which obeys a second-order algebraic, i.e., quadratic, equation [cf. Bieberbach, 1965, p. 204, (2.3.24)]:

$$\alpha_{z_k}\left(\alpha_{z_k}-1\right)+A_{z_k}\,\alpha_{z_k}+B_{z_k}=0, \quad k=1,\ldots,3,$$
$$\alpha_{\infty}\left(\alpha_{\infty}-1\right)+\left(2-\sum_{k=1}^{3}A_{z_k}\right)\alpha_{\infty}+\sum_{k=1}^{3}\left(B_{z_k}+C_{z_k}\,z_k\right)=0, \tag{3.2.11}$$

from which result four pairs of characteristic exponents:

$$\alpha_{z_k 1},\ \alpha_{z_k 2},\ \alpha_{\infty 1},\ \alpha_{\infty 2}, \quad k=1,\ldots,3. \tag{3.2.12}$$

As was already mentioned at the end of §2.3.1, the sum of all characteristic exponents of a Fuchsian differential equation is an invariant of linear transformations, Moebius as well as s-homotopic. In the case where infinity is an ordinary point of the differential equation (i.e., the coefficients $P(z)$ and $Q(z)$ are holomorphic there) this sum is $n-2$ where n is the number of finite singularities of the differential equation, and $n-1$ in the case where infinity is a (regular) singularity of the differential equation, where n is also the number of finite singularities (cf. Bieberbach, 1965, pp. 204–5).

Thus, the sum of all eight characteristic exponents of the general form (3.2.5) as well as of the natural form (3.2.9) is

$$\sum_{k=1}^{3}\left(\alpha_{z_k 1}+\alpha_{z_k 2}\right)+\alpha_{\infty 1}+\alpha_{\infty 2}=2. \tag{3.2.13}$$

It may be of some interest to have the relations between the coefficients of the Heun differential equation (3.2.9) and its characteristic exponents. These are given in the following.

By means of Moebius transformations it is always possible to put four of the singularities of (3.2.5) at $z=0$, $z=a=\frac{z_3-z_2}{z_3-z_1}\frac{z_4-z_1}{z_4-z_2}$, $z=1$ and $z=\infty$. Moreover, by means of s-homotopic transformations it is always possible to make the coefficients B_{z_k}, $k=1,2,3$ vanish, thus

$$B_{z_k}=0. \tag{3.2.14}$$

Differential equations in natural form for which two finite singularities are located at $z=0$ and $z=1$ and, moreover, for which (3.2.14) holds are said to be in **standard form**.

The Fuchsian differential equation having four singularities in its standard form looks like

$$\frac{d^2y}{dz^2}+\left(\frac{A_0}{z}+\frac{A_a}{z-a}+\frac{A_1}{z-1}\right)\frac{dy}{dz}+\left(\frac{C_0}{z}+\frac{C_a}{z-a}+\frac{C_1}{z-1}\right)y=0. \tag{3.2.15}$$

In order for the point at infinity to be a regular singularity, the condition

$$C_0+C_a+C_1=0$$

has to hold. In the literature, this differential equation is often written in the form in which the parameters are the non-zero characteristic exponents of its singularities:[3]

$$\frac{d^2y}{dz^2} + \left(\frac{1-\alpha_0}{z} + \frac{1-\alpha_a}{z-a} + \frac{1-\alpha_1}{z-1}\right)\frac{dy}{dz} + \frac{\alpha_{\infty 1}\,\alpha_{\infty 2}\,z + q}{z\,(z-a)\,(z-1)}\,y = 0 \tag{3.2.16}$$

or

$$\frac{d^2y}{dz^2} + \left(\frac{1-\alpha_0}{z} + \frac{1-\alpha_a}{z-a} + \frac{1-\alpha_1}{z-1}\right)\frac{dy}{dz}$$
$$+ \left[\frac{\alpha_{\infty 1}\,\alpha_{\infty 2} + q}{a\,z} + \frac{\alpha_{\infty 1}\,\alpha_{\infty 2} + q}{a\,(a-1)\,(z-a)} - \frac{\alpha_{\infty 1}\,\alpha_{\infty 2} + q}{(a-1)\,(z-1)}\right] y = 0.$$

Thus, the relations between the parameters of equation (3.2.15) and the characteristic exponents of the singularities of the differential equation (3.2.16) are

$$\begin{aligned}
A_0 &= 1 - \alpha_{01},\\
A_a &= 1 - \alpha_{a1},\\
A_1 &= 1 - \alpha_{11},\\
C_0 &= \frac{q}{a},\\
C_a &= \frac{a\,\alpha_{\infty 1}\,\alpha_{\infty 2} + q}{a\,(a-1)} = -\frac{q}{a} + \frac{\alpha_{\infty 1}\,\alpha_{\infty 2} + q}{a-1},\\
C_1 &= -\frac{\alpha_{\infty 1}\,\alpha_{\infty 2} + q}{a-1}.
\end{aligned}$$

As may be seen

$$C_0 + C_a + C_1 = 0$$

in order for the point at infinity to be a regular singularity; moreover

$$\begin{aligned}
\alpha_{01} &= 1 - A_0,\\
\alpha_{a1} &= 1 - A_a,\\
\alpha_{11} &= 1 - A_1,\\
\alpha_{\infty 1} &= \frac{1}{2}\Big[A_0 + A_a + A_1 - 1\\
&\quad + \sqrt{(A_0 + A_a + A_1 - 1)^2 - 4\,(B_0 + B_a + B_1 + C_a\,a + C_1)}\Big].
\end{aligned} \tag{3.2.17}$$

[3] Sometimes, the sign of the accessory parameter q is negative, thus it is written $-q$ instead of q (see, e.g., Ronveaux, 1995, p. 7); however, it is obvious that this is not as convenient as taking a positive sign, as done here.

whereas

$$\begin{aligned}
\alpha_{02} &= 0, \\
\alpha_{a2} &= 0, \\
\alpha_{12} &= 0, \\
\alpha_{\infty 2} &= \frac{1}{2}\Big[A_0 + A_a + A_1 - 1 \\
&\quad - \sqrt{(A_0 + A_a + A_1 - 1)^2 - 4\,(B_0 + B_a + B_1 + C_a\,a + C_1)}\Big].
\end{aligned} \tag{3.2.18}$$

Because there are no quadratic terms in the coefficients

$$B_0 = B_a = B_1 = 0,$$

thus the characteristic exponents at infinity simplify to

$$\alpha_{\infty 1} = \frac{1}{2}\left[A_0 + A_a + A_1 - 1 + \sqrt{(A_0 + A_a + A_1 - 1)^2 - 4\,(C_a\,a + C_1)}\right],$$

$$\alpha_{\infty 2} = \frac{1}{2}\left[A_0 + A_a + A_1 - 1 - \sqrt{(A_0 + A_a + A_1 - 1)^2 - 4\,(C_a\,a + C_1)}\right].$$

As a result, it may be verified that (cf. the last paragraph of §2.3.1 and equation (3.2.24))

$$\alpha_{01} + \alpha_{a1} + \alpha_{11} + \alpha_{\infty 1} + \alpha_{02} + \alpha_{a2} + \alpha_{12} + \alpha_{\infty 2} = 2. \tag{3.2.19}$$

3.2.1 Linear Transformations

Moebius Transformations

The Heun differential equation whose four singularities are located at $x = 0$, $x = a$, $x = 1$ and $x = \infty$ is given by

$$\begin{aligned}
&\frac{d^2y}{dx^2} + \left(\frac{A_0}{x} + \frac{A_a}{x-a} + \frac{A_1}{z-1}\right)\frac{dy}{dx} \\
&\quad + \left(\frac{B_0}{x^2} + \frac{B_a}{(x-a)^2} + \frac{B_1}{(x-1)^2} + \frac{C_0}{x} + \frac{C_a}{x-a} + \frac{C_1}{x-1}\right) y = 0.
\end{aligned} \tag{3.2.20}$$

This may be derived from equation (3.2.5) by means of the Moebius transformation (1.2.26):

$$x = \frac{z_3 - z_4}{z_3 - z_1}\,\frac{z - z_1}{z - z_4}, \tag{3.2.21}$$

mapping the four singularities z_1, z_2, z_3, z_4 according to the following:

$$\begin{array}{cccccc}
z: & z_1 & z_2 & z_3 & z_4 \\
\downarrow & \downarrow & \downarrow & \downarrow & \downarrow \\
x: & 0 & a & 1 & \infty
\end{array}$$

whereby the singularity at $z = a$ is given by

$$a = \frac{z_3 - z_4}{z_3 - z_1}\,\frac{z_2 - z_1}{z_2 - z_4}.$$

In order for the point at infinity to be a regular singularity of the differential equation (3.2.20), the condition

$$C_0 + C_a + C_1 = 0 \tag{3.2.22}$$

has to hold.

The values of the characteristic exponents of the singularities of the differential equation (3.2.20) are not touched by the Moebius transformation (3.2.21), thus

$$\begin{aligned}
\alpha_{z_1 1} &= \alpha_{01}, \ \alpha_{z_1 2} = \alpha_{02}, \\
\alpha_{z_2 1} &= \alpha_{a1}, \ \alpha_{z_2 3} = \alpha_{a2}, \\
\alpha_{z_3 1} &= \alpha_{11}, \ \alpha_{z_3 2} = \alpha_{12}, \\
\alpha_{z_4 1} &= \alpha_{\infty 1}, \ \alpha_{z_4 2} = \alpha_{\infty 2}.
\end{aligned} \tag{3.2.23}$$

Therefore, this sum is [cf. equation (3.2.19)]

$$\begin{aligned}
&\alpha_{z_1 1} + \alpha_{z_1 2} + \alpha_{z_2 1} + \alpha_{z_2 2}\alpha_{z_3 1} + \alpha_{z_3 2} + \alpha_{z_4 1} + \alpha_{z_4 2} \\
&\quad = \alpha_{01} + \alpha_{02} + \alpha_{a1} + \alpha_{a2}\alpha_{11} + \alpha_{12} + \alpha_{\infty 1} + \alpha_{\infty 2} = 2.
\end{aligned} \tag{3.2.24}$$

The Heun equation (3.2.20) with (3.2.22) is in its *natural form.*

s-Homotopic Transformations

In §1.2.2 the s-homotopic transformation of the differential equation (1.2.2), (1.2.1)

$$\frac{d^2y}{dz^2} + P(z)\,\frac{dy}{dz} + Q(z)\,y(z) = 0, \quad z \in \mathbb{C}, \tag{3.2.25}$$

with respect to a supposed singularity at the origin $z = 0$ was defined. It was shown that the quadratic pole in the coefficient $Q(z)$ of the differential equation of the underlying singularity thereby becomes zero. The reason for this is that in (1.2.30), (1.2.31),

$$f(z) = z^{\alpha_0} \tag{3.2.26}$$

was taken, i.e., one of the two characteristic exponents α_{01} or α_{02} of the singularity of the differential equation at the origin $z = 0$. This has the consequence that one of the two Frobenius solutions of the singularity at hand degenerates to a Taylor series, thus representing a particular solution of the differential equation that is holomorphic there. If $\alpha_0 = \alpha_{01}$ is taken, then the particular solution with just this characteristic exponent degenerates.

It is evident that s-homotopic transformations can be applied simultaneously with respect to several singularities. This effect then occurs with respect to all these Frobenius solutions.

So, if we apply such an s-homotopic transformation (1.2.30), (1.2.31) with respect to all the finite singularities of the Heun differential equation (3.2.20):

$$w(z) = z^{\alpha_{0i}}\,(z-a)^{\alpha_{aj}}\,(z-1)^{\alpha_{1k}}\,y(z), \tag{3.2.27}$$

whereby i, j, k may – in each of these three cases – take on the values 1 and 2.

This transformation results in the Heun differential equation in its standard form (cf. §1.2.3, above) and thus at least one of its characteristic exponents at all of its finite singularities vanishes:

$$\frac{d^2w}{dx^2} + \left(\frac{A_0}{x} + \frac{A_a}{x-a} + \frac{A_1}{x-1}\right)\frac{dw}{dx} + \left(\frac{C_0}{x} + \frac{C_a}{x-a} + \frac{C_1}{x-1}\right) w = 0 \qquad (3.2.28)$$

with

$$C_0 + C_a + C_1 = 0.$$

Moreover, the action of the s-homotopic transformation is the vanishing of all of the parameters B_0, B_a, B_1, which means the vanishing of one of the two characteristic exponents in (3.2.8) at each of the finite singularities of the differential equation (3.2.28).

As may be seen, the Heun differential equation (3.2.28) in its standard form as a Fuchsian differential equation having four singularities (3.2.5) may be obtained from the general form by means of a *Moebius transformation* (3.2.21) as well as an s-*homotopic transformation* (3.2.27).

As already mentioned in §1.2.3, the totality of Moebius as well as s-homotopic transformations of differential equations (2.1.2) creates the *set of forms* of a differential equation. This will be considered in more detail in the following.

3.2.2 Forms and Symbols

Forms

It is quite clear from what is written in §1.2.3 and in §3.2.1 that each Fuchsian differential equation having four singularities may be transformed by means of a Moebius and a subsequent s-homotopic transformation such that one of its singularities comes to lie on the origin $z = 0$, one comes to lie at unity $z = 1$ and one comes to lie at infinity $z = \infty$. Moreover, one of the two characteristic exponents of the singularity at hand becomes zero. Such a differential equation was defined to be in its standard form.

A consequence of this is that by means of Moebius and s-homotopic, thus linear, transformations, at least one of the two Frobenius solutions at each of the singularities at hand of these Fuchsian differential equations is holomorphic there.

The *form of a differential equation* distinguishes equation (3.1.1) according to the locations of its singularities, as well as to the concrete values of the pairs of characteristic exponents belonging to its singularities. As may be understood, this opens a rich variety of forms of equation (3.1.1). The totality of forms of a differential equation (3.1.1) is generated by means of Moebius as well as s-homotopic transformations, creating an innumerable set.

It is the merit of the theory to give incisive criteria for the substantial characteristics of equation (3.1.1). It is a matter of historical evolution that we have a few forms of equation (3.1.1) that have gained a certain attention and thus are introduced here and listed below.

If nothing is specified further, Heun differential equations (3.1.1) will be in a form that is characterised as follows:

- *General form.* No parameter of the differential equation is specified.
- *Natural form.* The s-rank of the singularity at infinity of the differential equation (3.1.1) is equal to or higher than the s-ranks of all of its finite singularities.
- *Standard form.* This is the natural form with in addition the following conditions met:
  - At least one of the characteristic exponents at each of the finite singularities is zero.
  - If the differential equation (3.1.1) has two singularities, then one of them is located at the origin $z = 0$.
  - If the differential equation (3.1.1) has three or more singularities, then one of them is located at unity $z = 1$.

There are three other forms of equation (3.1.1) mentioned in the following (cf. Slavyanov and Lay, 2000, pp. 4, 5, 13, 21):

- *Normal form.* The coefficient in front of the first derivative vanishes.
- *Self-adjoint form.* As is generally known, the self-adjoint form of a differential equation (2.1.1) is characterised by the condition

$$\frac{\mathrm{d}P_0(z)}{\mathrm{d}z} = P_1(z)$$

  (see Courant and Hilbert, 1968, p. 238).
- *Canonical form.* A differential equation for which all the zeros of $P_0(z)$ are simple ones is said to be in canonical form.

s-Rank and P-Symbols

It is of some use to have a symbol, attached to a differential equation, from which the number of its singularities as well as their s-ranks may be seen at a glance. Such a symbol is called the s-*rank symbol*. Since the s-rank is characteristic for irreducible singularities of equation (2.1.2), this restriction is to be assumed when an s-rank symbol is written down. Thus, the s-rank symbol is a horizontal list of the s-ranks of all the singularities of a differential equation (2.1.2), separated by a comma, while the singularity at infinity is separated at the end of the list by means of a semicolon. The singularities of the differential equation in the finite plane are not ordered. As a consequence, from the s-rank symbol it cannot be seen where the finite singularities of the differential equation are located.

Examples

- The s-rank symbol $\{1; 1\}$ represents a Fuchsian differential equation (2.1.2) having two non-elementary regular singularities, one of which is located at infinity and the other somewhere in the finite complex plane. Thus, it is in its natural form.
- The s-rank symbol $\{1, 1;\}$ represents a Fuchsian differential equation (2.1.2) having two non-elementary regular singularities, both of which are located somewhere in the finite complex plane. Thus, it is not in its natural form.
- The s-rank symbol $\{1; 2\}$ represents a differential equation (2.1.2) in natural form having two singularities, the non-elementary regular one of which is located somewhere in the finite complex plane. The s-rank of the irregular singularity of equation (2.1.2) is two.
- The s-rank symbol $\{1, 2;\}$ represents a differential equation (2.1.2) in non-natural form having two singularities, one of which is non-elementary regular and the other irregular with s-rank two. Both of these singularities are located somewhere in the finite complex plane.
- The s-rank symbol of equation (3.2.5) is

$$\{1, 1, 1, 1;\}$$

and that of equation (3.2.15) is

$$\{1, 1, 1; 1\}.$$

The s-rank symbol does not contain all the information of a differential equation (2.1.2), thus it is not possible to write down the differential equation explicitly, with only the information of the s-rank symbol available.

However, it may be quite useful to have a symbol that contains the full information about the underlying differential equation such that it is possible to write it down from this symbol. This idea is not new; already in the middle of the nineteenth century the German mathematician **Bernhard Riemann** realised it for the Fuchsian differential equation having three singularities. This is pointed out in the following.

Consider a Fuchsian differential equation (2.1.2) having three singularities. At each of its singularities z_i there exist two characteristic exponents $\alpha_{z_i 1}$, $\alpha_{z_i 2}$. These pairs of singularity parameters completely determine the Frobenius solutions of equation (2.1.2) at its singularities. Therefore, instead of writing the differential equation it is equivalent to write down a scheme in which all these parameters are displayed:

$$\boxed{P\begin{pmatrix} z_1 & z_2 & z_3 & \\ \alpha_{z_1 1} & \alpha_{z_2 1} & \alpha_{z_3 1} & z \\ \alpha_{z_1 2} & \alpha_{z_2 2} & \alpha_{z_3 2} & \end{pmatrix}.} \tag{3.2.29}$$

In the first line are written the locations of the three singularities of the differential equation. The second and third lines present the characteristic exponents. In the last row is written the independent variable of the differential equation at hand. This scheme is referred to as the *Riemann* P-*symbol*.

Two generalisations of the Riemann P-symbol are given, the first of which is introduced in the following and the second introduced in §3.5.1.

It is not a big step to generalise the Riemann P-symbol to a Fuchsian differential equation having more than three singularities, simply by adding as many rows as necessary and displaying the accessory parameters of the differential equation. This sort of symbol will be called the *conventional Riemann symbol*.[4]

The conventional Riemann symbol of the Fuchsian differential equation (3.2.5) having four finite singularities is given by

$$P\begin{pmatrix} 1 & 1 & 1 & 1 & \\ z_1 & z_2 & z_3 & z_4 & z \\ \alpha_{z_1 1} & \alpha_{z_2 1} & \alpha_{z_3 1} & \alpha_{z_4 1} & q \\ \alpha_{z_1 2} & \alpha_{z_2 2} & \alpha_{z_3 2} & \alpha_{z_4 2} & \end{pmatrix}. \tag{3.2.30}$$

Shifting the singularities at $z = z_1$, $z = z_2$, $z = z_3$ and $z = z_4$ to $x = 0$, $x = a$, $x = 1$ and $x = \infty$, respectively, the conventional Riemann symbol becomes

$$P\begin{pmatrix} 1 & 1 & 1 & 1 & \\ 0 & a & 1 & \infty & x \\ \alpha_{z_1 1} & \alpha_{z_2 1} & \alpha_{z_3 1} & \alpha_{z_4 1} & q \\ \alpha_{z_1 2} & \alpha_{z_2 2} & \alpha_{z_3 2} & \alpha_{z_4 2} & \end{pmatrix}.$$

Eventually, the s-homotopic transformation (3.2.27) yields the Heun equation (3.2.15) or (3.2.16), of which the conventional Riemann symbol is

$$P\begin{pmatrix} 1 & 1 & 1 & 1 & \\ 0 & a & 1 & \infty & x \\ 0 & 0 & 0 & \alpha_{z_4 1} + \alpha_{z_1 1} + \alpha_{z_2 1} + \alpha_{z_3 1} & q \\ \alpha_{z_1 2} - \alpha_{z_1 1} & \alpha_{z_2 2} - \alpha_{z_2 1} & \alpha_{z_3 2} - \alpha_{z_3 1} & \alpha_{z_4 2} + \alpha_{z_1 1} + \alpha_{z_2 1} + \alpha_{z_3 1} & \end{pmatrix}$$

or

$$P\begin{pmatrix} 1 & 1 & 1 & 1 & \\ 0 & a & 1 & \infty & x \\ 0 & 0 & 0 & \alpha_{\infty 1} & q \\ \alpha_0 & \alpha_a & \alpha_1 & \alpha_{\infty 2} & \end{pmatrix} \tag{3.2.31}$$

[4] For the sake of the second generalisation below, the s-ranks of the singularities are already written in the first line.

by means of the identities

$$\begin{aligned}
\alpha_0 &= \alpha_{z_12} - \alpha_{z_11},\\
\alpha_a &= \alpha_{z_22} - \alpha_{z_21},\\
\alpha_1 &= \alpha_{z_32} - \alpha_{z_31},\\
\alpha_{\infty 1} &= \alpha_{z_41} + \alpha_{z_11} + \alpha_{z_21} + \alpha_{z_31},\\
\alpha_{\infty 2} &= \alpha_{z_42} + \alpha_{z_11} + \alpha_{z_21} + \alpha_{z_31}.
\end{aligned}$$

Finally, writing the characteristic exponents of the Heun differential equation in terms of the singularity parameters, the conventional Riemann symbol takes the form

$$P\begin{pmatrix} 1 & 1 & 1 & 1 & \\ 0 & a & 1 & \infty & x \\ 0 & 0 & 0 & \alpha_{\infty 1} & C_0\, a \\ 1 - A_0 & 1 - A_a & 1 - A_1 & \alpha_{\infty 2} & \end{pmatrix}$$

with $\alpha_{\infty 1}, \alpha_{\infty 2}$ from (3.2.18).

3.2.3 Local Solutions

We start with Heun's differential equation (3.2.16) in its traditional form and notation

$$\frac{d^2y}{dz^2} + \left(\overbrace{\frac{1-\alpha_0}{z}}^{=\gamma} + \overbrace{\frac{1-\alpha_a}{z-a}}^{=\epsilon} + \overbrace{\frac{1-\alpha_1}{z-1}}^{=\delta} \right) \frac{dy}{dz} + \frac{\alpha_{\infty 1}\, \alpha_{\infty 2}\, z + q}{z\,(z-a)\,(z-1)}\, y = 0, \tag{3.2.32}$$

the modern form of which is

$$\frac{d^2y}{dz^2} + \left(\frac{\gamma}{z} + \frac{\epsilon}{z-a} + \frac{\delta}{z-1} \right) \frac{dy}{dz} + \left(\frac{\frac{q}{a}}{z} + \frac{\frac{a\,\alpha\beta + q}{a\,(a-1)}}{z-a} + \frac{\frac{\alpha\beta+q}{1-a}}{z-1} \right) y = 0 \tag{3.2.33}$$

with

$$\alpha_0 + \alpha_a + \alpha_1 + \alpha_{\infty 1} + \alpha_{\infty 2} = 2$$

and

$$\alpha_{\infty 1} = \alpha, \ \alpha_{\infty 2} = \beta,$$

which may also be written in dependence on α, β, γ, ϵ, δ as

$$\gamma + \delta + \epsilon = \alpha + \beta + 1.$$

This relation may be used to eliminate one of the parameters, ϵ, say.

Without losing generality, it is assumed in the following that $0 < a < 1$.

According to (3.2.31), there is a basic solution that is expandable in a Taylor series (cf. Ronveaux, 1995, p. 34):

$$y(z) = \sum_{n=0}^{\infty} a_n z^n, \quad a_0 \neq 0, \tag{3.2.34}$$

resulting in a three-term recurrence relation (i.e., a linear second-order homogeneous difference equation)

$$\begin{aligned} q\, a_0 + a\,\gamma\, a_1 &= 0, \\ P_n\, a_{n-1} - (Q_n - q)\, a_n + R_n\, a_{n+1} &= 0, \quad n \geq 1, \end{aligned} \tag{3.2.35}$$

with

$$\begin{aligned} P_n &= (n-1+\alpha)(n-1+\beta), \\ Q_n &= n\,[(n-1+\gamma)\,(1+a) + a\,\delta + \epsilon], \\ R_n &= (n+1)\,(n+\gamma)\,a \end{aligned} \tag{3.2.36}$$

or

$$\begin{aligned} \frac{P_n}{n^2} &= 1 + \frac{\alpha+\beta-2}{n} + \frac{(\alpha-1)(\beta-1)}{n^2}, \\ -\frac{Q_n - q}{n^2} &= -1 - a\,\frac{a(\delta+\gamma-1)+\epsilon+\gamma-1}{n} + \frac{q}{n^2}, \\ \frac{R_n}{n^2} &= a + \frac{a\,(\gamma+1)}{n} + \frac{a\,\gamma}{n^2}. \end{aligned} \tag{3.2.37}$$

If $\gamma \neq 0, -1, -2, \ldots$ the linear second-order difference equation (3.2.35), (3.2.36) turns to a three-term recurrence equation such that after having fixed a_0, all the subsequent terms a_n, $n = 1, 2, 3, \ldots$, are determined by the difference equation and may be explicitly calculated recursively.

The characteristic equation of (3.2.35), (3.2.36) is

$$a\,t^2 - (1+a)\,t + 1 = 0, \tag{3.2.38}$$

the solutions of which are

$$\begin{aligned} t_1 &= \frac{1}{a}, \\ t_2 &= 1. \end{aligned} \tag{3.2.39}$$

The series (3.2.34) represents what are generally known as the *Heun functions*. Under the application of appropriate Moebius and s-homotopic transformations, each particular solution about the singularities of the Fuchsian differential equation (3.2.32) having four singularities may be written by means of (3.2.34).

It should be mentioned that the radius of convergence of the series (3.2.34) generally is $r = a$ if $a < 1$, because it is a solution of the differential equation (3.2.32) expanded about the origin and thus ranges up to the neighbouring singularity of the equation, which is located at $z = a$.

The second-order linear difference equation (3.2.35) just has two non-trivial particular solutions that are linearly independent and thus may serve as constituents of a fundamental system. These two particular solutions are characterised by the two solutions (3.2.39) of its characteristic equation (3.2.38) and, moreover, are in a one-to-one relation with the particular solutions of the differential equation (3.2.32) via the ansatz (3.2.34). This linear second-order differential equation (3.2.32) also has two linearly independent particular solutions that are characterised by the two different solutions (3.2.39) of the characteristic equation (3.2.38) of the Frobenius ansatz (3.2.34). Since, in the case of differential equations, the solution of the characteristic equation determines the radius of convergence r of the series in (3.2.34), this means that if a_n via (3.2.35) admits the particular solution that is characterised by t_1 then the corresponding series in (3.2.34) has radius of convergence $r = a < 1$. If a_n via (3.2.35) admits the particular solution that is characterised by t_2 then the radius of convergence of the corresponding series in (3.2.34) increases to $r = 1$.

There is one, and only one, alternative to these two opportunities, namely if the difference equation breaks off, leading to the trivial solution (viz. having only a finite number of non-zero terms a_n, $n = 0, 1, 2, \ldots, N_0 - 1, N_0$) that generates a polynomial solution $y(z) = P_{N_0}(z)$ of the differential equation (3.2.32) via the Frobenius ansatz (3.2.34). This is quite understandable, since the singularity of the differential equation at infinity is regular. This polynomial solution is a global solution of the differential equation (3.2.32), meaning that it is holomorphic at all of the finite singularities of the differential equation (3.2.32).

A necessary condition for the difference equation (3.2.35), (3.2.36) to break off is $\alpha = -N_0$, where $N_0 = 0, 1, 2, \ldots$. In this case $P_{N_0+1} = 0$ in (3.2.36). If one of the remaining parameters takes on a value such that $a_{N_0+2} = 0$, then the subsequent relations yield $a_n = 0$, $n > N_0$. The series (3.2.34) thus breaks off, becoming the representation of a polynomial of degree $N_0 + 1$ via (3.2.34).

The sufficient condition is dealt with in §3.2.4. The resulting polynomials are higher ones in the sense that they do not stem from Fuchsian differential equations having at most three singularities.

3.2.4 Accessory Parameter and Polynomial Solutions

Accessory Parameter

The main difference between equation (3.2.15) and the Fuchsian differential equation having less than four singularities is that it has an accessory parameter

$$q = C_0\, a.$$

The accessory parameter does not influence the behaviour of the solutions of (3.2.15) at any of its singularities. Thus, there is a one-dimensional manifold of solutions of (3.2.15) having the selfsame behaviour at all four singularities of the differential equation (3.2.15).

Principally, there are two types of parameters of a differential equation (3.2.5): parameters that touch its singularities (called *singularity parameters*) and parameters that do not. The former may be distinguished by whether they determine the location of one of its singularities in the complex plane or whether they imply the characteristic exponents of its singularities. The latter may only imply the solutions of the differential equation. These parameters are called *accessory parameters*. The accessory parameters determine the dimensions of the manifold of solutions that behave in the same manner at all of the singularities of the underlying differential equation.

For Fuchsian differential equations it is possible to give the number m of accessory parameters as

$$m = n - 3,$$

where $n \geq 3$ is the number of singularities.

Polynomial Solutions

There are three types of solutions of the Heun differential equation (3.2.15):

- Frobenius solutions
- Heun functions
- Heun polynomials.

While the first of these have been treated in §2.3.1 and the second in §3.2.3, the last type is dealt with in the following.

Heun polynomials are solutions of the Heun differential equation (3.2.15)

$$\frac{d^2y}{dz^2} + \left(\frac{A_0}{z} + \frac{A_0}{z-a} + \frac{A_0}{z-1}\right)\frac{dy}{dz} + \left(\frac{C_0}{z} + \frac{C_a}{z-a} + \frac{C_1}{z-1}\right) y = 0, \quad z, a \in \mathbb{C},$$

with

$$C_0 + C_a + C_1 = 0$$

that are holomorphic over the complex plane $z \in \mathbb{C}$. This means that the ansatz[5]

$$y(z) = \sum_{n=0}^{\infty} a_n z^n$$

breaks, such that the solution may be written in the form

$$y(z) = \sum_{n=0}^{N_0} a_n z^n.$$

[5] This is also true for Frobenius solutions having a non-zero characteristic exponent. In this case we speak of generalised Heun polynomials.

This, in turn, means that the three-term recurrence relation (3.2.35), (3.2.36) for determining the coefficients a_n, $n \in \mathbb{N}^0$ of the series (3.2.4) breaks.

In order for the three-term recurrence relation (3.2.35), (3.2.36) to break, it is necessary that the one-sided infinite determinant (i.e., a determinant of a matrix that has an infinite number of rows and lines)

$$\begin{vmatrix} \tilde{Q}_0 & R_0 & 0 & 0 & \dots & 0 \\ P_1 & \tilde{Q}_1 & R_1 & 0 & \dots & 0 \\ 0 & P_2 & \tilde{Q}_2 & R_2 & \dots & 0 \\ \dots & \dots & \dots & \dots & \dots & \dots \\ 0 & 0 & \dots & P_{N_0-1} & \tilde{Q}_{N_0-1} & R_{N_0-1} \\ 0 & 0 & \dots & 0 & P_{N_0} & \tilde{Q}_{N_0} \end{vmatrix}$$

becomes zero:

$$\begin{vmatrix} \tilde{Q}_0 & R_0 & 0 & 0 & \dots & 0 \\ P_1 & \tilde{Q}_1 & R_1 & 0 & \dots & 0 \\ 0 & P_2 & \tilde{Q}_2 & R_2 & \dots & 0 \\ \dots & \dots & \dots & \dots & \dots & \dots \\ 0 & 0 & \dots & P_{N_0-1} & \tilde{Q}_{N_0-1} & R_{N_0-1} \\ 0 & 0 & \dots & 0 & P_{N_0} & \tilde{Q}_{N_0} \end{vmatrix} = 0. \tag{3.2.40}$$

This condition (3.2.40) was formulated by Felix Medland Arscott (1922–1996) in Ronveaux (1995, p. 43). It is a quite simple one that may be verified easily in a numerical calculation. The result, however, is astonishing and is discussed and displayed in Figures 3.1 to 3.6.

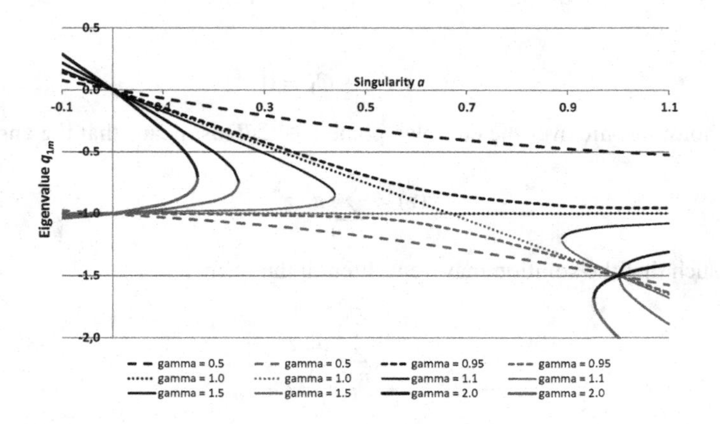

Figure 3.1 Eigenvalue curves of polynomial solutions of Heun's equation I.

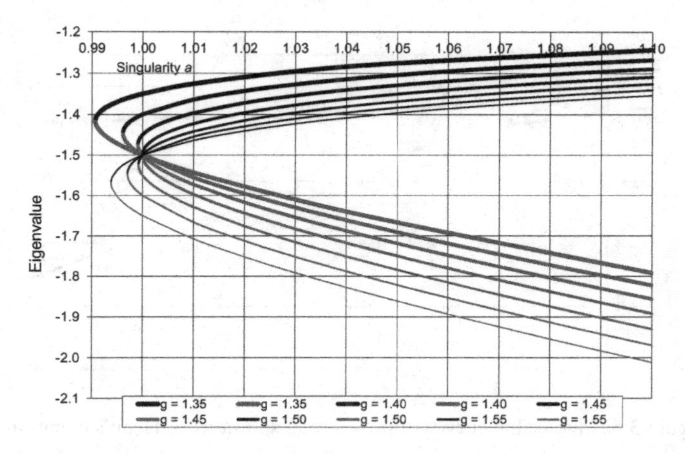

Figure 3.2 Eigenvalue curves of polynomial solutions of Heun's equation II.

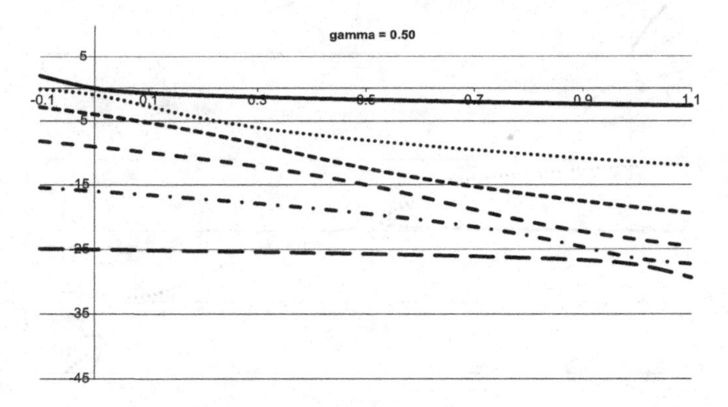

Figure 3.3 Eigenvalue curves of polynomial solutions of Heun's equation III.

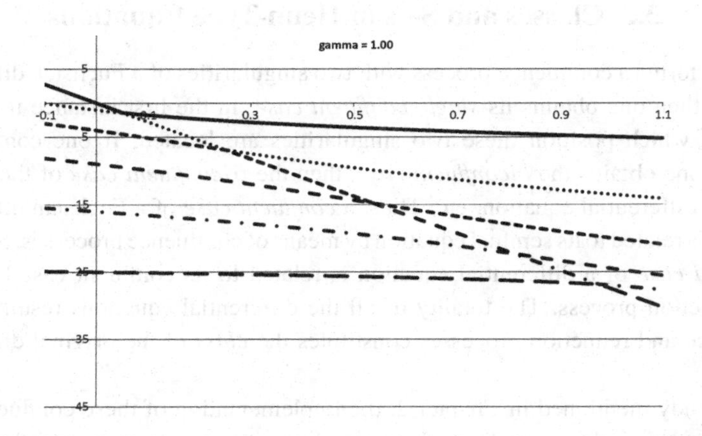

Figure 3.4 Eigenvalue curves of polynomial solutions of Heun's equation IV.

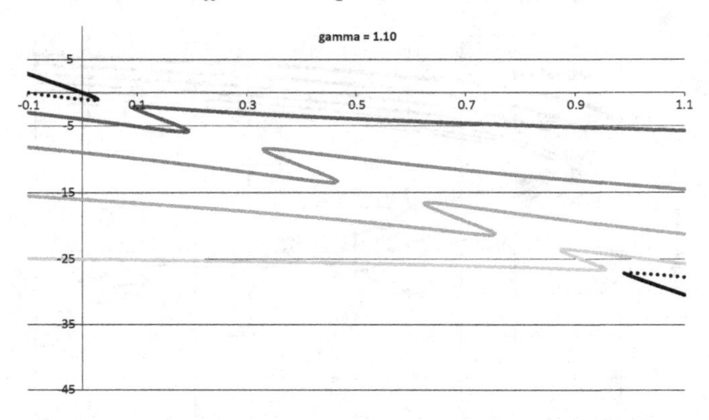

Figure 3.5 Eigenvalue curves of polynomial solutions of Heun's equation V.

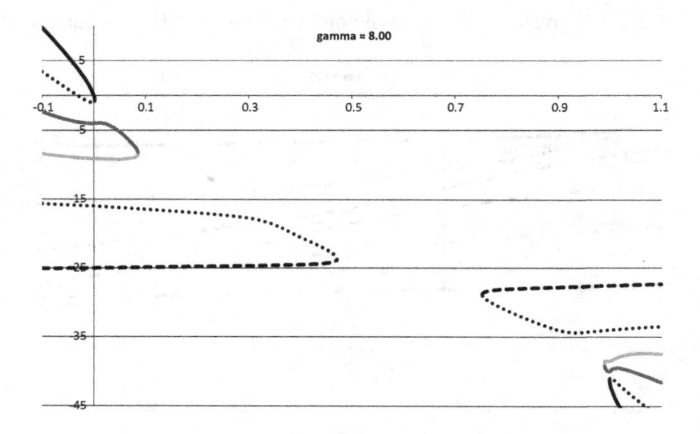

Figure 3.6 Eigenvalue curves of polynomial solutions of Heun's equation VI.

3.3 Classes and Sets of Heun-Type Equations

If one performs a confluence process with two singularities of a Fuchsian differential equation then one obtains its *single confluent case*. In the first instance it does not matter at which position these two singularities are located. If one continues in this way one obtains the *biconfluent case*, then the *triconfluent case* of the original Fuchsian differential equation, etc. Thus, a *confluent case* of a Fuchsian differential equation is related to its seminal equation by means of confluence processes. Similarly, a *reduced case* of a differential equation is related to its confluent case by means of a reduction process. The totality of all the differential equations resulting from confluence and reduction processes constitutes the *class* of the original differential equation.

As already mentioned in Chapter 2, the implementation of these confluence processes is not unambiguous. Thus, there are always two fundamentally different processes: strong and weak ones. Depending on whether the sum of the coalescing singularities of the differential equation is as large as the resulting singularity or whether it is smaller, one speaks of a *strong* or a *weak confluence process*, respectively.

The Fuchsian differential equations as seminal equations for their respective classes bear the names of great mathematicians.

The simplest Fuchsian differential equation is that with only one singularity. It bears the name of the French mathematician **Pierre-Simon Laplace** (1749–1827). As a consequence of having just one singularity there is no confluence process to consider here. The class of the Laplace differential equation consists of just this equation.

The Fuchsian differential equation having two singularities bears the name of the Swiss mathematician **Leonhard Euler** (1707–1783). The class of Euler differential equations is the totality of the Euler equation, the confluent and the reduced cases of this differential equation.

In an analogous way there are classes of differential equations that go back to Fuchsian differential equations with three and four singularities. These equations are named after the German mathematicians **Johann Carl Friedrich Gauss** (1777–1855) and Karl Heun. Accordingly, there are (cf. Figure 3.7) the *Laplace differential equation*, *differential equations of Euler class*, *differential equations of Gauss class* and *differential equations of Heun class*, respectively.

This chapter is devoted to the main properties of the Heun class of differential equations in its standard forms.

3.3.1 Heun Class: Confluent Cases

Regular and irregular singularities of linear ordinary second-order differential equations were defined in §2.1. Regular singularities of linear ordinary second-order differential equations are to be distinguished in elementary and non-elementary regular ones. The basic singularity of linear ordinary second-order differential equations is the non-elementary regular one, having s-rank unity.

Irregular singularities of linear ordinary second-order differential equations are thought to be composed ones. Carrying out this concept in a mathematically satisfactory way, confluence processes were defined in §2.2. Among these, a standard confluence process is emphasised that consists of a coalescence of two non-elementary regular singularities yielding an irregular singularity of the underlying differential equation, the s-rank of which is the sum of the two coalescing singularities, thus is two. In contrast to all the other singularities, elementary ones have been defined in a specific way by relating two of the singularity parameters of a non-elementary regular singularity to one another.

The topic of the underlying §3.3.1 in the following is the treatment of the *confluent cases* of the Heun differential equation. Consider a Heun differential equation (3.2.9). Suppose that two of its singularities undergo a confluence process. The result is a differential equation that is called a *confluent case* of the original differential equation.

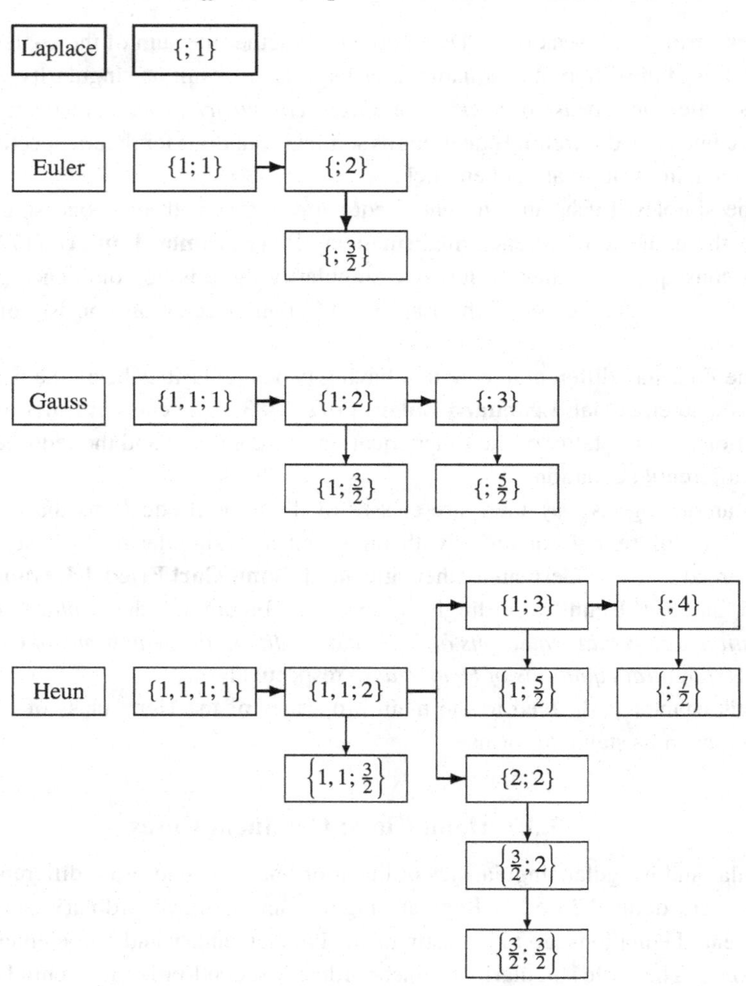

Figure 3.7 Confluent and reduced cases of Fuchsian differential equations.

A subsequent reduction process may reduce the s-rank of the resulting singularity. In this way a *reduced case* occurs, if the resulting singularity of the confluence process becomes reducible.

Generally speaking, the totality of differential equations generated by all possible confluence and reduction processes starting with a Fuchsian differential equation is called the *class of confluent and reduced cases.*

In order to get an overview of the Laplace, Euler, Gauss and Heun classes of confluent and reduced cases, we present the following scheme (cf. Figure 3.7) where all the equations (represented by their s-rank symbols) are displayed, connected by confluence and reduction processes:

- The starting differential equations are Fuchsian ones having an ascending number of singularities, the maximum of which here is taken to be four.
- Since, as has already been mentioned, there is no differential equation (2.1.2) without singularities, the minimum number of singularities of the starting Fuchsian differential equation is unity.
- Differential equations are considered that are thought to be in natural form, exclusively. This is always achievable by means of an appropriate Moebius transformation.
  - This means that the point at infinity is always a singularity of the underlying differential equation and that no finite singularity has an s-rank that is higher than the s-rank of the singularity that is located at infinity.
- The s-ranks of the singularities of the differential equations at hand may either be integer or odd half-integer. However, elementary singularities of the differential equation (2.1.2) are left completely aside because of their specific definition (2.1.9).
  - Therefore, the s-ranks of the singularities of all the differential equations (2.1.2), considered below, are at least unity.
  - Thus, singularities of differential equations having odd half-integer s-ranks are irregular ones, necessarily.
- Confluent cases of Fuchsian differential equations having singularities with integer s-ranks are thought to be generated by means of strong confluence processes.
- Confluent cases having singularities with odd half-integer s-ranks are thought to be generated by means of a confluence process followed by a reduction process of the resulting singularity, i.e., by weak confluence processes.
  - This means that – starting with Fuchsian differential equations in natural form – confluence as well as reduction processes are considered only in as far as they do create once again differential equations in natural form.

Under these conditions, as an example, the Gauss differential equation has two confluent cases, namely the single confluent case and the biconfluent case; this was already seen in Chapter 1. Each of these two confluent cases may undergo a reduction process, resulting in a reduced confluent case having a singularity with an odd half-integer s-rank. Because of the arrangement displayed above, i.e., considering exclusively natural forms, the irregular singularities of all of these four confluent cases are located at infinity. In the scheme shown in Figure 3.7 the s-rank symbols of all possible Fuchsian differential equations having at most four singularities are given, and all confluent and reduced cases (horizontal vectors mean confluence processes, vertical vectors mean reduction processes).

As may be seen, the *Heun class of confluent and reduced cases* comprises ten differential equations, five of which have singularities with exclusively integer s-ranks.

3.3.2 Heun Sets: Forms

§§1.2.3, 3.2.1 and 3.2.2 showed how linear transformations, i.e., Moebius and s-homotopic transformations, change any Fuchsian differential equation into other forms. Thus, the totality of linear transformations constitutes an (innumerable) set of differential equations, the seminal one of which may be thought of as being the *standard form* of the Fuchsian differential equation.

Generalised Riemann Symbol

Consider a Fuchsian differential equation (3.1.1)–(3.1.4). At each of its singularities z_k exist parameters A_{z_k}, B_{z_k} that determine two characteristic exponents: $\alpha_{z_k 1}, \alpha_{z_k 2}$, which themselves determine the Frobenius solutions

$$y_{z_k 1}(z) = (z - z_{k1})^{\alpha_{z_k 1}} \sum_{n=0}^{\infty} (z - z_{k1})^n,$$

$$y_{z_k 1}(z) = (z - z_{k1})^{\alpha_{z_k 1}} \sum_{n=0}^{\infty} (z - z_{k1})^n$$

of equation (3.1.1)–(3.1.4). Therefore, instead of writing the differential equation it is equivalent to write down a scheme in which all of its characteristic exponents, as well as its accessory parameters, are displayed.

This idea stemmed from Bernhard Riemann in 1857 for the Fuchsian differential equation with three singularities. Here, the idea is picked up and widened in order to display not only the location but also the s-rank of all the singularities of equation (3.1.1)–(3.1.4). Moreover, if the differential equation has accessory parameters (cf. §3.2.4), these are also displayed. Thus, for the Heun differential equation the scheme looks like (3.2.30) where one of the four singularities may also be placed at infinity. This scheme is called *conventional Riemann symbol.*

The first line shows the s-ranks of the singularities. For Fuchsian differential equations, all the quantities written there are unity or one half, since all the singularities of the differential equations are regular ones. The second line lists the locations of the singularities, as well as the independent variable; the entry z_n is ∞ if the point at infinity is a (regular) singularity of the differential equation. The third line contains the first of the two characteristic exponents at each of the singularities of the equation, as well as the accessory parameter, and in the last line are written the second characteristic exponents.

Thus, the *conventional Riemann symbol* (3.2.29) of a Fuchsian differential equation (3.1.1)–(3.1.4) is defined as a tableau in which are displayed all the determining quantities of the Fuchsian differential equation: the s-ranks of the singularities of the differential equation, their locations and the pairs of characteristic exponents (of first kind) for each singularity. Moreover, in the last row are stated the independent variable as well as the accessory parameters, if there are any. Conventional Riemann symbols are defined exclusively for Fuchsian differential equations.

Conventional Riemann symbols may become *generalised Riemann symbols* for differential equations having irregular singularities as well. This may be done by adding the pairs of characteristic exponents of the second kind below the characteristic exponents of the first kind in the row of the respective irregular singularities, put in sequence from higher to lower order.

In the following some examples are given.

(1) Differential equation: Confluent case of Euler equation in standard form

$$\frac{d^2y}{dz^2} + a\,\frac{dy}{dz} + (b\,z + c)\;y = 0.$$

Generalised Riemann symbol:

$$P\begin{pmatrix} 2 & \\ \infty & ;z \\ \alpha_{11} & \\ \alpha_{12} & \\ \alpha_{01} & \\ \alpha_{02} & \end{pmatrix}.$$

(2) Differential equation: Single confluent case of Gauss equation in standard form

$$\frac{d^2y}{dz^2} + \left(\frac{A}{z} + a\right)\frac{dy}{dz} + \left(\frac{B}{z^2} + \frac{C}{z} + b\,z + c\right)\;y = 0.$$

Generalised Riemann symbol:

$$P\begin{pmatrix} 1 & 2 & \\ 0 & \infty & ;z \\ \alpha_{0r1} & \alpha_{1i1} & \\ \alpha_{0r2} & \alpha_{1i2} & \\ & \alpha_{0i1} & \\ & \alpha_{0i2} & \end{pmatrix}.$$

The index r means 'regular', the index i means 'irregular'.

Here and in §3.2.2, three different symbols, all of them representing linear, ordinary differential equations of second order, have been defined and all of them are named after Riemann: Riemann P-symbol, conventional Riemann symbol and generalised Riemann symbol. As a generic term for these three notions, and without specifying any particular one of them, I suggest coining the notion of the *Riemannian scheme*.

3.4 Confluent Cases of Heun's Equation

3.4.1 Single Confluent Case

The *single confluent case* of the Fuchsian differential equation (3.2.5) having four singularities in its natural form is

$$\frac{\mathrm{d}^2y}{\mathrm{d}z^2} + \left[\frac{A_{z_1}}{z - z_1} + \frac{A_{z_2}}{z - z_2} + G_0\right]\frac{\mathrm{d}y}{\mathrm{d}z} + \left[\frac{B_{z_1}}{(z - z_1)^2} + \frac{B_{z_2}}{(z - z_2)^2} + \frac{C_{z_1}}{z - z_1} + \frac{C_{z_2}}{z - z_2} + D_0\right] y = 0. \tag{3.4.1}$$

In the standard form $z_1 = 0$ and $z_2 = 1$ as well as $B_{z_1} = B_{z_2} = 0$ holds, thus

$$\frac{\mathrm{d}^2y}{\mathrm{d}z^2} + \left[\frac{A_{z_1}}{z} + \frac{A_{z_2}}{z - 1} + G_0\right]\frac{\mathrm{d}y}{\mathrm{d}z} + \left[\frac{C_{z_1}}{z} + \frac{C_{z_2}}{z - 1} + D_0\right] y = 0. \tag{3.4.2}$$

Equation (3.4.1) may be thought of as being constructed by means of a confluence process of the differential equation (3.2.5) under which the finite singularities z_3 and z_4 of the differential equation are brought to infinity, forming an irregular singularity there, the s-rank of which is two.

Equally, it might be thought of as being constructed by an ad-hoc adding of a non-elementary regular singularity at $z = z_2$ to the single confluent case [cf. equation (1.2.45)] of the Fuchsian differential equation having three singularities in its natural form:

$$\frac{\mathrm{d}^2y}{\mathrm{d}z^2} + \left[\frac{A_0}{z - z_1} + G_0\right]\frac{\mathrm{d}y}{\mathrm{d}z} + \left[\frac{B_0}{(z - z_1)^2} + \frac{C_0}{z - z_1} + D_0\right] y = 0. \tag{3.4.3}$$

Remember that the differential equations (3.4.1) or (3.4.3) are in their natural form, meaning that the s-rank of the singularity at infinity is equal to or higher than the s-ranks of all of its finite singularities.

Local Solutions

Frobenius Solutions about the Finite Singularities

The local solutions of the two finite singularities of (3.4.1) at $z = z_1$ and at $z = z_2$ may be handled both at once, since the procedure for both of them is the same. In order to keep the formulae as simple as possible, this singularity, the one at $z = z_1$, say, is assumed to be at the origin $z = 0$. Thus, z_1 is taken to be zero, i.e., $z_1 = 0$ without losing generality. In the same sense, A_{z_1} and B_{z_1} are written as A and B, for the sake of simplicity.

If this equation is in its natural form, it looks like (3.4.1) on setting $z_1 = 0$:

$$\frac{\mathrm{d}^2y}{\mathrm{d}z^2} + \left[\frac{A}{z} + \frac{A_{z_2}}{z - z_2} + G_0\right]\frac{\mathrm{d}y}{\mathrm{d}z} + \left[\frac{B}{z^2} + \frac{C}{z} + \frac{B_{z_2}}{(z - z_2)^2} + \frac{C_{z_2}}{z - z_2} + D_0\right] y = 0. \tag{3.4.4}$$

According to (2.3.7) and (2.3.9), the Frobenius solutions about the regular singularity at the origin are given by

$$y(z) = z^{\alpha_{0r}}\, v(z) \tag{3.4.5}$$

with

$$v(z) = \sum_{n=0}^{\infty} a_n\, z^n. \tag{3.4.6}$$

According to (2.3.3)–(2.3.6), using the identification

$$\begin{aligned} p_i &= 0 \text{ for } i > 1,\\ p_1 &= G_1,\\ p_0 &= G_0,\\ q_i &= 0 \text{ for } i > 2,\\ q_2 &= D_2,\\ q_1 &= D_1,\\ q_0 &= D_0, \end{aligned}$$

the characteristic exponents α_{0ri}, $i = 1, 2$ at the regular singularity are given by the condition (2.3.13)

$$\alpha_{0r}^2 + (A-1)\,\alpha_{0r} + B = 0 \tag{3.4.7}$$

and thus are given by (2.3.14)

$$\alpha_{0r1,0r2} = \frac{1 - A \pm \sqrt{(1-A)^2 - 4\,B}}{2}. \tag{3.4.8}$$

The difference equation for the coefficients a_n, $n = 1, 2, 3, \ldots$, with $a_0 \neq 0$ is given by (2.3.12), the following five-term recurrence relation:

$$\begin{aligned} (z-z_0)^1 &: (1\cdot 0 + 1\,\hat{p}_{-1})\,a_1 + (0\,\hat{p}_0 + \hat{q}_{-1})\,a_0 = 0,\\ (z-z_0)^2 &: (2\cdot 1 + 2\,\hat{p}_{-1})\,a_2 + (1\,\hat{p}_0 + \hat{q}_{-1})\,a_1 + \hat{q}_0\,a_0 = 0,\\ (z-z_0)^3 &: (3\cdot 2 + 3\,\hat{p}_{-1})\,a_3 + (2\,\hat{p}_0 + \hat{q}_{-1})\,a_2 + (\hat{p}_1 + \hat{q}_0)\,a_1 + \hat{q}_1\,a_0 = 0,\\ (z-z_0)^4 &: (4\cdot 3 + 4\,\hat{p}_{-1})\,a_4 + (3\,\hat{p}_0 + \hat{q}_{-1})\,a_3 + (2\,\hat{p}_1 + \hat{q}_0)\,a_2\\ &\qquad + \hat{q}_1\,a_1 + \hat{q}_2\,a_0 = 0,\\ (z-z_0)^5 &: (5\cdot 4 + 5\,\hat{p}_{-1})\,a_5 + (4\,\hat{p}_0 + \hat{q}_{-1})\,a_4 + (3\,\hat{p}_1 + \hat{q}_0)\,a_3\\ &\qquad + \hat{q}_1\,a_2 + \hat{q}_2\,a_1 = 0,\\ \ldots &\qquad \ldots \end{aligned}$$

or

$$
\boxed{
\begin{aligned}
&(1 \cdot 0 + 1\,\hat{p}_{-1})\,a_1 + (0\,\hat{p}_0 + \hat{q}_{-1})\,a_0 = 0,\\
&(2 \cdot 1 + 2\,\hat{p}_{-1})\,a_2 + (1\,\hat{p}_0 + \hat{q}_{-1})\,a_1 + \hat{q}_0\,a_0 = 0,\\
&(3 \cdot 2 + 3\,\hat{p}_{-1})\,a_3 + (2\,\hat{p}_0 + \hat{q}_{-1})\,a_2 + (\hat{p}_1 + \hat{q}_0)\,a_1 + \hat{q}_1\,a_0 = 0,\\
&\\
&(n+2)\,[n+1+\hat{p}_{-1}]\;a_{n+2} + [(n+1)\,\hat{p}_0 + \hat{q}_{-1}]\;a_{n+1} + [n\,\hat{p}_1 + \hat{q}_0]\;a_n\\
&\qquad + \hat{q}_1\,a_{n-1} + \hat{q}_2\,a_{n-2} = 0, \qquad n = 2,3,4,\ldots,
\end{aligned}
}
\tag{3.4.9}
$$

with

$$
\begin{aligned}
\hat{p}_{-1} &= A + 2\,\alpha_{0r},\\
\hat{p}_i &= p_i,\; i = 0,1,2,\ldots,\\
\hat{q}_{-2} &= \alpha_{0r}\,(\alpha_{0r} - 1) + B + A\,\alpha_{0r},\\
\hat{q}_{-1} &= C + \alpha_{0r}\,p_0,\\
\hat{q}_i &= q_i + \alpha_{0r}\,p_{i+1},\; i = 0,1,2,\ldots.
\end{aligned}
$$

Thomé Solution at the Infinite Singularity

The s-rank symbol of the single confluent case (3.4.1)

$$
\begin{aligned}
&\frac{d^2y}{dz^2} + \left[\frac{A_{z_1}}{z-z_1} + \frac{A_{z_2}}{z-z_2} + G_0\right]\frac{dy}{dz}\\
&\qquad + \left[\frac{B_{z_1}}{(z-z_1)^2} + \frac{B_{z_2}}{(z-z_2)^2} + \frac{C_{z_1}}{z-z_1} + \frac{C_{z_2}}{z-z_2} + D_0\right] y = 0
\end{aligned}
$$

of the Heun differential equation is

$$
\{1,1;2\}\,.
$$

Thus, according to (2.3.25), (2.3.28), the Thomé solution ansatz about the irregular singularity at infinity is given by

$$
y(z) = \exp(\alpha_1\,z)\;z^{\alpha_{0i}}\;w(z) \tag{3.4.10}
$$

with

$$
w(z) = \sum_{n=0}^{\infty} C_n\,z^{-n}. \tag{3.4.11}
$$

Because

$$
\frac{A_{z_1}}{z-z_1} + \frac{A_{z_2}}{z-z_2} = \frac{1}{z}\left[\frac{A_{z_1}}{1-\frac{z_1}{z}} + \frac{A_{z_2}}{1-\frac{z_2}{z}}\right] \sim \frac{1}{z}\,\left(A_{z_1} + A_{z_2}\right)
$$

as $z \to \infty$, according to (2.3.26) the following identification holds:

$$
\begin{aligned}
p_{-1} &= A_{z_1} + A_{z_2},\\
p_0 &= G_0,\\
p_i &= 0 \text{ for } i > 1.
\end{aligned}
$$

Because

$$\frac{B_{z_1}}{(z-z_1)^2}+\frac{B_{z_2}}{(z-z_2)^2}=\frac{1}{z^2}\left[\frac{B_{z_1}}{(1-\frac{z_1}{z})^2}+\frac{B_{z_2}}{(1-\frac{z_2}{z})^2}\right]\sim\frac{1}{z^2}\left(B_{z_1}+B_{z_2}\right)$$

as $z\to\infty$, according to (2.3.27) the following identification holds:

$$q_{-2}=B_{z_1}+B_{z_2}.$$

Because

$$\frac{C_{z_1}}{z-z_1}+\frac{C_{z_2}}{z-z_2}=\frac{1}{z}\left[\frac{C_{z_1}}{1-\frac{z_1}{z}}+\frac{C_{z_2}}{1-\frac{z_2}{z}}\right]\sim\frac{1}{z}\left(C_{z_1}+C_{z_2}\right)$$

as $z\to\infty$, according to (2.3.27) the following identification holds:

$$q_{-1}=C_{z_1}+C_{z_2},$$
$$q_0=D_0,$$
$$q_i=0 \text{ for } i>1.$$

According to (2.3.29) the characteristic exponents α_{1j}, $j=1,2$, at the irregular singularity at infinity are given by

$$\alpha_{11}=\frac{1}{2}\left(-G_0+\sqrt{G_0^2-4\,D_0}\right),$$

$$\alpha_{12}=\frac{1}{2}\left(-G_0-\sqrt{G_0^2-4\,D_0}\right).$$

According to (2.3.30) the characteristic exponents α_{0ij}, $j=1,2$, at the irregular singularity at infinity are given by

$$\alpha_{0i1}=-\frac{(A_{z_1}+A_{z_2})\,\alpha_{11}+C_{z_1}+C_{z_2}}{G_0+2\,\alpha_{11}},$$

$$\alpha_{0i2}=-\frac{(A_{z_1}+A_{z_2})\,\alpha_{12}+C_{z_1}+C_{z_2}}{G_0+2\,\alpha_{12}}.$$

3.4.2 Biconfluent Case

The *biconfluent case* of the Fuchsian differential equation (3.2.20) having four singularities in its natural form is

$$\boxed{\begin{aligned}&\frac{d^2y}{dz^2}+\left[\frac{A_{z_1}}{z-z_1}+G_0+G_1\,z\right]\frac{dy}{dz}\\&+\left[\frac{B_{z_1}}{(z-z_1)^2}+\frac{C_{z_1}}{z-z_1}+D_0+D_1\,z+D_2\,z^2\right]y=0\end{aligned}} \qquad (3.4.12)$$

where z_1 may also be zero, such that the finite singularity of equation (3.4.12) is located at the origin.

Equation (3.4.12) may be thought of as being constructed by means of a confluence process of the differential equation (3.4.1) under which one of the finite singularities (z_2, say) of the differential equation is brought to infinity, forming an irregular singularity there, the s-rank of which is three.

Equally, it might be thought of as being constructed by adding a finite, non-elementary regular singularity to the biconfluent case (1.2.22)

$$\frac{d^2y}{dz^2} + (G_0 + G_1\, z)\,\frac{dy}{dz} + \left(D_0 + D_1\, z + D_2\, z^2\right)\, y = 0 \tag{3.4.13}$$

of the Fuchsian differential equation having three singularities (1.2.2), (1.2.3) in its natural form. The s-rank symbol of the biconfluent case (3.4.12) of the Fuchsian differential equation having four singularities is

$$\{1; 3\}.$$

Local Solutions

Frobenius Solution

The biconfluent case of the Fuchsian differential equation having four singularities has two singularities: one regular and the other irregular, the s-rank of which is three.

If this equation is in its natural form, it looks like (3.4.12), where the location of the finite singularity $z = z_1$ may be put to the origin $z = 0$ without losing generality by means of setting $z_1 = 0$:

$$\frac{d^2y}{dz^2} + \left[\frac{A}{z} + G_0 + G_1\, z\right]\frac{dy}{dz} + \left[\frac{B}{z^2} + \frac{C}{z} + D_0 + D_1\, z + D_2\, z^2\right]\, y = 0.$$

According to (2.3.7) and (2.3.9), the Frobenius solutions about the regular singularity at the origin are given by

$$y(z) = z^{\alpha_{0r}}\, w(z) \tag{3.4.14}$$

with

$$w(z) = \sum_{n=0}^{\infty} a_n\, (z)^n. \tag{3.4.15}$$

Putting $z_0 = 0$ there, and according to (2.3.3)–(2.3.6), using the identification

$$\begin{aligned}
p_i &= 0 \text{ for } i > 1,\\
p_1 &= G_1,\\
p_0 &= G_0,\\
q_i &= 0 \text{ for } i > 2,\\
q_2 &= D_2,\\
q_1 &= D_1,\\
q_0 &= D_0,
\end{aligned}$$

the characteristic exponents α_{0ri}, $i = 1, 2$, at the irregular singularity are given by the condition (2.3.13)

$$\alpha_{0r}^2 + (A-1)\,\alpha_{0r} + B = 0 \tag{3.4.16}$$

and thus by (2.3.14),

$$\alpha_{0r1,0r2} = \frac{1 - A \pm \sqrt{(1-A)^2 - 4\,B}}{2}.$$

The difference equation for the coefficients a_n, $n = 1, 2, 3, \ldots$, with $a_0 \neq 0$ is given by (2.3.12), the following five-term recurrence relation:

$$\begin{aligned}
(z-z_0)^1 &: (1\cdot 0 + 1\,\hat{p}_{-1})\,a_1 + (0\,\hat{p}_0 + \hat{q}_{-1})\,a_0 = 0,\\
(z-z_0)^2 &: (2\cdot 1 + 2\,\hat{p}_{-1})\,a_2 + (1\,\hat{p}_0 + \hat{q}_{-1})\,a_1 + \hat{q}_0\,a_0 = 0,\\
(z-z_0)^3 &: (3\cdot 2 + 3\,\hat{p}_{-1})\,a_3 + (2\,\hat{p}_0 + \hat{q}_{-1})\,a_2 + (\hat{p}_1 + \hat{q}_0)\,a_1 + \hat{q}_1\,a_0 = 0,\\
(z-z_0)^4 &: (4\cdot 3 + 4\,\hat{p}_{-1})\,a_4 + (3\,\hat{p}_0 + \hat{q}_{-1})\,a_3 + (2\,\hat{p}_1 + \hat{q}_0)\,a_2\\
&\qquad + \hat{q}_1\,a_1 + \hat{q}_2\,a_0 = 0,\\
(z-z_0)^5 &: (5\cdot 4 + 5\,\hat{p}_{-1})\,a_5 + (4\,\hat{p}_0 + \hat{q}_{-1})\,a_4 + (3\,\hat{p}_1 + \hat{q}_0)\,a_3\\
&\qquad + \hat{q}_1\,a_2 + \hat{q}_2\,a_1 = 0,\\
\ldots &\qquad \ldots
\end{aligned}$$

or

$$\boxed{\begin{aligned}
&(1\cdot 0 + 1\,\hat{p}_{-1})\,a_1 + (0\,\hat{p}_0 + \hat{q}_{-1})\,a_0 = 0,\\
&(2\cdot 1 + 2\,\hat{p}_{-1})\,a_2 + (1\,\hat{p}_0 + \hat{q}_{-1})\,a_1 + \hat{q}_0\,a_0 = 0,\\
&(3\cdot 2 + 3\,\hat{p}_{-1})\,a_3 + (2\,\hat{p}_0 + \hat{q}_{-1})\,a_2 + (\hat{p}_1 + \hat{q}_0)\,a_1 + \hat{q}_1\,a_0 = 0,\\
&(n+2)\,[n+1+\hat{p}_{-1}]\,a_{n+2} + [(n+1)\,\hat{p}_0 + \hat{q}_{-1}]\,a_{n+1} + [n\,\hat{p}_1 + \hat{q}_0]\,a_n\\
&\qquad + \hat{q}_1\,a_{n-1} + \hat{q}_2\,a_{n-2} = 0, \qquad n = 2, 3, 4, \ldots,
\end{aligned}} \tag{3.4.17}$$

with

$$\begin{aligned}
\hat{p}_{-1} &= A + 2\,\alpha_{0r},\\
\hat{p}_i &= p_i,\ i = 0, 1, 2, \ldots,\\
\hat{q}_{-2} &= \alpha_{0r}\,(\alpha_{0r} - 1) + B + A\,\alpha_{0r},\\
\hat{q}_{-1} &= C + \alpha_{0r}\,p_0,\\
\hat{q}_i &= q_i + \alpha_{0r}\,p_{i+1},\ i = 0, 1, 2, \ldots.
\end{aligned}$$

Thomé Solution

According to (2.3.32), the Thomé solutions about the irregular singularity at infinity are given by

$$y(z) = \exp\left(\frac{\alpha_2}{2}\,z^2 + \alpha_1\,z\right)\,z^{\alpha_{0i}}\,w(z) \tag{3.4.18}$$

with

$$w(z) = \sum_{n=0}^{\infty} C_n \, z^{-n}. \tag{3.4.19}$$

Putting $z_0 = 0$, and according to (2.3.31), using the identification

$$\begin{aligned}
p_i &= 0 \text{ for } i > 1, \\
p_1 &= G_1, \\
p_0 &= G_0, \\
p_{-1} &= A, \\
q_i &= 0 \text{ for } i > 2, \\
q_2 &= D_2, \\
q_1 &= D_1, \\
q_0 &= D_0, \\
q_{-1} &= C, \\
q_{-2} &= B,
\end{aligned}$$

and according to (2.3.33), the characteristic exponents $\alpha_{2j}, \alpha_{1j}, \alpha_{0ij}$, $j = 1, 2$, at the irregular singularity are given by

$$\begin{aligned}
\alpha_{2i} &= \frac{1}{2}\left(-G_1 \pm \sqrt{G_1^2 - 4\,D_2}\right), \\
\alpha_{1i} &= -\frac{D_1 + G_0\,\alpha_{2j}}{G_1 + 2\,\alpha_{2j}}, \\
\alpha_{0i} &= -\frac{\alpha_{2j}\,(A+1) + \alpha_{1j}\,\left(G_0 + \alpha_{1j}\right) + D_0}{G_1 + 2\,\alpha_{2j}}.
\end{aligned}$$

The difference equation for the coefficients C_n is given by

$$\boxed{\begin{array}{lll}
C_0 \neq 0 \text{ given,} & & \\
z^{-1} & : & -1\,\hat{p}_1\,C_1 + \hat{q}_{-1}\,C_0 = 0 \rightsquigarrow C_1, \\
z^{-2} & : & -2\,\hat{p}_1\,C_2 + (-1\,\hat{p}_0 + \hat{q}_{-1})\,C_1 + \hat{q}_{-2}\,C_0 = 0 \rightsquigarrow C_2, \\
z^{-3} & : & -3\,\hat{p}_1\,C_3 + (-2\,\hat{p}_0 + \hat{q}_{-1})\,C_2 + (1\cdot 2 - 1\,\hat{p}_{-1} \\
 & & \quad + \hat{q}_{-2})\,C_1 + \hat{q}_{-3}\,C_0 = 0 \rightsquigarrow C_3, \\
z^{-4} & : & -4\,\hat{p}_1\,C_4 + (-3\,\hat{p}_0 + \hat{q}_{-1})\,C_3 + (2\cdot 3 - 2\,\hat{p}_{-1} + \hat{q}_{-2})\,C_2 \\
 & & \quad + (-1\,\hat{p}_{-2} + \hat{q}_{-3})\,C_1 + \hat{q}_{-4}\,C_0 = 0 \rightsquigarrow C_4, \\
\ldots & &
\end{array}} \tag{3.4.20}$$

Since

$$\begin{aligned}
\hat{p}_1 &= G_1 + 2\,\alpha_2,\\
\hat{p}_0 &= G_0 + 2\,\alpha_1,\\
\hat{p}_{-1} &= A + 2\,\alpha_0,\\
\hat{p}_{-i} &= 0 \text{ for } i > 1
\end{aligned}$$

and

$$\begin{aligned}
\hat{q}_{-1} &= C + G_0\,\alpha_1 + A\,\alpha_0 + p_{-1}\,\alpha_1 + 2\,\alpha_0\,\alpha_1,\\
\hat{q}_{-2} &= B + A\,\alpha_0 + \alpha_0\,(\alpha_0 - 1),\\
\hat{q}_{-i} &= 0 \text{ for } i > 2,
\end{aligned}$$

this triangular-shaped difference equation simplifies to a diagonal-shaped recursively solvable second-order regular difference equation of Poincaré–Perron type:

$$\begin{aligned}
&C_0 \neq 0 \text{ given},\\
&z^{-1} \;:\; -1\,\hat{p}_1\,C_1 + \hat{q}_{-1}\,C_0 = 0 \rightsquigarrow C_1,\\
&z^{-2} \;:\; -2\,\hat{p}_1\,C_2 + (-1\,\hat{p}_0 + \hat{q}_{-1})\,C_1 + \hat{q}_{-2}\,C_0 = 0 \rightsquigarrow C_2,\\
&z^{-3} \;:\; -3\,\hat{p}_1\,C_3 + (-2\,\hat{p}_0 + \hat{q}_{-1})\,C_2 + (1\cdot 2 - 1\,\hat{p}_{-1} + \hat{q}_{-2})\,C_1 = 0 \rightsquigarrow C_3,\\
&z^{-4} \;:\; -4\,\hat{p}_1\,C_4 + (-3\,\hat{p}_0 + \hat{q}_{-1})\,C_3 + (2\cdot 3 - 2\,\hat{p}_{-1} + \hat{q}_{-2})\,C_2 = 0 \rightsquigarrow C_4,\\
&\ldots
\end{aligned}$$

or

$$\boxed{\begin{aligned}
&C_0 \neq 0 \text{ given},\\
&z^{-1} \;:\; -1\,\hat{p}_1\,C_1 + \hat{q}_{-1}\,C_0 = 0 \rightsquigarrow C_1,\\
&z^{-2} \;:\; -2\,\hat{p}_1\,C_2 + (-1\,\hat{p}_0 + \hat{q}_{-1})\,C_1 + \hat{q}_{-2}\,C_0 = 0 \rightsquigarrow C_2,\\
&z^{-3} \;:\; -(n+1)\,\hat{p}_1\,C_{n+1} + (-n\,\hat{p}_0 + \hat{q}_{-1})\,C_n\\
&\qquad + [(n-1)\,n - (n-1)\,\hat{p}_{-1} + \hat{q}_{-2}]\,C_{n-1} = 0 \rightsquigarrow C_{n+1}
\end{aligned}} \tag{3.4.21}$$

with $n = 2, 3, 4, \ldots$.

3.4.3 Triconfluent Case

The pendant of the triconfluent case of the Heun differential equation is the biconfluent case of the Gauss differential equation. This triconfluent case is a cornerstone of the theory of the central two-point connection problem, presented in this book. Therefore, this differential equation is treated in more detail than the foregoing equations of this class.

If the finite singularity of the biconfluent case of the Fuchsian differential equation having four singularities is supposed to be mobile, then the coefficients generally have to be dependent on ϵ:

$$
\begin{aligned}
P(z;\epsilon) &= \frac{A(\epsilon)}{z-\epsilon} + G_0(\epsilon) + G_1(\epsilon)\,z \\
&= \frac{1}{\epsilon}\frac{A(\epsilon)}{\frac{z}{\epsilon}-1} + G_0(\epsilon) + G_1(\epsilon)\,z \\
&= \frac{A(\epsilon) + G_0(\epsilon)(z-\epsilon) + G_1(\epsilon)\,z\,(z-\epsilon)}{\epsilon\,(\frac{z}{\epsilon}-1)} \\
&= \frac{A(\epsilon) - G_0(\epsilon)\epsilon + (G_0(\epsilon) - G_1(\epsilon))\,z + G_1\,z^2}{\epsilon\,(\frac{z}{\epsilon}-1)} \\
&= \frac{1}{\epsilon}\,\frac{G_1\,z^2 + [G_0(\epsilon) - G_1(\epsilon)]\,z + A(\epsilon) - G_0(\epsilon)\epsilon}{\frac{z}{\epsilon}-1}
\end{aligned}
\tag{3.4.22}
$$

and

$$
\begin{aligned}
&Q(z;\epsilon) \\
&= \frac{B(\epsilon)}{(z-\epsilon)^2} + \frac{C(\epsilon)}{z-\epsilon} + D_0(\epsilon) + D_1(\epsilon)\,z + D_2(\epsilon)\,z^2 \\
&= \frac{D_2(\epsilon)\,z^4 + [D_1(\epsilon) - 2\,\epsilon\,D_2(\epsilon)]\,z^3 + [D_0(\epsilon) - \epsilon\,(2\,D_1(\epsilon) - \epsilon\,D_2(\epsilon))]\,z^2}{(z-\epsilon)^2} \\
&+ \frac{[C(\epsilon) - \epsilon\,(2\,D_0(\epsilon) - \epsilon\,D_1(\epsilon))]\,z + B(\epsilon) - \epsilon\,(C(\epsilon) - D_0(\epsilon)\,\epsilon)}{(z-\epsilon)^2} \\
&= \frac{1}{\epsilon^2}\,\frac{D_2(\epsilon)\,z^4 + [D_1(\epsilon) - 2\,\epsilon\,D_2(\epsilon)]\,z^3 + \{D_0(\epsilon) - \epsilon\,[2\,D_1(\epsilon) - \epsilon\,D_2(\epsilon)]\}\,z^2}{(\frac{z}{\epsilon}-1)^2} \\
&+ \frac{1}{\epsilon^2}\,\frac{\{C(\epsilon) - \epsilon\,[2\,D_0(\epsilon) - \epsilon\,D_1(\epsilon)]\}\,z + B(\epsilon) - \epsilon\,[C(\epsilon) - D_0(\epsilon)\,\epsilon]}{(\frac{z}{\epsilon}-1)^2}.
\end{aligned}
\tag{3.4.23}
$$

From this expression it becomes obvious that the coefficients need to have the following dependencies on ϵ as $\epsilon \longrightarrow \infty$:

$$
\begin{aligned}
G_1(\epsilon) &= \ldots + G_{1,0} + G_{1,1}\,\epsilon, \\
G_0(\epsilon) &= \ldots + G_{0,0} + G_{0,1}\,\epsilon, \\
A(\epsilon) &= \ldots + A_0 + A_1\,\epsilon, \\
D_2(\epsilon) &= \ldots + D_{2,0}, \\
D_1(\epsilon) &= \ldots + D_{1,0}, \\
D_0(\epsilon) &= \ldots + D_{0,0}, \\
C(\epsilon) &= \ldots + C_1\,\epsilon + C_2\,\epsilon^2, \\
B(\epsilon) &= \ldots + B_1\,\epsilon + B_2\,\epsilon^2.
\end{aligned}
$$

Carrying out the confluence process $\epsilon \longrightarrow \infty$ in (3.4.22) and (3.4.23) results in the coefficients $P(z)$ and $Q(z)$ having the form

$$\lim_{\epsilon \to \infty} P(z;\epsilon) = P(z) = G_0 + G_1\,z + G_2\,z^2,$$
$$\lim_{\epsilon \to \infty} Q(z;\epsilon) = Q(z) = D_0 + D_1\,z + D_2\,z^2 + D_3\,z^3 + D_4\,z^4.$$

These coefficients in turn result in a differential equation having just one singularity. It is located at infinity, is of irregular type and its s-rank is four, viz. $R(\infty) = 4$:

$$\boxed{\begin{aligned}\frac{\mathrm{d}^2y}{\mathrm{d}z^2} &+ \left(G_0 + G_1\,z + G_2\,z^2\right)\frac{\mathrm{d}y}{\mathrm{d}z}\\ &+ \left(D_0 + D_1\,z + D_2\,z^2 + D_3\,z^3 + D_4\,z^4\right)\,y = 0\end{aligned}} \tag{3.4.24}$$

with

$$\begin{aligned}
G_2 &= -G_{1,1},\\
G_1 &= -G_{0,1} + G_{1,1},\\
G_0 &= -A_1 + G_{0,1},\\
D_4 &= -D_{2,2},\\
D_3 &= -D_{1,2} + 2\,D_{2,1},\\
D_2 &= -D_{0,2} + 2\,D_{1,1} + D_{2,0},\\
D_1 &= -C_2 + 2\,D_{0,1} - D_{1,0},\\
D_0 &= -B_2 + C_1 - D_{0,0}.
\end{aligned}$$

Thus, it is in its natural form and the s-rank symbol is

$$\{;4\}.$$

The differential equation (3.4.24) is called the *triconfluent case of the Fuchsian differential equation having four singularities*, since it may be thought of as being constructed by means of three confluence processes in which all the finite singularities of the Fuchsian equation having four singularities coalesce with the singularity at infinity. It is the most interesting differential equation treated in this book.

Local Solutions

The fact that the triconfluent case of the Fuchsian differential equation having four singularities in its natural form only has one singularity, namely an irregular singularity of s-rank $R(\infty) = 4$ that is placed at infinity, means that a local solution about the origin is a pure convergent power series, the radius of convergence of which is infinity. The asymptotic behaviour of this local solution is given by the asymptotic factor (2.3.34) of the Thomé series (2.3.25).

Since the origin $z = 0$ is an ordinary point of the differential equation (3.4.24), all its particular solutions are holomorphic there; hence, all its particular solutions are entire functions. Thus, they may be represented by means of a power series ansatz

$$y(z) = \sum_{n=0}^{\infty} a_n \, z^n, \tag{3.4.25}$$

being convergent on the whole complex plane. Here, in order to determine the particular solution by means of a Taylor series about the origin $z = 0$, it is necessary not only to fix the first coefficient a_0 but the first two a_0 and a_1 in order to get a recurrence relation for calculating the coefficients a_n, $n = 2, 3, 4, \ldots$, from the difference equation obtained by inserting (3.4.25) into (3.4.24):

$$\boxed{\begin{aligned}
&2 \cdot 1 \, a_2 + 1 \, p_0 \, a_1 + (0 \, p_1 + q_0) \, a_0 = 0,\\
&3 \cdot 2 \, a_3 + 2 \, p_0 \, a_2 + (1 \, p_1 + q_0) \, a_1 + (0 \, p_2 + q_1) \, a_0 = 0,\\
&4 \cdot 3 \, a_4 + 3 \, p_0 \, a_3 + (2 \, p_1 + q_0) \, a_2 + (1 \, p_2 + q_1) \, a_1 + (0 \, p_3 + q_2) \, a_0 = 0,\\
&\\
&5 \cdot 4 \, a_5 + 4 \, p_0 \, a_4 + (3 \, p_1 + q_0) \, a_3 + (2 \, p_2 + q_1) \, a_2\\
&\qquad + (1 \, p_3 + q_2) \, a_1 + (0 \, p_4 + q_3) \, a_0 = 0,\\
&6 \cdot 5 \, a_6 + 5 \, p_0 \, a_5 + (4 \, p_1 + q_0) \, a_4 + (3 \, p_2 + q_1) \, a_3\\
&\qquad + (2 \, p_3 + q_2) \, a_2 + (1 \, p_4 + q_3) \, a_1 = 0,\\
&7 \cdot 6 \, a_7 + 6 \, p_0 \, a_6 + (5 \, p_1 + q_0) \, a_5 + (4 \, p_2 + q_1) \, a_4\\
&\qquad + (3 \, p_3 + q_2) \, a_3 + (2 \, p_4 + q_3) \, a_2 = 0,\\
&\ldots
\end{aligned}} \tag{3.4.26}$$

Determining a_0 and a_1, this fifth-order difference equation changes into a six-term recurrence relation having three initial equations.

The Point at Infinity

The Thomé solutions for the triconfluent case of the Fuchsian differential equation having four singularities in its natural form (3.4.24) are given by [cf. (2.3.25), (2.3.34)]

$$y(z) = \exp\left(\frac{\alpha_3}{3} \, z^3 + \frac{\alpha_2}{2} \, z^2 + \alpha_1 \, z\right) \, z^{\alpha_{0i}} \left(1 + \sum_{n=0}^{\infty} \frac{C_n}{z^{-n}}\right). \tag{3.4.27}$$

The characteristic exponents α_3, α_2, α_1 and α_{0i} may be taken from (2.3.35) by means of the following identifications:

$$\begin{aligned}
&p_0 = G_0, \quad p_1 = G_1, \quad p_2 = G_2,\\
&q_0 = D_0, \quad q_1 = D_1, \quad q_2 = D_2, \quad q_3 = D_3, \quad q_4 = D_4.
\end{aligned}$$

The result is given by

$$\begin{aligned}
\alpha_{31} &= \frac{1}{2}\left(-G_2 + \sqrt{G_2^2 - 4\,D_4}\right), \\
\alpha_{32} &= \frac{1}{2}\left(-G_2 - \sqrt{G_2^2 - 4\,D_4}\right), \\
\alpha_{2j} &= -\frac{\alpha_{3j}\,G_1 + D_3}{G_2 + 2\,\alpha_{3j}}, \\
\alpha_{1j} &= -\frac{\alpha_{3j}\,G_0 + \alpha_{2j}\,(G_1 + \alpha_{2j}) + D_2}{G_2 + 2\,\alpha_{3j}}, \\
\alpha_{0ij} &= -\frac{2\,(\alpha_{2j}\,\alpha_{1j} + \alpha_{3j}) + \alpha_{2j}\,G_0 + \alpha_{1j}\,G_1 + D_1}{G_2 + 2\,\alpha_{3j}},
\end{aligned} \tag{3.4.28}$$

with $j = 1, 2$. The coefficients C_n obey the difference equation

$$-1\,\hat{p}_2\,C_1 + (-0\,\hat{p}_1 + \hat{q}_0)\,C_0 = 0 \rightsquigarrow C_1,$$
$$-2\,\hat{p}_2\,C_2 + (-1\,\hat{p}_1 + \hat{q}_0)\,C_1 + (-0\,\hat{p}_0 + \hat{q}_{-1})\,C_0 = 0 \rightsquigarrow C_2,$$

$$\begin{aligned}
&-(n+1)\,\hat{p}_2\,C_{n+1} \\
&\quad + (-n\,\hat{p}_1 + \hat{q}_0)\,C_n \\
&\quad\quad + [-(n-1)\,\hat{p}_0 + \hat{q}_{-1}]\,C_{n-1} \\
&\quad\quad\quad + [(n-2)\cdot(n-1) - (n-2)\,\hat{p}_{-1} + \hat{q}_{-2}]\,C_{n-2} = 0 \rightsquigarrow C_{n+1}
\end{aligned}$$

for $n = 2, 3, 4, 5, 6, \ldots$ with the following identifications:

$$\hat{p}_2 = G_2 + 2\,\alpha_3, \quad \hat{p}_1 = G_1 + 2\,\alpha_2, \quad \hat{p}_0 = G_0 + 2\,\alpha_1, \quad \hat{p}_{-1} = 2\,\alpha_0$$

and

$$\begin{aligned}
\hat{q}_0 &= \alpha_2\,(2\,\alpha_0 + 1) + \alpha_1\,(G_0 + \alpha_1) + G_1\,\alpha_0 + D_0, \\
\hat{q}_{-1} &= \alpha_1\,(G_0 + 2\,\alpha_0), \\
\hat{q}_{-2} &= \alpha_0\,(\alpha_0 - 1).
\end{aligned}$$

It is important to mention that – under the condition $\alpha_3 \neq 0$ – it is just the value of α_3 that determines the asymptotic behaviour of the solution $y(z)$ as $z \to \infty$. Assuming that $D_4 > 0$ holds, there is one Thomé solution the asymptotic factor of which increases exponentially as $z \to \infty$ according to

$$y_{1\infty}(z) \sim \exp\left(\frac{\alpha_{31}}{3}\,z^3\right), \; \alpha_{31} > 0$$

and one further Thomé solution that decreases exponentially as $z \to \infty$ according to

$$y_{2\infty}(z) \sim \exp\left(\frac{\alpha_{32}}{3}\,z^3\right), \; \alpha_{32} < 0. \tag{3.4.29}$$

For both of the values α_{31} and α_{32}, the ansatzes (3.4.27) represent a particular solution $y_{1\infty}(z)$ and $y_{2\infty}(z)$ of the differential equation (3.4.24) (cf. Bieberbach, 1965, p. 196). The general solution $y_g(z)$ of the differential equation (3.4.24) is the linear hull

$$y_g(z) = C_{1\infty}\, y_{1\infty}(z) + C_{2\infty}\, y_{2\infty}(z), \quad C_{1\infty}, C_{2\infty} \in \mathbb{C},$$

where the factors $C_{1\infty}$ and $C_{2\infty}$ do not depend on z.

From what is written above it is quite clear now that generally, the solution of the differential equation (3.4.24) increases exponentially as $z \to \infty$, as this behaviour dominates the exponentially decreasing one. However, if there are parameters of the differential equation for which

$$C_{1\infty} = 0$$

holds, then the solution of the differential equation (3.4.24) decreases exponentially as $z \to \infty$ according to (3.4.29).

There remains the question of how to find those parameters or parameter combinations for which this happens. This is just the two-point connection problem and is at the centre of this book, the solution of which is given in Chapter 4 along the lines presented in Chapter 1.

3.4.4 Double Confluent Case

The *double confluent case* of the Fuchsian differential equation (3.2.20) having four singularities is the most peculiar differential equation treated in this book. It does not have an analogue in Gauss-type differential equations. It has two singularities, both of which have s-rank two and thus are irregular singularities of the differential equation.

If the two singularities of the double confluent case of the Fuchsian differential equation (3.2.20) having four singularities are located at the origin and at infinity, then the differential equation admits a remarkable symmetry, namely a symmetry with respect to inversion. This symmetry is heavily taken advantage of when solving its central two-point connection problem, see below (cf. §4.2.3).

In the form where its singularities are located at $z = \epsilon$ and at infinity, it looks like

$$\boxed{\begin{aligned} &\frac{d^2y}{dz^2} + \left[\frac{G_{-2}}{(z-\epsilon)^2} + \frac{G_{-1}}{z-\epsilon} + G_0\right]\frac{dy}{dz} \\ &\quad + \left[\frac{D_{-4}}{(z-\epsilon)^4} + \frac{D_{-3}}{(z-\epsilon)^3} + \frac{D_{-2}}{(z-\epsilon)^2} + \frac{D_{-1}}{z-\epsilon} + D_0\right] y = 0 \end{aligned}} \tag{3.4.30}$$

where ϵ might also be zero such that the finite singularity of equation (3.4.30) is located at the origin. In this case, the above-mentioned symmetry occurs.

The double confluent case (3.4.30) of the Fuchsian differential equation (3.2.20) having four singularities may be thought of as being constructed by means of a confluence process of the two regular singularities of the single confluent case (3.4.1) of the Fuchsian differential equation (3.2.20) having four singularities or by an ad-hoc increasing of the s-rank of the single confluent case (1.2.13) of the finite singularity of the Gauss differential equation (1.2.1).

Local Solutions

The double confluent case (3.4.30) of the Fuchsian differential equation with four singularities has two singularities, both of which have s-rank two, i.e., $R(\epsilon) = R(\infty) = 2$, thus, the s-rank symbol of the differential equation is $\{2; 2\}$, whereby the finite singularity may be thought of as being located at the origin $z = 0$. Thus, the coefficients are

$$P(z) = \frac{G_{-2}}{z^2} + \frac{G_{-1}}{z} + G_0,$$
$$Q(z) = \frac{D_{-4}}{z^4} + \frac{D_{-3}}{z^3} + \frac{D_{-2}}{z^2} + \frac{D_{-1}}{z} + D_0. \tag{3.4.31}$$

Therefore, a Thomé series exists at both of its singularities, asymptotically representing the particular solutions on approaching the singularity radially, i.e., along the positive real axis, say. At the point at infinity, this is given by (2.3.28)

$$y_\infty(z) = \exp(\alpha_{1\infty}\, z)\, z^{\alpha_{0i\infty}}\, w_\infty(z).$$

Accordingly, at the origin the ansatz takes the form

$$y_0(z) = \exp\left(\frac{\alpha_{10}}{z}\right) z^{\alpha_{0i0}}\, w_0(z).$$

The characteristic exponents $\alpha_{1\infty j}$ and $\alpha_{0i\infty j}$, $j = 1, 2$ of the Thomé solution at infinity are given by (2.3.29), (2.3.30):

$$\alpha_{1\infty 1} = \frac{1}{2}\left(-G_0 + \sqrt{G_0^2 - 4\, D_0}\right),$$

$$\alpha_{1\infty 2} = \frac{1}{2}\left(-G_0 - \sqrt{G_0^2 - 4\, D_0}\right)$$

and

$$\alpha_{0i\infty 1} = -\frac{G_{-1}\,\alpha_{11} + D_{-1}}{G_0 + 2\,\alpha_{11}},$$

$$\alpha_{0i\infty 2} = -\frac{G_{-1}\,\alpha_{12} + D_{-1}}{G_0 + 2\,\alpha_{12}}.$$

The characteristic exponents α_{10j} and α_{0i0j}, $j = 1, 2$, of the Thomé solution at the origin are given analogously by

$$\alpha_{101} = -\frac{G_{-2}}{2} + \frac{1}{2}\sqrt{G_{-2}^2 - 4\, D_{-4}},$$

$$\alpha_{102} = -\frac{G_{-2}}{2} - \frac{1}{2}\sqrt{G_{-2}^2 - 4\, D_{-4}}$$

and

$$\alpha_{0i01} = \frac{D_{-3} + 2\sqrt{-D_{-4}}}{G_{-2} - 2\sqrt{-D_{-4}}},$$

$$\alpha_{0i02} = \frac{D_{-3} - 2\sqrt{-D_{-4}}}{G_{-2} + 2\sqrt{-D_{-4}}}.$$

The differential equation (2.1.2) having the coefficients (3.4.31) does not have a regular singularity; therefore, it is a non-convergent, one-sided particular solution about non-ordinary points of the differential equation (3.4.31).

3.5 The Land Beyond Heun

3.5.1 Generalisations

In the previous sections and chapters, the terms *class of the Heun differential equations* and *set of the Heun differential equations* were coined. In this system of Heun-type differential equations, the Fuchsian differential equation having four singularities plays the seminal role and it is straightforward to generalise it to Fuchsian differential equations having an arbitrary number of singularities.

Classes and Sets of Equations

A *class* of Fuchsian differential equations consists of its confluent and reduced cases. Here, one case of a Fuchsian differential equation emerges from another one by means of a confluence or a reduction process. A class of Fuchsian differential equations is thus the totality of all differential equations that have emerged from each other by means of a confluence or a reduction process.

A *set* of a case of a class of Fuchsian differential equations consists of all of its forms. Thus, the elements of such a set carry a specific form. One element of such a set emerges from another one by means of a Moebius or an s-homotopic transformation. A set of a case of a class of Fuchsian differential equations is thus the totality of all differential equations that have emerged from each other by a Moebius or an s-homotopic transformation.

The totality of all differential equations that comprise a class of Fuchsian differential equations, as well as all of its sets, are said to be of its *type*. For example, the differential equations of Heun's type are the totality of all the equations of the class of Heun's differential equations, as well as of all the sets of this class.

These relationships are graphically illustrated in Figure 3.8.

Generalised Riemann Symbol

The conventional Riemann symbols may be generalised to characterise differential equations not only beyond the Fuchsian ones with four singularities but also confluent and reduced cases having an arbitrary number of regular as well as irregular

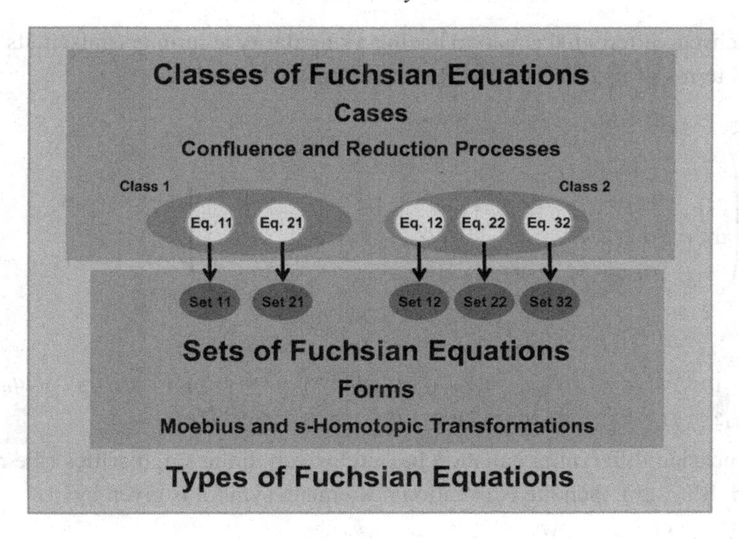

Figure 3.8 Sets and classes.

singularities. This may be done in a straightforward manner and is to be discussed in the following, starting with Fuchsian equations.

Consider a Fuchsian differential equation (2.1.2). At each of its singularities z_i there exist two characteristic exponents $\alpha_{z_i 1}$ and $\alpha_{z_i 2}$. These pairs of singularity parameters determine the Frobenius solutions of equation (2.1.2) at its singularities. Therefore, instead of writing the differential equation, it is equivalent to write down its conventional Riemann symbol. For Fuchsian equations having n singularities, the conventional Riemann symbol looks like

$$P\begin{pmatrix} 1 & 1 & \dots & 1 & 1 & \\ z_1 & z_2 & \dots & z_{n-1} & z_n & z \\ \alpha_{z_1 1} & \alpha_{z_2 1} & \dots & \alpha_{z_{n-1} 1} & \alpha_{z_n 1} & q_1, q_2, \dots, q_{n-3} \\ \alpha_{z_1 2} & \alpha_{z_2 2} & \dots & \alpha_{z_{n-1} 2} & \alpha_{z_n 2} & \end{pmatrix}. \tag{3.5.1}$$

The first line shows the s-ranks of the singularities. The second line lists the locations of the singularities, as well as the independent variable; the entry z_n is ∞ if the point at infinity is a (regular) singularity of the differential equation. The third line contains the first of the two characteristic exponents at each of the singularities of the equation, as well as all the accessory parameters, and the last line contains the second characteristic exponents.

Conventional Riemann symbols are applicable to Fuchsian equations and are rather useful tools, to get an immediate survey about the Fuchsian differential equation at hand, but also the locations and s-ranks of its singularities, the characteristic exponents and the accessory parameters when carrying out s-homotopic transformations; Moebius transformations (1.2.26) become trivial. So,the s-homotopic transformation for a Fuchsian differential equation having a singularity at infinity (thus, in its natural form) in terms of its conventional Riemann symbol looks like

$$
\begin{aligned}
&(z-z_1)^{\gamma_1}(z-z_2)^{\gamma_2}\cdots(z-z_{n-1})^{\gamma_{n-1}}\\
&\times P\begin{pmatrix} 1 & 1 & \cdots & 1 & 1 & \\ z_1 & z_2 & \cdots & z_{n-1} & \infty & z\\ \alpha_{z_1 1} & \alpha_{z_2 1} & \cdots & \alpha_{z_{n-1} 1} & \alpha_{z_\infty 1} & q_1, q_2, \ldots, q_{n-3}\\ \alpha_{z_1 2} & \alpha_{z_2 2} & \cdots & \alpha_{z_{n-1} 2} & \alpha_{z_\infty 2} & \end{pmatrix} =\\
&P\begin{pmatrix} 1 & \cdots & 1 & 1 & \\ z_1 & \cdots & z_{n-1} & \infty & z\\ \alpha_{z_1 1}-\gamma_1 & \cdots & \alpha_{z_{n-1} 1}-\gamma_{n-1} & \alpha_{z_\infty 1}+\gamma_1+\cdots+\gamma_{n-1} & q_1, q_2, \ldots, q_{n-3}\\ \alpha_{z_1 2}-\gamma_1 & \cdots & \alpha_{z_{n-1} 2}-\gamma_{n-1} & \alpha_{z_\infty 2}+\gamma_1+\cdots+\gamma_{n+1} & \end{pmatrix}.
\end{aligned}
$$

If the Fuchsian differential equation has exclusively finite singularities (the number of which is n, say), then the conventional Riemann symbol is given by

$$
\begin{aligned}
&\left(\frac{z-z_1}{z-z_n}\right)^{\rho_1}\left(\frac{z-z_2}{z-z_n}\right)^{\rho_2}\cdots\left(\frac{z-z_{n-1}}{z-z_n}\right)^{\rho_{n-1}}.\\
&= P\begin{pmatrix} 1 & 1 & \cdots & 1 & 1 & \\ z_1 & z_2 & \cdots & z_{n-1} & z_n & z\\ \alpha_1'' & \alpha_2'' & \cdots & \alpha_{n-1}'' & \alpha_n'' & q_1, q_2, \ldots, q_n\\ \alpha_1' & \alpha_2' & \cdots & \alpha_{ne1}' & \alpha_n' & \end{pmatrix}\\
&P\begin{pmatrix} 1 & 1 & \cdots & 1 & 1 & \\ z_1 & z_2 & \cdots & z_{n-1} & z_n & z\\ \alpha_1''-\rho_1 & \alpha_2''-\rho_2 & \cdots & \alpha_{n-1}''-\rho_{n-1} & \alpha_n''+\sum_{i=1}^{n-1}\rho_i & q_1, q_2, \ldots, q_n\\ \alpha_1'-\rho_1 & \alpha_2'-\rho_2 & \cdots & \alpha_{ne1}'-\rho_{n-1} & \alpha_n'+\sum_{i=1}^{n-1}\rho_i & \end{pmatrix}.
\end{aligned}
$$

Turning now to the generalised Riemann symbol, each of the two characteristic exponents of second kind and of highest order creates a sequence of characteristic exponents of second kind and lower order and of first kind. The position within the groups of pairs of characteristic exponents in each order is the same as the pair of exponents of highest order.

According to the definition, a column in a generalised Riemann symbol related to a singularity at $z = z_1$ having s-rank R includes $2R + 2$ numbers or algebraic symbols, respectively, of which a number R are characteristic exponents from which one pair is always of first order.

Such a generalised Riemann symbol for any differential equation may thus look like the following. If a differential equation has l regular and m irregular singularities located at

$$z_1, \ldots, z_l, z_{l+1}, \ldots, z_m$$

then the s-ranks of the irregular ones are

$$R_{z_{l+1}}, \ldots, R_{z_{n-1}}, R_{z_m},$$

respectively, and its generalised Riemann symbol is given by

$$P\begin{pmatrix} 1 & \dots & 1 & R_{z_{l+1}} & \dots & R_{z_{n-1}} & R_{z_m} & \\ z_1 & \dots & z_l & z_{l+1} & \dots & z_{m-1} & z_m & z \\ \alpha_{z_{11}} & \dots & \alpha_{z_{l1}} & \alpha_{z_{(l+1)1}}^{(R_{z_{l+1}}-1)} & \dots & \alpha_{z_{m-1\,1}} & \alpha_{z_{m,1}} & q_1, q_2, \dots, q_{k-3} \\ \alpha_{z_{12}} & \dots & \alpha_{z_{l2}} & \alpha_{z_{(l+1)2}}^{(R_{z_{l+1}}-1)} & \dots & \alpha_{z_{m-1\,2}} & \alpha_{z_{m,2}} & \\ & & & \dots & \dots & \dots & \dots & \\ & & & \alpha_{z_{(l+1)}\,1}^{(1)} & \dots & \alpha_{z_{m-1}\,1}^{(1)} & \alpha_{z_m\,1}^{(1)} & \\ & & & \alpha_{z_{(l+1)}\,2}^{(1)} & \dots & \alpha_{z_{m-1}\,2}^{(1)} & \alpha_{z_m\,2}^{(2)} & \\ & & & \alpha_{z_{(l+1)}\,1}^{(0)} & \dots & \alpha_{z_{m-1}\,1}^{(0)} & \alpha_{z_m\,1}^{(1)} & \\ & & & \alpha_{z_{(l+1)}\,2}^{(0)} & \dots & \alpha_{z_{m-1}\,2}^{(0)} & \alpha_{z_m\,2}^{(2)} & \end{pmatrix}$$

where

$$k = l + R_{z_{l+1}} + \cdots + R_{z_{n-1}} + R_{z_m}.$$

Here, the left-hand rows list the s-ranks, the locations as well as the pairs of characteristic exponents of the regular singularities. All other columns contain the data of irregular singularities of the underlying differential equation.

Eventually, it should be mentioned that Riemann symbols are sometimes called *Riemannian schemes*.

3.5.2 Arscott and Ince Equations

It is the special feature of the Jaffé method that it is not limited to such singular boundary eigenvalue problems that lead to two-term recurrence relations for the calculation of the coefficients of their Taylor series. This means that it is also not limited to solving singular boundary eigenvalue problems whose differential equation has only a certain maximum number of singularities. Nor is the method limited to singular boundary eigenvalue problems whose irregular singularities have only a certain maximum of its s-rank. Thus, from a methodological point of view, there is no reason to abstain from an application of the Jaffé method to differential equations beyond the Heun type.

Since the Fuchsian differential equations are determined by just one parameter, namely the number of their singularities, which are all regular ones, it is obvious to exceed the set of Heun differential equations by considering Fuchs equations with more than four singularities. As already stated in the introductory text to this chapter, I refer to the Fuchs equation with five singularities as the *Arscott differential equation*, because he was the first to treat this differential equation (Arscott et al., 1983). In its natural form it looks like

$$\frac{d^2y}{dz^2} + \sum_{i=1}^{4} \frac{A_{z_i}}{z - z_i} \frac{dy}{dz} + \sum_{i=1}^{4} \left[\frac{B_{z_i}}{(z - z_i)^2} + \frac{C_{z_i}}{z - z_i} \right] y = 0$$

with

$$\sum_{i=1}^{4} C_{z_i} = 0.$$

I suggest calling this equation, together with the totality of all its confluent cases, the *class of Arscott differential equations*. Analogously, the totality of forms of a confluent case can be called the *set of Arscott differential equations*.

Similarly, I propose naming the Fuchs equation with six singularities after Edward Lindsay Ince, thus calling it the *Ince differential equation*, because it was he who seems to have mentioned it first, in Ince (1956, pp. 501–505). In its natural form it looks like

$$\frac{\mathrm{d}^2 y}{\mathrm{d}z^2} + \sum_{i=1}^{5} \frac{A_{z_i}}{z - z_i} \frac{\mathrm{d}y}{\mathrm{d}z} + \sum_{i=1}^{5} \left[\frac{B_{z_i}}{(z - z_i)^2} + \frac{C_{z_i}}{z - z_i} \right] y = 0$$

with

$$\sum_{i=1}^{5} C_{z_i} = 0.$$

Moreover, I suggest calling this equation, together with the totality of all its confluent cases, the *class of Ince differential equations*. Analogously, the totality of forms of a confluent case can be called the *set of Ince differential equations*.

The Jaffé method for solving singular boundary eigenvalue problems and their confluent and reduced cases can be applied mutatis mutandis to the class of Arcott as well as to the class of Ince differential equations, and even to classes of Fuchsian differential equations having more than six singularities. In §5.6 a concrete example is treated.

4

Central Two-Point Connection Problems and Higher Special Functions

According to the way of thinking here, the totality of special functions (viz. solutions of singular boundary eigenvalue problems of linear second-order differential equations having polynomial coefficients) is divided into *classical special functions* and *higher special functions*. While the classical special functions are particular solutions of differential equations of Gauss type, the special functions are called higher in those cases where the underlying differential is of Heun type or higher. Types of differential equations beyond the Heun one are those whose seminal Fuchsian differential equation has more than four singularities.

The singular boundary eigenvalue problems producing the classical special functions are not denoted in a particular manner, while the singular boundary eigenvalue problems producing higher special functions are called *central two-point connection problems* (CTCP for short).

Solving a CTCP means determining the parameters of the underlying differential equation for which a particular solution obeys the singular boundary eigenvalue conditions.

In this chapter the method to solve the CTCP is developed in detail. This method does not have limits that lie in the type or form of the differential equation, as was the case with the infinite continued fraction method relating to second-order difference equations, but the restrictions lie only in numerical reasons. Moreover, since there is no approximation made, the method may be assigned to be exact in a strong mathematical sense. This means that all the applications which may be solved by means of this method can be considered as being solved exactly. For calculatory reasons, the method is developed for the Heun class.

From a scientific point of view, Chapter 4 is the main mathematical part of this book. Here, the power of what I call the *Jaffé method* is shown with respect to generalising the central two-point connection problem for the confluent cases of Fuchsian differential equations, shown at the differential equations of the Heun class. The asymptotic behaviour of the power-series part on the one side and the asymptotic factor on the other is balanced in such a way that just the eigenfunction behaviour appears to become apparent when the eigenvalue parameter admits an eigenvalue. This is the ingenious aspect of Jaffé's ansatz.

Although this chapter deals with more fundamental mathematical issues, it fills, in a certain view, a missing gap, which helps a great deal to understand in what manner the Jaffé approach to the central two-point connection problem works when applied to Fuchsian differential equations and their confluent cases.

The considerations are split into a section dealing with the basic concept and another one presenting the calculatory procedure of carrying out the aforementioned concept. It is just the latter section that is new, since the eigenvalue condition has to be solved by means of the Jaffé approach. In all the confluent cases of Heun's differential equation, it leads to a linear difference equation of Poincaré–Perron type for the coefficients of the ansatz. The eigenvalue condition then requires a singling out of the particular solution belonging to the algebraically smaller root of the characteristic equation, viz. the minimum solution.

For the Heun differential equation itself, this singling out, however, was achieved already in the 1920s by the Danish mathematician **Niels Erik Nörlund** (1885–1981), applying infinite continued fraction methods (cf. §4.2.1) to become the first transcendental eigenvalue condition for singular boundary eigenvalue problems in the field of higher special functions. The necessity of developing Jaffé's approach occurred because Nörlund's method only works for second-order difference equations. However, there are confluent cases within the Heun class for which the Jaffé method leads to difference equations of higher than second order. The fact that the Jaffé approach is not restricted to the order of the underlying difference equation is a comfortable aspect.

Carrying out, as well as reasoning, the eigenvalue conditions of the central two-point connection problems of the Heun class are the main topic of this chapter.

4.1 Basic Mathematical Concept: Unmasking Recessive Solutions

Given a linear, second-order ordinary differential equation with polynomial coefficients (2.1.2),

$$\frac{\mathrm{d}^2 y}{\mathrm{d}z^2} + P(z,\underline{c})\,\frac{\mathrm{d}y}{\mathrm{d}z} + Q(z,\underline{c})\,y(z) = 0, \quad z \in \mathbb{C}, \tag{4.1.1}$$

the parameters of which are denoted $\underline{c}$ and are supposed to be real-valued. It is supposed that there is an ordinary point or a regular singularity of (4.1.1) at the origin $z = 0$ and an irregular singularity at infinity $z = \infty$, and that there is no other singularity of this differential equation on the positive real axis (that is – with respect to the possibility of carrying out appropriate Moebius transformations – no substantial restriction). If the s-rank of the irregular singularity at infinity is denoted $R(\infty) = s$ then – according to the Thomé solutions – the asymptotic behaviour of the solutions of (4.1.1) may be either

$$y_{\max}(z) \sim \exp\left(+k_{\max}\,\frac{z^{s-1}}{s-1}\right)$$

or

$$y_{\min}(z) \sim \exp\left(-k_{\min}\frac{z^{s-1}}{s-1}\right)$$

as $z \longrightarrow \infty$. If the discriminant in (1.2.16) is non-zero, it holds that

$$\Re(k_{\max}) > 0 \in \mathbb{R},$$
$$\Re(k_{\min}) > 0 \in \mathbb{R}.$$

For the sake of calculatory simplicity, in the following we assume $k_{\max}$ and $k_{\min}$ to be real-valued. $y_{\max}$ is called the *maximum solution*, while $y_{\min}$ is the *minimum solution*. Moreover, it is supposed that $y_{\max}(z)$ and $y_{\min}(z)$ are linearly independent particular solutions of the differential equation (4.1.1). Then, the general solution of (4.1.1) is given by

$$y_{gen}(z) = c_1\, y_{\max}(z) + c_2\, y_{\min}(z), \quad c_1, c_2 \in \mathbb{C}. \tag{4.1.2}$$

This means that each particular solution $y_p(z)$ of the differential equation (4.1.1) may be written as

$$y_p(z) = c_1(\underline{c})\, y_{\max}(z) + c_2(\underline{c})\, y_{\min}(z). \tag{4.1.3}$$

Definition 4.1. The determination of those values of the parameters $\underline{c}$ of the differential equation (4.1.1) for which a certain particular solution $y_{\text{ctcp}}(z)$ is just a multiple of the minimum solution $y_{\min}$ at a given particular solution at the origin is called the *central two-point connection problem* of the differential equation (4.1.1), or CTCP for short.

This means

$$y_{\text{ctcp}}(z) = c_2(\underline{c})\, y_{\min}(z) \tag{4.1.4}$$

or

$$c_1(\underline{c}) = 0, \tag{4.1.5}$$

which is a determining equation for the parameters of the underlying differential equation (4.1.1).

This definition of a central two-point connection problem is illustrated in Figure 4.1.

The CTCP is normally neither a local problem nor a global one of the underlying differential equation, but an intermediate one. Generally speaking, it looks for particular solutions that connect two singularities of a differential equation (4.1.1) on a (straight) line between them lying in the complex plane, not bothering about the behaviour of the solution outside this connecting line. Without loss of generality this line is the positive real axis, called the *relevant interval*.

The asymptotic behaviour of the solution for z tending to infinity along the positive real axis is at the centre of the problem.

The crucial thing for the application of the method presented here is the existence of two linearly independent partial solutions of the differential equation that differ

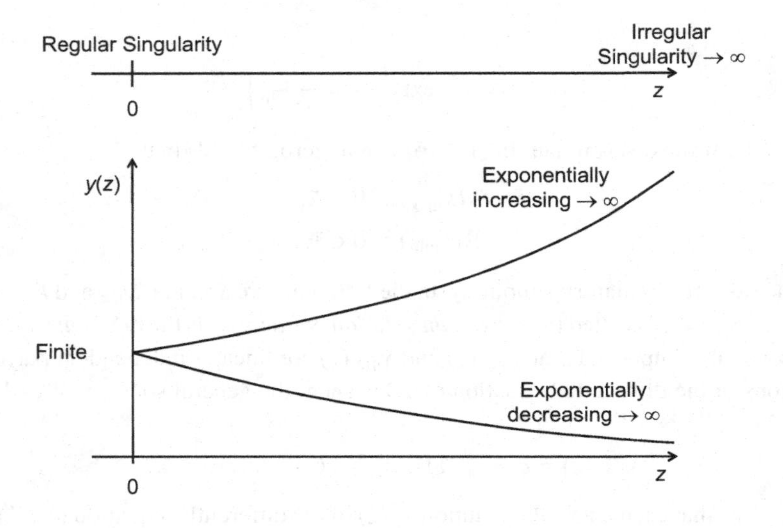

Figure 4.1 The central two-point connection problem.

significantly in their asymptotic behaviour for $z \to \infty$ along the positive real axis. The other end of the relevant interval is the origin of the coordinate system. Either there is a regular singularity of the differential equation or an ordinary point. In both cases, two linearly independent partial solutions of the differential equation exist there as well, which differ significantly in their asymptotic behaviour for $z \to 0$ along the positive real axis. In the first case it is the two Frobenius solutions, in the second the ratio of the function value and the first derivative. In both cases, one chooses one of the two particular solutions. A word about the nature of the parameters: basically, a differential equation knows two types of parameters – those, whose value influences the behaviour of its solutions at their singularities and those whose value does not influence the behaviour of its solutions at its singularities. In the first case we speak of *singularity parameters*, in the second of *accessory parameters*. In principle, both types of parameters can serve as boundary eigenvalue parameters, but it is obvious that their influence on the singular boundary eigenvalue problem is of a different nature. Therefore, the singular boundary eigenvalue problem may be defined for differential equations (4.1.1) having no accessory parameter, but such a problem is of a totally different character, since the asymptotic behaviour in this case is dependent on the eigenvalue parameter(s).

Concerning the solution of the CTCP for the Heun differential equation, as well as for its confluent cases, we request an exact solution in an algebraic as well as an analytic sense, meaning that neither algebraic approximations nor non-convergent analytic (e.g., non-convergent asymptotic) solutions are accepted. This means that we require a solution which in principle is able to solve the CTCP to an arbitrary accuracy. Such a solution is presented in the following; it consists of two steps. First, the CTCP of the differential equation is solved formally by means of an ansatz consisting of an asymptotic factor and a *convergent* series of power type, the coefficients of which are solvable by means of a difference equation in the form of a recurrence relation. The radius of convergence of this series comprises the whole relevant interval of the CTCP, viz. both of the singularities between which the CTCP is defined. In a second step, the asymptotic behaviour of the solutions of the difference equation for large values of the index is interpreted.

The problem of the first step is to find an ansatz that yields a convergent representation of its solution, although the representations of the particular local solutions (i.e., the Thomé-type expansions) are, in general, not convergent, but only non-convergent asymptotic on radially approaching the singularity of the differential equation. This problem has already been tackled by George Jaffé in 1933. As far as I know, he solved this problem – following a hint by **Wolfgang Pauli** and **Friedrich Hund** – for the first time (see Jaffé, 1933) on calculating the eigensolutions and eigenvalues of the hydrogen molecule ion by quantum theoretical means.[1] In honour of having had the idea to solve this special CTCP of the single confluent case of the Heun differential equation in this (exact) manner, I call this sort of solution the *Jaffé solution*, consisting of Jaffé transformations as well as Jaffé expansions (see below). This is all the more worthwhile since it is generalisable significantly in that, according to this method, all CTCPs – not only of the Heun differential equation itself, but of all confluent cases of Fuchsian differential equations – may be solved as well. Regarding the relevance of these differential equations with respect to applied sciences, it is rather a great step.

The resolution of the second step is, on the one hand, based on the *Euler–McLaurin summation formula* as well as on *Laplace's method* of integration, and, on the other hand, on a large-scale investigation of the Canadian mathematicians **George D. Birkhoff** (1884–1944) and **Waldemar J. Trjitzinsky** (1901–1973) on the asymptotics of linear difference equations. Applying the results of Birkhoff and Trjitzinsky to the irregular difference equations of Poincaré–Perron type of arbitrary order stemming from the central two-point connection problems of Fuchsian differential equations and its confluent cases by means of a thorough analytical investigation of asymptotically interpreting irregular difference equations reveals the meaning of its asymptotic solutions.

What is remarkable, and the main reason for interpreting the difference equations, is that the linear differential equation is of second order while the convergent ansatzes result in difference equations that are still linear, but the order of which is not dependent on the order of the original differential equation and thus not necessarily of second order. Moreover, the order of the resulting difference equation is a consequence of the s-rank of the irregular singularity of the CTCP. Thus, generally, the above-written ansatz yields a linear difference equation of Poincaré–Perron type, the order

[1] It was Jaffé's last work as a Jewish professor at the University of Gießen before he had to go into exile to Batôn Rouge in the US state of Louisiana. Unfortunately, he never came back to develop further his marveilous idea on calculating boundary eigenvalue problems of differential equations, the solutions of which have irregular singularities.

of which is by and large just the s-rank of the irregular singularity of the underlying differential equation that is involved in the CTCP.

It is quite clear that a linear difference equation of Poincaré–Perron type is more complicated in theory, as well as in numerical calculations, the higher its order is. This was the main obstacle up to now in treating singular boundary eigenvalue problems of differential equations of Fuchsian type with more than three singularities and their confluent cases, since a theory was lacking as well as methods to treat this sort of difference equations, the order of which is higher than two.

As long as boundary eigenvalue value problems of linear, second-order differential equations are breakable down to two-term recurrence relations, the according eigenvalue conditions are of algebraic type. This is always the case when boundary eigenvalue problems originate from Fuchsian differential equations having at most three singularities. As soon as Fuchsian differential equations have more than three singularities, the eigenvalue conditions become transcendental. The first transcendental eigenvalue condition involved infinite continued fractions. However, only three-term recurrence relations are describable by infinite continued fractions. Therefore, only the Heun differential equation and its single confluent case are treatable by this approach. For all the other confluent cases of the Heun differential equation and for Fuchsian differential equations having more than four singularities, there is no method to describe the transcendental eigenvalue condition coming out of power-type ansatzes in order to solve their CTCPs.

This was the reason for dividing the whole field of special functions into classical and higher special functions. By and large, classical special functions are restricted in that the underlying difference equations for the eigensolutions are restricted to second order, and thus treatable by either algebraic methods or by infinite continued fraction methods.[2] However, it has to be recognised that there are fundamental effects occurring for higher special functions which are not seen in the field of classical special functions.

In the following a theory is given that solves this problem not only on a theoretical but also on a calculatory level for all confluent cases of the Heun differential equation and thus for the complete Heun class. The order of the resulting irregular difference equations of Poincaré–Perron type for this class of equtions is at most four. However, the difference from all treatments of this sort of boundary eigenvalue problem up to now is that the theory presented here is generalisable to the case of arbitrary s-ranks and thus to arbitrary order of the difference equation. As a consequence, the theory of higher special functions stemming from a CTCP may be considered essentially as generally solved.

[2] This is not completely correct. Arscott et al. (1983) have treated a four-term recurrence relation by means of backward recurrence relation methods; however, this approach is applicable only to regular difference equations.

4.2 Methods of Solution for the Heun Class

4.2.1 Heun Equation

The Heun differential equation is a Fuchsian equation having four singularities. In its standard form, three of them are finite and one of the characteristic exponents at each of the three finite singularities of the differential equation is zero. Thus, it is given by (3.2.15)

$$\frac{d^2y}{dz^2} + \left\{\frac{A_0}{z} + \frac{A_a}{z-a} + \frac{A_1}{z-1}\right\}\frac{dy}{dz} + \left\{\frac{C_0}{z} + \frac{C_a}{z-a} + \frac{C_1}{z-1}\right\} y = 0. \tag{4.2.1}$$

In order for the singularity at infinity to be a regular singularity, the relation

$$C_0 + C_a + C_1 = 0 \tag{4.2.2}$$

has to hold.

Traditionally, this differential equation is written in the form

$$\frac{d^2y}{dz^2} + \left\{\frac{A_0}{z} + \frac{A_a}{z-a} + \frac{A_1}{z-1}\right\}\frac{dy}{dz} + \frac{\alpha\,\beta\,z + q}{z\,(z-a)\,(z-1)}\,y = 0, \tag{4.2.3}$$

whereby (4.2.2) adopts the form

$$A_0 + A_a + A_1 = \alpha + \beta + 1.$$

Therefore, the relations between the parameters in (4.2.1) and (4.2.3) are given by

$$\alpha\,\beta = -\left[(a+1)(C_0 + C_1) + C_0\,a\right],$$
$$q = C_0\,a$$

or

$$C_0 = \frac{q}{a},$$
$$C_a = \frac{\alpha\,\beta + \frac{q}{a}}{a-1},$$
$$C_1 = -\frac{\alpha\,\beta + q}{a-1}.$$

The characteristic exponents of the singularity at infinity are given by [cf. (3.2.17), (3.2.18)]

$$\alpha = -\frac{1}{2}\,(1 - A_0 - A_a - A_1) + \frac{1}{2}\,\sqrt{(1 - A_0 - A_a - A_1)^2 - 4\,(C_a\,a + C_1)},$$
$$\beta = -\frac{1}{2}\,(1 - A_0 - A_a - A_1) - \frac{1}{2}\,\sqrt{(1 - A_0 - A_a - A_1)^2 - 4\,(C_a\,a + C_1)}.$$

This may be verified by using the parameters of equation (4.2.1):

$$\left.\begin{matrix}\alpha\\ \beta\end{matrix}\right\} = -\frac{1}{2}\underbrace{(1 - A_0 - A_a - A_1)}_{=-(\alpha+\beta)} \pm \frac{1}{2}\underbrace{\sqrt{\underbrace{(1 - A_0 - A_a - A_1)^2}_{(\alpha+\beta)^2} - 4\underbrace{(C_a\, a + C_1)}_{=\alpha\beta}}}_{=\alpha-\beta}$$

$$= \frac{1}{2}(\alpha + \beta) \pm \frac{1}{2}(\alpha - \beta) = \begin{cases} \alpha, \\ \beta. \end{cases}$$

The conventional Riemann symbol of the Heun differential equation in the form (4.2.1) is

$$\begin{pmatrix} 1 & 1 & 1 & 1 & \\ 0 & a & 1 & \infty & z \\ 0 & 0 & 0 & \alpha & q \\ 1 - A_0 & 1 - A_a & 1 - A_1 & \beta & \end{pmatrix}. \tag{4.2.4}$$

The standard particular solution of (4.2.1) is given by means of a pure power series (because the related characteristic exponent is zero)

$$y(z) = \sum_{n=0}^{\infty} a_n\, z^n, \tag{4.2.5}$$

the coefficients of which are determined by a regular linear second-order difference equation of Poincaré–Perron type (cf. Ronveaux, 1995, p. 34)

$$\begin{aligned} a\, A_0\, a_1 + q\, a_0 &= 0, \\ P_n\, a_{n+1} - (Q_n - q)\, a_n + R_n\, a_{n-1} &= 0, \quad n \geq 1, \end{aligned} \tag{4.2.6}$$

where

$$\begin{aligned} P_n &= (n + 1)(n + A_0)\, a = n^2\, a\left(1 + \frac{1 + A_0}{n} + \frac{A_0}{n^2}\right), \\ Q_n &= n\,[(n - 1 + A_0)(1 + a) + a\, A_1 + A_a] \\ &= n^2\left\{1 + a + \frac{A_0\,(a + 1) + a\,(A_1 - 1) + A_a - 1}{n}\right\}, \\ R_n &= (n - 1 + \alpha)(n - 1 + \beta) = n^2\left(1 + \frac{\alpha + \beta - 2}{n} + \frac{(\alpha - 1)(\beta - 1)}{n^2}\right). \end{aligned} \tag{4.2.7}$$

Considering the limiting processes

$$\lim_{n\to\infty}\left(\frac{P_n}{n^2}\right) = a,$$
$$\lim_{n\to\infty}\left(\frac{Q_n}{n^2}\right) = 1+a,$$
$$\lim_{n\to\infty}\left(\frac{R_n}{n^2}\right) = 1,$$

the characteristic equation of the difference equation (4.2.6) is

$$a\,t^2 - (1+a)\,t + 1 = 0 \tag{4.2.8}$$

with

$$t = \lim_{n\to\infty} t_n = \lim_{n\to\infty}\left(\frac{a_{n+1}}{a_n}\right).$$

The two solutions of its characteristic equation (4.2.8) are

$$\begin{aligned} t_1 &= \frac{1}{a}, \\ t_2 &= 1. \end{aligned} \tag{4.2.9}$$

Without loss of generality, the parameter a may be considered as being real-valued and placed between the origin $z = 0$ and $z = 1$:

$$0 \le a \le 1$$

so that

$$t_1 \ge t_2.$$

For $a \to 0$ as well as for $a \to 1$ a confluence process occurs, from which results a single confluent case of the Fuchsian differential equation having four singularities in non-natural form.

According to (4.2.9), the radius of convergence r of the series (4.2.5) locally representing a particular solution of (4.2.1) may adopt two values. Generally (i.e., for no specific choice of the parameters) it takes the inverse of the larger value t_1 of (4.2.8), thus $r = a < 1$, viz. it ranges from the origin to the neigbouring singularity of the differential equation (4.2.1) at $z = a$.

As can be seen from the conventional Riemann symbol (4.2.4), a particular solution of (4.2.1) that is determined by an initial condition $y(0) = y_0$ at the origin $z = 0$ generally yields a linear combination

$$c_1\, y_1(a) + c_2\, y_2(a), \quad c_1, c_2 \text{ arbitrary},$$

of two fundamental solutions $y_1(z)$, $y_2(z)$ of equation (4.2.1) at the neigbouring singularity at $z = a$. However, there is a second solution $t_2 = 1$ of equation (4.2.8) since the difference equation (4.2.6) is of second order. This solution $\{a_n\}$, $n = 0, 1, \ldots$, generated via the ansatz (4.2.4) produces a particular solution $y = y(z)$

of the differential equation (4.2.1). The asymptotic behaviour of the coefficients a_n of the power series (4.2.6) is different, since the ratio of two consecutive numbers a_{n+1} and a_n tends to 1 for $n \to \infty$ here, and no longer to $1/a$. Thus, for this particular solution of the differential equation (4.2.8), the radius of convergence of the series (4.2.4) becomes 1, instead of $a \leq 1$. This, however, means that the solution is holomorphic at $z = a$. This is just possible if the Frobenius solution at $z = 0$ that adopts the characteristic exponent $\alpha_{00} = 0$ is the Frobenius solution at $z = a$ that adopts the characteristic exponent $\alpha_{0a} = 0$. These solutions are called *Heun functions* (cf. Ronveaux, 1995, p. 39). They are holomorphic at the two singularities of the differential equation (4.2.1) at $z = 0$, as well as at $z = a$ simultaneously.

It is quite clear that Heun functions as particular solutions of the Heun differential equation only occur for special values of the parameters of the differential equation (4.2.1). This raises the question of how to get just these values of the parameters.

Infinite Continued Fractions

A Frobenius ansatz

$$y(z) = z^{\alpha_{0j}} \sum_{n=0}^{\infty} a_{nj}\, z^n, \tag{4.2.10}$$

where α_{0j}, $j = 1, 2$, is one of the two characteristic exponents of the singularity at the origin for solving the CTCP of the Heun differential equation (3.2.15) with $0 \leq a \leq 1$ on the interval $0 \leq z \leq a$ of the positive real axis leads to a linear, regular, homogeneous, second-order *difference equation of Poincaré–Perron type* (4.2.6), (4.2.7). Moreover, there is an initial condition in (4.2.6) such that after having determined the term a_0, all the infinitely many subsequent terms a_n, $n = 1, 2, 3, \ldots$, are fixed by means of solving the difference equation (3.2.35) recursively. Therefore, (4.2.6), (4.2.7) may be called a three-term *recurrence relation*.

The characterisation of a difference equation to be of Poincaré–Perron type means that the ratios of two subsequent terms tend to finite numbers as the index n tends to infinity, or, what is the same, all of its coefficients tend to finite numbers, which means that there is a characteristic equation like (3.2.38), all of the roots of which are finite numbers.

The characterisation of a difference equation to be regular means that all the roots of its characteristic equation are simple ones.

The difference equation (4.2.6) for the coefficients a_n of the Frobenius ansatz (4.2.10) may be written in the form

$$a_{n+1} + p_n\, a_n - q_n\, a_{n-1} = 0, \; n = 1, 2, \ldots, \tag{4.2.11}$$

with

$$p_n = -\frac{Q_n - q}{P_n},$$

$$q_n = -\frac{R_n}{P_n},$$

where P_n, Q_n, R_n, q are from (3.2.36) and are dependent on the parameters, denoted $\underline{c}$, of the underlying differential equation (3.2.20). The roots of its characteristic equation are [cf. (4.2.9)]

$$t_1 = \frac{1}{a} \quad \text{and} \quad t_2 = 1. \tag{4.2.12}$$

Generally, the ratio

$$\frac{a_{n+1}}{a_n}$$

of two consecutive terms of (4.2.11) tends to $\frac{1}{a}$:

$$\lim_{n\to\infty} \frac{a_{n+1}}{a_n} = \frac{1}{a}$$

as $n \to \infty$, meaning that the inverse a of this is also the radius of convergence $r = r_1 = a$ of the series (4.2.10). For special values of the parameters of (3.2.20) there are particular solutions of (4.2.11) that behave like

$$\lim_{n\to\infty} \frac{a_{n+1}}{a_n} = 1 < \frac{1}{a},$$

resulting in a larger radius of convergence $r = r_2 = 1 > a$. This means that the series (4.2.10) representing a particular solution of the differential equation (3.2.20) is a holomorphic function at the singularity $z = a$ of the differential equation (3.2.20). This holomorphicity indeed solves the CTCP of the Heun differential equation, since it incorporates the asymptotic behaviour of the particular solutions as the singularity is approached, radially, along the positive real axis, determined by the recessive particular solution of the differential equation.

The question occurs – and it is, indeed, the solution of the CTCP for the Heun differential equation – how to determine these parameters or parameter combinations $\underline{c}$. The criterion that determines these parameters is called the *boundary eigenvalue condition.*

There is a classical answer to this question that actually leads to the first transcendental eigenvalue condition for the differential equations generating special functions. This condition is discussed in the following.

Consider the difference equation (4.2.11). This may be written as

$$\frac{a_{n+1}}{a_{n-1}}\frac{a_n}{a_n} + p_n \frac{a_n}{a_{n-1}} - q_n = 0, \; n = 1, 2, \ldots,$$

or

$$\frac{a_n}{a_{n-1}} \left[\frac{a_{n+1}}{a_n} + p_n \right] - q_n = 0$$

or

$$\frac{a_n}{a_{n-1}} = \frac{q_n}{p_n + \dfrac{a_{n+1}}{a_n}}. \tag{4.2.13}$$

Using the consecutive equation of (4.2.13) (i.e., replacing n by $n+1$)

$$\frac{a_{n+1}}{a_n} = \frac{q_{n+1}}{p_{n+1} + \dfrac{a_{n+2}}{a_{n+1}}}$$

and inserting this into (4.2.13) yields the finite continued fraction

$$\frac{a_n}{a_{n-1}} = \frac{q_n}{p_n + \dfrac{q_{n+1}}{p_{n+1} + \dfrac{a_{n+2}}{a_{n+1}}}}.$$

This may be continued to an arbitrary number $m \geq 2$, yielding

$$\frac{a_n}{a_{n-1}} = \frac{q_n}{p_n + \dfrac{q_{n+1}}{p_{n+1} + \dfrac{q_{n+2}}{p_{n+2} + \dfrac{q_{n+3}}{p_{n+3} + \dfrac{\cdots}{p_{n+m-1} + \dfrac{q_{n+m}}{p_{n+m} + \dfrac{a_{n+m}}{a_{n+m-1}}}}}}}} \tag{4.2.14}$$

with $m = 2, 3, 4, \ldots$.

For $n = 0$ the finite continued fraction (4.2.14) yields

$$\frac{a_1}{a_0} = \frac{q_1}{p_1 + \dfrac{q_2}{p_2 + \dfrac{q_3}{p_3 + \dfrac{q_4}{p_4 + \dfrac{\cdots}{p_{m-1} + \dfrac{q_m}{p_m + \dfrac{a_m}{a_{m-1}}}}}}}} \tag{4.2.15}$$

with $m \in \mathbb{N} \geq 2$. This formula needs interpretation:

- The linear difference equation (4.2.11) is transformed into a nonlinear relation (4.2.15). Thus, at this very point, the method presented here is surmounting or even leaving the region of linearity.
- The equation (4.2.15) incorporating a finite continued fraction constitutes fixed algebraic relations between the ratios a_1/a_0 and a_m/a_{m-1}; $m = 2, 3, 4, \ldots$.
- The linear difference equation (4.2.11) is of second order. Therefore, its general solution may be written as

$$a_n = L_1\, a_n^{(1)} + L_2\, a_n^{(2)},\ n \in \mathbb{N}^0$$

  where L_1 and L_2 may be dependent on the parameters of the difference equation (but are not dependent on n).
- The two particular fundamental solutions $a_n^{(1)}$ and $a_n^{(2)}$ of the difference equation (4.2.11) may be distinguished through their asymptotic behaviour for $n \to \infty$.

– According to the characteristic equation (4.2.8), the ratio of two consecutive terms of the particular solution $a_n^{(1)}$ behaves like

$$\lim_{n\to\infty} \frac{a_{n+1}^{(1)}}{a_n^{(1)}} = \frac{1}{a} > 1$$

while the ratio of two consecutive terms of the other particular solution $a_n^{(2)}$ behaves like

$$\lim_{n\to\infty} \frac{a_{n+1}^{(2)}}{a_n^{(2)}} = 1. \tag{4.2.16}$$

- This raises the question: Which of these two asymptotic behaviours is adopted in the formula (4.2.15) for

$$\frac{a_{m+1}}{a_m}$$

as $m \to \infty$?

– The answer is a classical one. As the Italian mathematician **Salvatore Pincherle** already showed in 1894 (cf. Pincherle, 1894) and the Danish mathematician Niels Erik Nörlund elaborated in his famous book (Nörlund, 1924), it is the solution that belongs to the lower value of t, thus $t_2 = 1$, which is adopted in the limit $m \to \infty$. The proof of this key argument is given on page 438 of Nörlund's book.

- Thus, equation (4.2.16) may be written as

$$\lim_{m\to\infty} C_m = C = \frac{a_1^{(2)}}{a_0^{(2)}}$$

with

$$C_m = \cfrac{q_1}{p_1 + \cfrac{q_2}{p_2 + \cfrac{q_3}{p_3 + \cfrac{q_4}{p_4 + \cfrac{\cdots}{p_{m-1} + \cfrac{q_m}{p_m + \cfrac{a_m}{a_{m-1}}}}}}}} \tag{4.2.17}$$

whereby $m = 2, 3, 4, \ldots$. Carrying out the limiting process $m \to \infty$ yields the infinite continued fraction

$$C = \cfrac{q_1}{p_1 + \cfrac{q_2}{p_2 + \cfrac{q_3}{p_3 + \cfrac{q_4}{p_4 + \ldots}}}}. \tag{4.2.18}$$

- This means that the continued fraction (4.2.18) has a rather peculiar property that is crucial for the role it plays in this problem: it selects just the minimum solution $\{a_n^{(2)}\}$ of the difference equation (4.2.11).
- A consequence of this is that for $a_n = a_n^{(2)}$, from (4.2.11) the radius of convergence r of the corresponding power series in the ansatz (4.2.10) that represents a particular solution of the differential equation (3.2.20) is enlarged from generally being $r = a < 1$ to $r = 1$.
- In a last step, the initial condition in (3.2.35) comes into play. If the infinite continued fraction (4.2.18) is convergent (cf. Perron, 1913), then it adopts the value of the ratio

$$C = \frac{a_1^{(2)}}{a_0^{(2)}}. \tag{4.2.19}$$

  This ratio

$$\frac{a_1}{a_0},$$

  however, is not arbitrary in the eigenvalue problem at hand. It is determined by the initial condition (3.2.35):

$$\frac{a_1}{a_0} = -\frac{q}{a\,\gamma}. \tag{4.2.20}$$

  Generally, this will not be consistent with the value coming out of (4.2.6). However, as already mentioned, this ratio as well as the infinite continued fraction (4.2.18) are dependent on the parameters $\underline{c}$ of the difference equation (4.2.11), which are also the parameters of the differential equation (4.2.1), (4.2.2) of the eigenvalue problem. Thus, it is the criterion for the eigenvalue problem to vary these parameters $\underline{c}$ such that the value of (3.2.35) meets that of equation (4.2.19).
- This yields the eigenvalue condition of the eigenvalue problem:

$$C = -\frac{q}{a\,A_0}. \tag{4.2.21}$$

- This transcendental condition (4.2.21) may be written in concise form as an eigenvalue condition of the problem:

$$F(\underline{c}) = C - \frac{q}{a\,A_0} = 0, \tag{4.2.22}$$

  with C being the infinite continued fraction (4.2.17)

$$C = \cfrac{q_1}{p_1 + \cfrac{q_2}{p_2 + \cfrac{q_3}{p_3 + \cfrac{q_4}{p_4 + \cdots}}}}. \tag{4.2.23}$$

- This condition is remarkable since it is a transcendental one, in contrast to all the conditions coming out of eigenvalue problems of the Fuchsian differential equations having at most three singularities. In these latter cases the eigenvalue condition is always of algebraic nature, see (1.2.41), (1.2.42), as is the case for Fuchsian differential equations having at most three singularities, which may always be converted into the standard forms of the Laplace, Euler or Gauss differential equation by means of Moebius as well as s-homotopic transformations, thus by algebraic manipulation.

- The eigenvalue condition (4.2.21) was the first one of transcendental nature. Unfortunately, it only works for second-order difference equations. This is the reason why a more general theory of singular eigenvalue problems is needed, which still works in the case of difference equations being of higher than second order. Such a theory is presented in this book.
- It should be mentioned that the numerical properties of continued fractions are excellent, since there occurs what is called 'superexponential convergence'. This means that if the boundary eigenvalue problem can be reduced to the condition (4.2.21), the resulting special functions may be obtained of good quality at low calculatory expenditure. A numerical procedure was elaborated by Salvatore Pincherle (1894), as already mentioned above.
- The price that has to be payed for the new method, presented below, is a moderately slower rate of convergence. Naturally, this new method also works in the case of second-order difference equations, thus in the underlying one here, viz. in the case of boundary eigenvalue problems for Fuchsian differential equations having four singularities.

4.2.2 An Integrated Procedure of Jaffé's Method

The CTCP of the Heun differential equation (4.2.1) is solvable by means of a three-term recurrence relation (4.2.6), (4.2.7) that yields an eigenvalue condition in the form of an infinite continued fraction (4.2.18). This eigenvalue condition was the first transcendental one being developed by Salvatore Pincherle and Niels Erik Nörlund in the late nineteenth and early twentieth century. The CTCP of the Heun differential equation is in between the generation of classical and higher special functions: it is classical in a historical sense and higher in the mathematical sense that the eigenvalue condition is no longer algebraic but of transcendental nature, viz. the zeros of the secular function (4.2.22) is no longer algebraic but a transcendental function.

This is no longer so for the confluent cases of the Heun differential equation. Because the recurrence relations are not restricted to three-term ones and because the infinite continued fraction methods are restricted to three-term recurrence relations, it becomes necessary for new methods to be elaborated, as is done in this book.

It is an aesthetic aspect of the method presented below that it may be developed in a concise manner with respect to all three confluent cases of the Heun differential equation: the single confluent case, the biconfluent case and the triconfluent case. The double confluent case, however, is to be treated separately below. The s-rank symbols of these differential equations are given by

$$\begin{aligned}\text{Single confluent case: } & \{1,1;2\},\\ \text{Biconfluent case: } & \{1;3\},\\ \text{Triconfluent case: } & \{;4\}.\end{aligned}$$

To be developed in a concise manner means that the *form* of the mathematical objects is the same for all three confluent cases, while only the parameters are different. This helps in writing down formulae in the general case, where the s-rank s of the irregular singularity of the CTCP is arbitrary.

So, the single confluent, biconfluent and triconfluent cases are treated here, where an irregular singularity of the differential equation is placed at infinity and a regular singularity (in the single and biconfluent cases) or an ordinary point (in the triconfluent case) is located at the origin. The relevant interval is the positive real axis.

In order not to confuse formulae between the different confluent cases of the Heun differential equation, in the following, indices are written at the parameters: s stands for the single confluent case, b the biconfluent case and t the triconfluent case of the Heun differential equation.

In order to avoid extended presentations, the main steps and main results of the computations are shown in the following, without a detailed carrying out of the calculations. These are completely analogous to the single confluent case of the Gauss equation, treated in Chapter 1. Moreover, it should be mentioned that for the sake of not having too many indices, the notation of the characteristic exponents is changed.

The starting point is the *Heun differential equation* (3.2.9) in its natural form:

$$\frac{\mathrm{d}^2 y}{\mathrm{d}z^2} + \left[\frac{A_0}{z} + \frac{A_{z_1}}{z - z_1} + \frac{A_{z_2}}{z - z_2}\right]\frac{\mathrm{d}y}{\mathrm{d}z} + \left[\frac{C_0}{z} + \frac{C_{z_1}}{z - z_1} + \frac{C_{z_2}}{z - z_2} + \frac{B_0}{z^2} + \frac{B_{z_1}}{(z - z_1)^2} + \frac{B_{z_2}}{z - z_{z_2}}\right] y = 0,$$

with

$$C_0 + C_{z_1} + C_{z_2} = 0.$$

Ansatzes and Characteristic Exponents

From this are derived the three confluent cases discussed

Single confluent case:

$$\frac{\mathrm{d}^2 y_s}{\mathrm{d}z^2} + \left[G_{0s} + \frac{A_{1s}}{z} + \frac{A_{2s}}{z + z_0}\right]\frac{\mathrm{d}y_s}{\mathrm{d}z} + \left[D_{0s} + \frac{C_{1s}}{z} + \frac{C_{2s}}{z + z_0} + \frac{B_{1s}}{z^2} + \frac{B_{2s}}{(z + z_0)^2}\right] y_s = 0. \tag{4.2.24}$$

Biconfluent case:

$$\frac{d^2 y_b}{dz^2} + \left[G_{1b}\, z + G_{0b} + \frac{A_b}{z}\right]\frac{dy_b}{dz} + \left[D_{2b}\, z^2 + D_{1b}\, z + D_{0b} + \frac{C_b}{z} + \frac{B_b}{z^2}\right] y_b = 0.$$

Triconfluent case:

$$\frac{d^2 y_t}{dz^2} + \left[G_{2t}\, z^2 + G_{1t}\, z + G_{0t}\right]\frac{dy_t}{dz} + \left[D_{4t}\, z^4 + D_{3t}\, z^3 + D_{2t}\, z^2 + D_{1t}\, z + D_{0t}\right] y_t = 0.$$

As mentioned above, the central two-point connection problem for these differential equations is defined on the positive real axis (i.e., between the origin $z = 0$ and infinity). This means that it looks for those parameters of the differential equations for which there are particular solutions, the asymptotic behaviour of which for $z \to \infty$ is decreasing while the behaviour at the origin $z = 0$ is prescribed, according to the characteristic exponents of the Frobenius solutions in the single and biconfluent cases, and the functional behaviour in the triconfluent case.

According to Jaffé, the ansatzes consist of one of the asymptotic factors of the Thomé solutions at the irregular singularity at infinity (viz. at one of the boundary points) as well as of a convergent, generalised power-type series about the origin at the origin. The ansatzes for solving the central two-point connection problems for these confluent cases of Heun's differential equation are given explicitly in the following.

Single confluent case:

$$\boxed{y_s(z) = \exp\left(\nu_s\, z\right)\, z^{\mu_{1s}}\, (z + z_{0s})^{\mu_{2s}}\, w(z)} \tag{4.2.25}$$

whereby z_{0s} is supposed to be real-valued and positive and the singularity parameters are given by

$$\nu_s^2 + G_{0s}\,\nu_s + D_{0s} = 0 \rightsquigarrow \nu_s =_{1,2} \nu_s,$$
$$\mu_{1s}^2 + (A_{1s} - 1)\,\mu_{1s} + B_{1s} = 0 \rightsquigarrow \mu_{1s} =_{1,2} \mu_{1s},$$
$$\mu_{2s} = -\mu_{1s} - \frac{(A_{1s} + A_{2s})\,\nu_s + C_{1s} + C_{2s}}{G_{0s} + 2\,\nu_s}.$$

It should be recognised that μ_{1s} takes into account the correct asymptotic behaviour of the Frobenius solution at the origin and $\mu_{1s} + \mu_{2s}$ takes into account the correct asymptotic behaviour of the Thomé solution on approaching the singularity of the differential equation (4.2.24) at infinity.

Biconfluent case:

$$\boxed{y_b(z) = \exp\left(\nu_b\, z + \tfrac{\kappa_b}{2}\, z^2\right)\, z^{\mu_b}\, (z + z_{0b})^{\alpha_b}\, w(z)} \tag{4.2.26}$$

whereby z_{0b} is supposed to be real-valued and positive and the singularity parameters are given by

$$\begin{aligned}
&\kappa_b^2 + G_{1b}\,\kappa_b + D_{2b} = 0 \rightsquigarrow \kappa_b =_{1,2} \kappa_b,\\
&\nu_b = -\frac{D_{1b} + G_{0b}\,\kappa_b}{G_{1b} + 2\,\kappa_b},\\
&\mu_b^2 + (A_b - 1)\,\mu_b + B_b = 0 \rightsquigarrow \mu_b =_{1,2} \mu_b,\\
&\alpha_b = -\frac{\nu_b^2 + G_{0b}\,\nu_b + D_{0b} + \kappa_b\,(A_b + 1)}{G_{1b} + 2\,\kappa_b} - \mu_b.
\end{aligned}$$

Triconfluent case:

$$\boxed{y_t\,(z) = \exp\left(\tfrac{\eta_t}{3}\, z^3 + \tfrac{\kappa_t}{2}\, z^2 + \nu_t\, z\right)\, (z + z_{0t})^{\alpha_t}\, w\,(z),} \tag{4.2.27}$$

$$\eta_t^2 + G_{2t}\,\eta_t + D_{4t} = 0 \rightsquigarrow \eta_t =_{1,2} \eta_t = -\frac{G_{2t}}{2} \pm \frac{1}{2}\sqrt{G_{2t}^2 - 4\,D_{4t}}$$

and thus there are – in dependence on the chosen value for η_t – two values of all the following quantities:

$$\begin{aligned}
\kappa_t &= -\frac{D_{3t} + \eta_t\,G_{1t}}{G_{2t} + 2\,\eta_t},\\
\nu_t &= -\frac{D_{2t} + \eta_t\,G_{0t} + \kappa_t\,(G_{1t} + \kappa_t)}{G_{2t} + 2\,\eta_t},\\
\alpha_t &= -\frac{D_{1t} + \nu_t\,G_{1t} + \kappa_t\,G_{0t} + 2\,\kappa_t\,\nu_t + 2\,\eta_t}{G_{2t} + 2\,\eta_t}.
\end{aligned}$$

A consequence of the Jaffé ansatzes above is a differential equation for the function $w(z)$ in all the treated confluent cases of the Heun differential equation.

Differential equation for $w(z)$:

$$\begin{aligned}
\Xi(z)\,\frac{d^2 w}{dz^2} &+ \left[g_2\,z^2 + g_1\,z + g_0 + \frac{g_{-1}}{z + z_0}\right]\frac{dw}{dz}\\
&+ \left[d_0 + \frac{d_{-1}}{z + z_0} + \frac{d_{-2}}{(z + z_0)^2}\right] w(z) = 0,
\end{aligned}$$

whereby the coefficients are given in the following.

Single confluent case:

$$
\begin{aligned}
\Xi(z) &= P_s(z) = z^2,\\
g_2 &= g_{2s} = G_{0s} + 2\,\nu_s,\\
g_1 &= g_{1s} = A_{1s} + A_{2s} + 2\,(\mu_{1s} + \mu_{2s}),\\
g_0 &= g_{0s} = -(A_{2s} + 2\,\mu_{2s})\,z_{0s},\\
g_{-1} &= g_{-1s} = (A_{2s} + 2\,\mu_{2s})\,z_{0s}^2 = -g_{0s}\,z_0,\\
d_0 &= d_{0s} = A_{1s}\,\mu_{2s} + A_{2s}\,(\mu_{1s} + \mu_{2s} - \nu_s\,z_{0s})\\
&\qquad + B_{2s} - C_{2s}\,z_{0s} - G_{0s}\,\mu_{2s}\,z_{0s}\\
&\qquad + 2\,\mu_{1s}\,\mu_{2s} + \mu_{2s}^2 - \mu_{2s}\,(2\,\nu_s\,z_{0s} + 1),\\
d_{-1} &= d_{-1s} = -z_{0s}\,\{A_{1s}\,\mu_{2s} + A_{2s}\,(\mu_{1s} + 2\,\mu_{2s} - \nu_s\,z_{0s}) + 2\,B_{2s}\\
&\qquad - C_{2s}\,z_{0s} - \mu_{2s}\,[G_{0s}\,z_{0s} - 2\,(\mu_{1s} + \mu_{2s} - \nu_s\,z_{0s} - 1)]\},\\
d_{-2} &= d_{-2s} = [B_{2s} + \mu_{2s}\,(A_{2s} + \mu_{2s} - 1)]\,z_{0s}^2.
\end{aligned}
$$

Biconfluent case:

$$
\begin{aligned}
\Xi(z) &= P_b(z) = z,\\
g_2 &= g_{2b} = G_{1b} + 2\,\kappa_b,\\
g_1 &= g_{1b} = G_{0b} + 2\,\nu_b,\\
g_0 &= g_{0b} = A_b + 2\,(\mu_b + \alpha_b),\\
g_{-1} &= g_{-1b} = -2\,\alpha_b\,z_{0b},\\
d_0 &= d_{0b} = C_b + \nu_b\,A_b + \mu_b\,G_{0b} + 4\,\mu_b\,\nu_b\\
&\qquad + \alpha_b\,[G_{0b} + 2\,\nu_b - (G_{1b} + 2\,\kappa_b)\,z_{0b}] + 2\,\alpha_b\,\frac{\mu_b}{z_{0b}},\\
d_{-1} &= d_{-1b}\\
&= \alpha_b\,\left[(G_{1b} + 2\,\kappa_b)\,z_{0b}^2 - (G_{0b} + 2\,\nu_b)\,z_{0b} + A_b + 2\,\mu_b + \alpha_b - 1\right],\\
d_{-2} &= d_{-2b} = -\alpha_b\,(\alpha_b - 1)\,z_{0b}.
\end{aligned}
$$

Triconfluent case:

$$
\begin{aligned}
\Xi(z) &= P_t(z) \equiv 1,\\
g_2 &= g_{2t} = G_{2t} + 2\,\eta_t,\\
g_1 &= g_{1t} = G_{1t} + 2\,\kappa_t,\\
g_0 &= g_{0t} = G_{0t} + 2\,\nu_t,\\
g_{-1} &= g_{-1t} = 2\,\alpha_t,\\
d_0 &= d_{0t} = D_{0t} + G_{0t}\,\nu_t + \nu_t^2 + \kappa_t - \alpha_t\,[(G_{2t} + 2\,\eta_t)\,z_{0t} - (G_{1t} + 2\,\kappa_t)],\\
d_{-1} &= d_{-1t} = \alpha_t\,\left\{(G_{2t} + 2\,\eta_t)\,z_{0t}^2 - [(G_{1t} + 2\,\kappa_t)\,z_{0t} - (G_{0t} + 2\,\nu_t)]\right\},\\
d_{-2} &= d_{-2t} = \alpha_t\,(\alpha_t - 1).
\end{aligned}
$$

Jaffé Transformations

The relevant interval of the CTCP is the positive real axis. This interval is now transformed onto the unit interval $[0, 1]$ by means of a Moebius transformation. This is for two reasons: first, in the single confluent case the further finite singularity is put outside the region of convergence covering the relevant interval and second, the asymptotic behaviour of the coefficients a_n is influenced in a definite way.

Moebius Transformation

$$x = \frac{z}{z + z_0} \tag{4.2.28}$$

with

Type of equation	z_0
Single confluent case	z_{0s}
Biconfluent case	z_{0b}
Triconfluent case	z_{0t}

where z_0 is supposed to be real-valued and positive.

In all the treated confluent cases, one and the same differential equation appears for the function $w(x)$. The distinction between the different confluent cases occurs in the coefficients of this final differential equation, as given below.

Differential Equation

$$P_4(x)\,\frac{\mathrm{d}^2 w}{\mathrm{d}x^2} + \sum_{i=0}^{3} \Gamma_i\, x^i\, \frac{\mathrm{d}w}{\mathrm{d}x} + \sum_{j=0}^{2} \Delta_j\, x^j\, w(x) = 0. \tag{4.2.29}$$

Coefficients

Single confluent case:

$$P_4(x) = P_{4s}(x) = x^4 - 2\,x^3 + x^2 = x^2\,(x-1)^2,$$

$$\Gamma_3 = \Gamma_{3s} = 2 - \frac{g_{-1s}}{z_{0s}^2},$$

$$\Gamma_2 = \Gamma_{2s} = g_{2s}\,z_{0s} - g_{1s} - 2 + \frac{g_{0s}}{z_{0s}} + \frac{3\,g_{-1s}}{z_{0s}^2},$$

$$\Gamma_1 = \Gamma_{1s} = g_{1s} - \frac{2\,g_{0s}}{z_{0s}} - \frac{3\,g_{-1s}}{z_{0s}^2},$$

$$\Gamma_0 = \Gamma_{0s} = \frac{g_{0s}}{z_{0s}} + \frac{g_{-1s}}{z_{0s}^2} \equiv 0,$$

$$\Delta_2 = \Delta_{2s} = \frac{d_{-2s}}{z_{0s}^2},$$

$$\Delta_1 = \Delta_{1s} = -\frac{2\,d_{-2s}}{z_{0s}^2} - \frac{d_{-1s}}{z_{0s}},$$

$$\Delta_0 = \Delta_{0s} = \frac{d_{-2s}}{z_{0s}^2} + \frac{d_{-1s}}{z_{0s}} + d_{0s} \equiv 0.$$

Biconfluent case:

$$P_4(x) = P_{4b}(x) = x^4 - 3\,x^3 + 3\,x^2 - x = x\,(x-1)^3,$$

$$\Gamma_3 = \Gamma_{3b} = 2 + \frac{g_{-1b}}{z_{0b}},$$

$$\Gamma_2 = \Gamma_{2b} = -4 - z_{0b}\,(g_{2b}\,z_{0b} - g_{1b}) - g_{0b} - 3\,\frac{g_{-1b}}{z_{0b}},$$

$$\Gamma_1 = \Gamma_{1b} = 2 - g_{1b}\,z_{0b} + 2\,g_{0b} + 3\,\frac{g_{-1b}}{z_{0b}},$$

$$\Gamma_0 = \Gamma_{0b} = -g_{0b} - \frac{g_{-1b}}{z_{0b}},$$

$$\Delta_2 = \Delta_{2b} = -\frac{d_{-2b}}{z_{0b}},$$

$$\Delta_1 = \Delta_{1b} = \frac{2\,d_{-2b}}{z_{0b}} + d_{-1b},$$

$$\Delta_0 = \Delta_{0b} = -\frac{d_{-2b}}{z_{0b}} - d_{-1b} - d_{0b}\,z_{0b}.$$

Triconfluent case:

$$P_4(x) = P_{4t}(x) = x^4 - 4\,x^3 + 6\,x^2 - 4\,x + 1 = (x-1)^4,$$

$$\Gamma_3 = \Gamma_{3t} = 2 - g_{-1t},$$

$$\Gamma_2 = \Gamma_{2t} = -6 + z_{0t}\,\left\{g_{2t}\,z_{0t}^2 - g_{1t}\,z_{0t} + g_{0t}\right\} + 3\,g_{-1t},$$

$$\Gamma_1 = \Gamma_{1t} = 6 + z_{0t}\,\{g_{1t}\,z_{0t} - 2\,g_{0t}\} - 3\,g_{-1t},$$

$$\Gamma_0 = \Gamma_{0t} = -2 + g_{0t}\,z_{0t} + g_{-1t},$$

$$\Delta_2 = \Delta_{2t} = d_{-2t},$$

$$\Delta_1 = \Delta_{1t} = -2d_{-2t} - d_{-1t}\,z_{0t},$$

$$\Delta_0 = \Delta_{0t} = d_{-2t} + d_{-1t}\,z_{0t} + d_{0t}\,z_{0t}^2.$$

There is an even more concise calculation. To see this, define the following quantities:

$$\tilde{\Gamma}_0 = g_0\,z_0 - g_{-1},$$

$$\tilde{\Gamma}_1 = g_1\,z_0^2 - 2\,g_0\,z_0 - 3\,g_{-1},$$

$$\tilde{\Gamma}_2 = g_2\,z_0^3 - g_1\,z_0^2 + g_0\,z_0 + 3\,g_{-1},$$

$$\begin{aligned}
\tilde{\Gamma}_3 &= -g_{-1}, \\
\tilde{\Delta}_0 &= d_0\, z_0^2 + d_{-1}\, z_0 + d_{-2}, \\
\tilde{\Delta}_1 &= 2\, d_{-2} - d_{-1}\, z_0, \\
\tilde{\Delta}_2 &= d_{-2}.
\end{aligned}$$

Then the coefficients may be written in the following strict manner.

Single confluent case:

$$\begin{aligned}
\Gamma_{0s} &= \frac{\tilde{\Gamma}_i}{z_{0s}^2} + k_j,\ j = 0, 1, 2, 3;\ k_0 = k_1 = 0,\ k_2 = -2,\ k_3 = +2, \\
\Delta_{is} &= \frac{\tilde{\Delta}_i}{z_{0s}^2},\ i = 0, 1, 2.
\end{aligned}$$

Biconfluent case:

$$\begin{aligned}
\Gamma_{0b} &= -\frac{\tilde{\Gamma}_i}{z_{0b}^2} + k_j,\ j = 0, 1, 2, 3;\ k_0 = 0,\ k_1 = -2,\ k_2 = +4,\ k_3 = -2, \\
\Delta_{ib} &= -\frac{\tilde{\Delta}_i}{z_{0b}^2},\ i = 0, 1, 2.
\end{aligned}$$

Triconfluent case:

$$\begin{aligned}
\Gamma_{0t} &= -\frac{\tilde{\Gamma}_i}{z_{0t}^2} + k_j,\ j = 0, 1, 2, 3;\ k_0 = -2,\ k_1 = +6,\ k_2 = -6,\ k_3 = +2, \\
\Delta_{it} &= -\frac{\tilde{\Delta}_i}{z_{0t}^2},\ i = 0, 1, 2.
\end{aligned}$$

Difference Equations

The reasoning of the ansatzes (4.2.25)–(4.2.27) and the Moebius transformations (4.2.28) is to obtain a differential equation (4.2.29) for $w(x)$, the solution of which is a pure power series in x about $x = z = 0$. This series I call the *Jaffé expansion*:

$$w(z) = \sum_{n=0}^{\infty} a_n \left(\frac{z}{z+1}\right)^n = \sum_{n=0}^{\infty} a_n\, x^n. \tag{4.2.30}$$

The result in all the confluent cases are irregular difference equations of Poincaré–Perron type for the coefficients a_n of the series (4.2.30), the order of which is just the s-rank of the irregular singularities of the underlying differential equations.

Single confluent case

$$\begin{aligned}
&a_{0,s} \text{ arbitrary}, \\
&\Gamma_{1s}\, a_{1,s} + \Delta_{1s}\, a_{0,s} = 0,
\end{aligned}$$

$$\left(1+\frac{\alpha_{1s}}{n}+\frac{\beta_{1s}}{n^2}\right)a_{n+1,s}+\left(-2+\frac{\alpha_{0s}}{n}+\frac{\beta_{0s}}{n^2}\right)a_{n,s}$$
$$+\left(1+\frac{\alpha_{-1s}}{n}+\frac{\beta_{-1s}}{n^2}\right)a_{n-1,s}=0,\ n\ge 1. \tag{4.2.31}$$

Coefficients of the Difference Equation

$$\begin{aligned}\alpha_{1s}&=\Gamma_{1s}+1, & \beta_{1s}&=\Gamma_{1s},\\ \alpha_{0s}&=\Gamma_{2s}+2, & \beta_{0s}&=\Delta_{1s},\\ \alpha_{-1s}&=\Gamma_{3s}-3, & \beta_{-1s}&=\Delta_{2s}-\Gamma_{3s}+2.\end{aligned} \tag{4.2.32}$$

Biconfluent case:

$$\begin{aligned}&a_{0,b}\ \text{arbitrary},\\ &\Gamma_{0b}\,a_{1,b}+\Delta_{0b}\,a_{0,b}=0,\\ &2\,(\Gamma_{0b}-1)\,a_{2,b}+(\Gamma_{0b}+\Delta_{0b})\,a_{1,b}+(\Delta_{1b}-\Gamma_{2b})\,a_{0,b}=0,\\ &\left(-1+\frac{\alpha_{1b}}{n}+\frac{\beta_{1b}}{n^2}\right)a_{n+1,b}+\left(3+\frac{\alpha_{0b}}{n}+\frac{\beta_{0b}}{n^2}\right)a_{n,b}\\ &\quad+\left(-3+\frac{\alpha_{-1b}}{n}+\frac{\beta_{-1,b}}{n^2}\right)a_{n-1,b}+\left(1+\frac{\alpha_{-2,b}}{n}+\frac{\beta_{-2,b}}{n^2}\right)a_{n-2,b}=0,\ n\ge 2.\end{aligned} \tag{4.2.33}$$

Coefficients of the Difference Equation

$$\begin{aligned}\alpha_{1b}&=\Gamma_{0b}-1, & \beta_{1b}&=\Gamma_{0b},\\ \alpha_{0b}&=\Gamma_{1b}-3, & \beta_{0b}&=\Delta_{0b},\\ \alpha_{-1b}&=\Gamma_{2b}+9, & \beta_{-1b}&=\Delta_{1b}-\Gamma_{2b}-6,\\ \alpha_{-2b}&=\Gamma_{3b}-5, & \beta_{-2b}&=\Delta_{2b}-2\,\Gamma_{3b}+6.\end{aligned} \tag{4.2.34}$$

Triconfluent case:

$$\begin{aligned}&a_0,\ a_1\ \text{arbitrary},\\ &2\,a_2+\Gamma_0\,a_1+\Delta_0\,a_0=0,\\ &6\,a_3+(2\,\Gamma_0-8)\,a_2+(\Gamma_1+\Delta_0)\,a_1+\Delta_1\,a_0=0,\end{aligned}$$

$$\begin{aligned}&\left(1+\frac{\alpha_2}{n}+\frac{\beta_2}{n^2}\right)a_{n+2}+\left(-4+\frac{\alpha_1}{n}+\frac{\beta_1}{n^2}\right)a_{n+1}\\ &\quad+\left(6+\frac{\alpha_0}{n}+\frac{\beta_0}{n^2}\right)a_n+\left(-4+\frac{\alpha_{-1}}{n}+\frac{\beta_{-1}}{n^2}\right)a_{n-1}\\ &\quad+\left(1+\frac{\alpha_{-2}}{n}+\frac{\beta_{-2}}{n^2}\right)a_{n-2}=0,\ n\ge 2.\end{aligned}$$

Coefficients of the Difference Equation

$$\begin{aligned}
\alpha_2 &= 3, \quad \beta_2 = 2,\\
\alpha_1 &= \Gamma_0 - 4, \quad \beta_1 = \Gamma_0,\\
\alpha_0 &= \Gamma_1 - 6, \quad \beta_0 = \Delta_0,\\
\alpha_{-1} &= \Gamma_2 + 12, \quad \beta_{-1} = \Delta_1 - \Gamma_2 - 8,\\
\alpha_{-2} &= \Gamma_3 - 5, \quad \beta_{-2} = \Delta_2 - 2\,\Gamma_3 + 6.
\end{aligned} \tag{4.2.35}$$

Second-Order Characteristic Equations

Single confluent case:

$$\left(-1 + \frac{\alpha_{1s}}{n}\right) t_b^2 + \left(2 + \frac{\alpha_{0s}}{n}\right) t_b + \left(-1 + \frac{\alpha_{-1s}}{n}\right) = 0.$$

Solutions:

$$\begin{aligned}
t_{s1} &= 1 + \sqrt[2]{\frac{-\sum_{i=-1}^{i=1} \alpha_{is}}{n}} + O\left(\frac{1}{n^{\frac{2}{2}}}\right),\\
t_{s2} &= 1 - \sqrt[2]{\frac{-\sum_{i=-1}^{i=1} \alpha_{is}}{n}} + O\left(\frac{1}{n^{\frac{2}{2}}}\right).
\end{aligned}$$

Graphics:

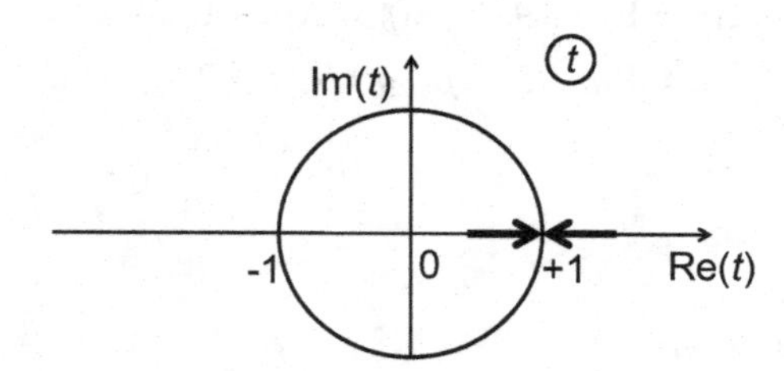

Figure 4.2 First-order asymptotics of irregular difference equations (single confluent case).

Biconfluent case:

$$\left(-1 + \frac{\alpha_{1b}}{n}\right) t_b^3 + \left(3 + \frac{\alpha_{0b}}{n}\right) t_b^2 + \left(-3 + \frac{\alpha_{-1b}}{n}\right) t_b + \left(1 + \frac{\alpha_{-2b}}{n}\right) = 0.$$

Solutions:

$$
\begin{aligned}
t_{b1} &= 1 - \sqrt[3]{\frac{-\sum_{i=-2}^{i=1} \alpha_{ib}}{n}} + O\left(\frac{1}{n^{\frac{2}{3}}}\right), \\
t_{b2} &= 1 + \frac{1}{2}\sqrt[3]{\frac{-\sum_{i=-2}^{i=1} \alpha_{ib}}{n}} \left(1 - \sqrt{3}\,\imath\right) + O\left(\frac{1}{n^{\frac{2}{3}}}\right), \\
t_{b3} &= 1 + \frac{1}{2}\sqrt[3]{\frac{-\sum_{i=-2}^{i=1} \alpha_{ib}}{n}} \left(1 + \sqrt{3}\,\imath\right) + O\left(\frac{1}{n^{\frac{2}{3}}}\right).
\end{aligned}
\tag{4.2.36}
$$

Graphics:

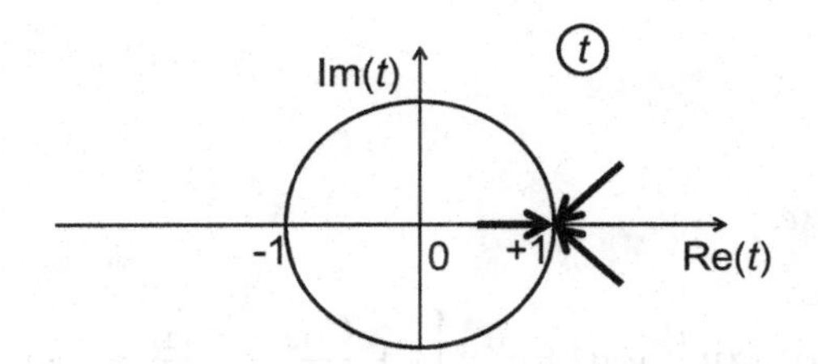

Figure 4.3 First-order asymptotics of irregular difference equations (biconfluent case).

Triconfluent case:

$$
\left(1 + \frac{\alpha_{2t}}{n}\right) t_t^4 + \left(-4 + \frac{\alpha_{1t}}{n}\right) t_t^3 + \left(6 + \frac{\alpha_{0t}}{n}\right) t_t^2 + \left(-4 + \frac{\alpha_{-1t}}{n}\right) t_t + \left(1 + \frac{\alpha_{-2t}}{n}\right) = 0.
$$

Solutions:

$$
t_{t1} = 1 + \sqrt[4]{\frac{-\sum_{i=-2}^{i=2} \alpha_{it}}{n}} + O\left(\frac{1}{n^{\frac{1}{2}}}\right), \tag{4.2.37}
$$

$$
t_{t2} = 1 - \sqrt[4]{\frac{-\sum_{i=-2}^{i=2} \alpha_{it}}{n}} + O\left(\frac{1}{n^{\frac{1}{2}}}\right), \tag{4.2.38}
$$

$$
t_{t3} = 1 + \imath\sqrt[4]{\frac{-\sum_{i=-2}^{i=2} \alpha_{it}}{n}} + O\left(\frac{1}{n^{\frac{1}{2}}}\right), \tag{4.2.39}
$$

$$
t_{t4} = 1 - \imath\sqrt[4]{\frac{-\sum_{i=-2}^{i=2} \alpha_{it}}{n}} + O\left(\frac{1}{n^{\frac{1}{2}}}\right). \tag{4.2.40}
$$

Graphics:

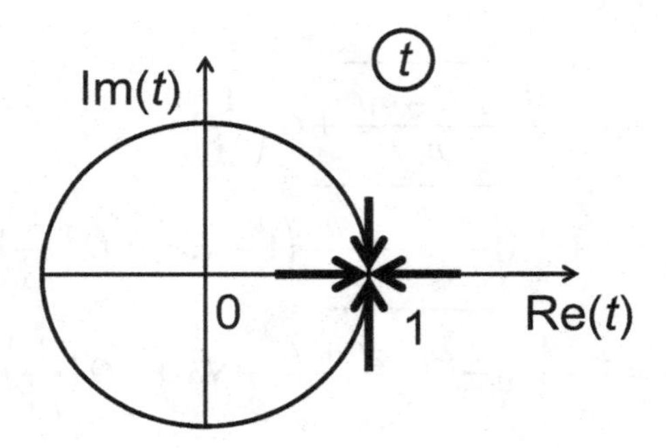

Figure 4.4 First-order asymptotics of irregular difference equations (triconfluent case).

Birkhoff Sets

Single confluent case:

$$\begin{aligned} s_{1s}(n) &= \varrho_s^n \exp\left(\gamma_{1s}\, n^{\frac{1}{2}}\right) n^{r_{1s}} \left[1 + \frac{C_{11s}}{n^{\frac{1}{2}}} + \frac{C_{12s}}{n^{\frac{2}{2}}} + \cdots\right], \\ s_{2s}(n) &= \varrho_s^n \exp\left(\gamma_{2s}\, n^{\frac{1}{2}}\right) n^{r_{2s}} \left[1 + \frac{C_{21s}}{n^{\frac{1}{2}}} + \frac{C_{22s}}{n^{\frac{2}{2}}} + \cdots\right]. \end{aligned} \tag{4.2.41}$$

Parameters:

$$\begin{aligned} \varrho_s &= 1, \\ \gamma_{1s} &= 2\sqrt{-\sum_{i=-1}^{i=1} \alpha_{is}}, \\ \gamma_{2s} &= 6 - \gamma_{1s}, \\ r_{1s} &= r_{2s} = r_s = -\frac{1}{4}\,(2\,\alpha_{1s} - 2\,\alpha_{-1s} - 1)\,. \end{aligned}$$

Thus, the general solution of the difference equation in (4.2.31) may be written in the form

$$a_n^{(g)} = L_1\, s_{1s}(n) + L_2\, s_{2s}(n),$$

where $L_i = L_i(\underline{c})$, $i = 1 - 4$ are (in general complex-valued) constants in n that still may be dependent on the parameters of the differential equation in (4.2.29).

Biconfluent case:

$$s_{1b}(n) = \varrho_b^n \exp\left(\gamma_{11b}\, n^{\frac{2}{3}} + \gamma_{12b}\, n^{\frac{1}{3}}\right) n^{r_{1b}} \left[1 + \frac{C_{11b}}{n^{\frac{1}{3}}} + \frac{C_{12b}}{n^{\frac{2}{3}}} + \cdots\right],$$

$$s_{2b}(n) = \varrho_b^n \exp\left(\gamma_{21b}\, n^{\frac{2}{3}} + \gamma_{22b}\, n^{\frac{1}{3}}\right) n^{r_{2b}} \left[1 + \frac{C_{21b}}{n^{\frac{1}{3}}} + \frac{C_{22b}}{n^{\frac{2}{3}}} + \cdots\right],$$

$$s_{3b}(n) = \varrho_b^n \exp\left(\gamma_{31b}\, n^{\frac{2}{3}} + \gamma_{32b}\, n^{\frac{1}{3}}\right) n^{r_{3b}} \left[1 + \frac{C_{31b}}{n^{\frac{1}{3}}} + \frac{C_{32b}}{n^{\frac{2}{3}}} + \cdots\right].$$

Parameters:

$$\gamma_{11b} = -\frac{2}{3} \sqrt[3]{-\sum_{i=-2}^{i=1} \alpha_{ib}} \neq 0,$$

$$\gamma_{21b} = \frac{3}{4} \sqrt[3]{-\sum_{i=-2}^{i=1} \alpha_{ib}} \left(1 - \sqrt{3}\,\imath\right),$$

$$\gamma_{31b} = \frac{3}{4} \sqrt{-\sum_{i=-2}^{i=1} \alpha_{ib}} \left(1 + \sqrt{3}\,\imath\right),$$

$$\gamma_{12b} = \frac{3}{4\,\gamma_{11b}} \left(3\,\alpha_{1b} + \alpha_{0b} - \alpha_{-1b} - 3\,\alpha_{-2b}\right),$$

$$\gamma_{22b} = -\frac{1}{2}\,\gamma_{12} \left(1 + \sqrt{3}\,\imath\right),$$

$$\gamma_{32b} = -\frac{1}{2}\,\gamma_{12} \left(1 - \sqrt{3}\,\imath\right),$$

$$r_{1b} = r_{2b} = r_{3b} = r_b = \frac{1}{3} \left(\alpha_{1b} + \alpha_{-2b} + 1\right).$$

Thus, the general solution of the difference equation in (4.2.33) may be written in the form

$$a_n^{(g)} = L_1\, s_{1b}(n) + L_2\, s_{2b}(n) + L_3\, s_{3b}(n),$$

where $L_i = L_i(\underline{c})$, $i = 1 - 4$ are (in general complex-valued) constants in n that still may be dependent on the parameters of the difference equation (4.2.33). It is seen that in the case where the parameters of the difference equation (4.2.33) are exclusively real-valued, it holds that

$$L_2 = L_3$$

since $a_n^{(g)}$ in this case is real-valued as well.

Triconfluent case

Encountering the difference equation in (4.2.35), (4.2.35) by means of the ansatz

$$s_{n1} = \exp\left(\gamma_{11}\, n^{\frac{3}{4}} + \gamma_{12}\, n^{\frac{2}{4}} + \gamma_{13}\, n^{\frac{1}{4}}\right) n^{r} \left[1 + \frac{C_{11}}{n^{\frac{1}{4}}} + \frac{C_{12}}{n^{\frac{2}{4}}} + \cdots\right],$$

$$
\begin{aligned}
s_{n2} &= \exp\left(\gamma_{21}\, n^{\frac{3}{4}} + \gamma_{22}\, n^{\frac{2}{4}} + \gamma_{23}\, n^{\frac{1}{4}}\right) n^r \left[1 + \frac{C_{21}}{n^{\frac{1}{4}}} + \frac{C_{22}}{n^{\frac{2}{4}}} + \cdots\right],\\
s_{n3} &= \exp\left(\gamma_{31}\, n^{\frac{3}{4}} + \gamma_{32}\, n^{\frac{2}{4}} + \gamma_{33}\, n^{\frac{1}{4}}\right) n^r \left[1 + \frac{C_{31}}{n^{\frac{1}{4}}} + \frac{C_{32}}{n^{\frac{2}{4}}} + \cdots\right],\\
s_{n4} &= \exp\left(\gamma_{41}\, n^{\frac{3}{4}} + \gamma_{42}\, n^{\frac{2}{4}} + \gamma_{43}\, n^{\frac{1}{4}}\right) n^r \left[1 + \frac{C_{41}}{n^{\frac{1}{4}}} + \frac{C_{42}}{n^{\frac{2}{4}}} + \cdots\right]
\end{aligned}
\tag{4.2.42}
$$

yields the following parameters:

$$
\begin{aligned}
\gamma_{11} &= \frac{4}{3}\sqrt[4]{-\sum_{i=-2}^{i=+2}\alpha_{it}} = \frac{4}{3}\sqrt[4]{-\tilde{g}_2} \neq 0,\\
\gamma_{12} &= -\frac{2\alpha_{2t} + \alpha_{1t} - \alpha_{-1t} - 2\alpha_{-2t}}{2\sqrt{-\sum_{i=-2}^{i=+2}\alpha_{it}}} = \frac{g_2 - g_1}{2\sqrt{-g_2}},\\
\gamma_{13} &= -\frac{C}{24\left(-\sum_{i=-2}^{i=+2}\alpha_{it}\right)^{\frac{5}{4}}},\\
\gamma_{21} &= -\gamma_{11}, \quad \gamma_{22} = \gamma_{12}, \quad \gamma_{23} = -\gamma_{13},\\
\gamma_{31} &= \imath\,\gamma_{11}, \quad \gamma_{32} = -\gamma_{12}, \quad \gamma_{33} = -\imath\,\gamma_{13},\\
\gamma_{41} &= -\imath\,\gamma_{11}, \quad \gamma_{42} = -\gamma_{12}, \quad \gamma_{43} = \imath\,\gamma_{13},\\
r &= -\frac{1}{8}\,(2\,\alpha_2 - 2\,\alpha_{-2} - 3)
\end{aligned}
\tag{4.2.43}
$$

where

$$
\begin{aligned}
C = &-32\,\alpha_2^2 - 16\,\alpha_2\,(4\,\alpha_{-1} + 7\,\alpha_{-2})\\
&-5\,\alpha_1^2 - 2\,\alpha_1\,(20\,\alpha_2 + 11\,\alpha_{-1} + 32\,\alpha_{-2})\\
&+4\,\alpha_0^2 - 4\,\alpha_0\,(\alpha_1 + 10\,\alpha_2 + \alpha_{-1} + 10\,\alpha_{-2})\\
&-5\,\alpha_{-1}^2 - 40\,\alpha_{-1}\,\alpha_{-2} - 32\,\alpha_{-2}^2.
\end{aligned}
$$

Thus, the general solution of the difference equation in (4.2.35), (4.2.35) may be written in the form

$$
a_n^{(g)} = L_1\, a_{n,1} + L_2\, a_{n,2} + L_3\, a_{n,3} + L_4\, a_{n,4},
$$

where $L_i = L_i(\underline{c})$, $i = 1 - 4$ are (in general complex-valued) constants in n that still may be dependent on the parameters of the differential equation in (4.2.35), (4.2.35).

Eventually, it is mentioned that there are relations between the relevant quantities of the difference and underlying differential equations:

$$
-\sum_{i=-1}^{+1}\alpha_{is} = -\sum_{i=0}^{+3}\Gamma_{is} = -g_{2s}\, z_{0s},
$$

$$-\sum_{i=-1}^{+1} \alpha_{ib} = -\sum_{i=0}^{+3} \Gamma_{ib} = +g_{2b}\, z_{0b}^2,$$

$$-\sum_{i=-1}^{+1} \alpha_{it} = -\sum_{i=0}^{+3} \Gamma_{it} = -g_{2t}\, z_{0t}^3,$$

$$-\sum_{i=-1}^{+1} \alpha_{id} = -\sum_{i=0}^{+3} \Gamma_{id} = -8\, g_{2d}\, z_{0d},$$

$$-\sum_{i=-1}^{+1} (-1)^i \alpha_{id} = -\sum_{i=0}^{+3} (-1)^i \Gamma_{id} = \frac{8}{z_{0d}} \left(\frac{g_{-1d}}{z_{0d}} + g_{0d} \right),$$

$$g_{2s} = \pm\sqrt{G_{0s}^2 - 4\, D_{0s}},$$

$$g_{2b} = \pm\sqrt{G_{1b}^2 - 4\, D_{2b}},$$

$$g_{2t} = \pm\sqrt{G_{2t}^2 - 4\, D_{4t}},$$

$$g_{2d} = \pm\sqrt{G_{-2d}^2 - 4\, D_{-4d}},$$

$$\gamma_{12t} = -\frac{1}{2}\sqrt{-g_{2t}\, z_{0t}^3} - \frac{g_{1t}\, z_{0t}^2}{2\sqrt{-g_{2t}\, z_{0t}^3}}.$$

4.2.3 The Double Confluent Case

The central two-point connection problem[3] of the *double confluent case* of the Fuchsian differential equation having four singularities is the only equation of the Heun class, connecting two irregular singularities. Thus, it looks for particular solutions of the differential equation

$$\frac{d^2 y_d}{dz^2} + \left[G_0 + \frac{G_{-1}}{z} + \frac{G_{-2}}{z^2} \right] \frac{dy_d}{dz} + \left[D_0 + \frac{D_{-1}}{z} + \frac{D_{-2}}{z^2} + \frac{D_{-3}}{z^3} + \frac{D_{-4}}{z^4} \right] y_d = 0 \tag{4.2.44}$$

that behave in a prescribed manner simultaneously on approaching the two singularities radially along the real axis. This requires the following *Jaffé ansatz*:

$$\boxed{y_d(z) = \exp\left(\alpha_{1\infty}\, z - \frac{\alpha_{10}}{z} \right) z^{\alpha_{00}} (z+1)^{\mu}\, w(z)} \tag{4.2.45}$$

whereby the singularity parameters are given by

[3] The two-point connection problem that is treated in the following must not be confused with the circular connection problem, dealt with in Ronveaux (1995, Part C, §2.5).

$$
\begin{aligned}
\alpha_{10}^2 + G_{-2}\,\alpha_{10} + D_{-4} &= 0,\\
\alpha_{00} &= -\frac{D_{-3} + \alpha_{10}\,(G_{-1} - 2)}{G_{-2} + 2\,\alpha_{10}},\\
\alpha_{1\infty}^2 + G_0\,\alpha_{1\infty} + D_0 &= 0,\\
\mu &= -\frac{D_{-1} + G_0\,\alpha_{00} + \alpha_{1\infty}\,(G_{-1} + 2\,\alpha_{00})}{G_0 + 2\,\alpha_{1\infty}}.
\end{aligned}
\tag{4.2.46}
$$

The first equation yields (in general) two values α_{101} and α_{102} of the characteristic exponent of the second kind of singularity at the origin and, on the basis of these, the second equation yields two values α_{001} and α_{002} of the characteristic exponent of the first kind.

The third equation yields (in general) two values $\alpha_{1\infty 1}$ and $\alpha_{1\infty 2}$ of the characteristic exponent of the second kind of singularity at the origin and, on the basis of these, the fourth equation yields two values μ_1 and μ_2 of the characteristic exponent of the first kind.

Inserting the ansatz (4.2.45) and using (4.2.46) yields an *intermediary differential equation* having the form

$$
z^2\,\frac{d^2w}{dz^2} + \left[g_2\,z^2 + g_1\,z + g_0 + \frac{g_{-1}}{z+1}\right]\frac{dw}{dz} + \left[d_0 + \frac{d_{-1}}{z+1} + \frac{d_{-2}}{(z+1)^2}\right] w(z) = 0
\tag{4.2.47}
$$

with

$$
\begin{aligned}
g_2 &= G_0 + 2\,\alpha_{1\infty},\\
g_1 &= G_{-1} + 2\,(\alpha_{00} + \mu),\\
g_0 &= G_{-2} + 2\,(\alpha_{10} - \mu),\\
g_{-1} &= 2\,\mu,\\
d_0 &= D_{-2} + \alpha_{1\infty}\,G_{-2} + \alpha_{10}\,G_0 + 2\,\alpha_{1\infty}\,\alpha_{10} + \alpha_{00}\,G_{-1}\\
&\qquad + \alpha_{1\infty}^2 - \alpha_{00} + \mu\,[G_{-1} - (G_0 + 2\,\alpha_{1\infty}) + 2\,\alpha_{00} + \mu - 1],\\
d_{-1} &= \alpha_d\,[(G_0 + 2\,\alpha_{1\infty}) - (G_{-1} + 2\,\alpha_{00} + 2\,\mu - 2) + G_{-2} + 2\,\alpha_{10}],\\
d_{-2} &= \mu\,(\mu - 1).
\end{aligned}
\tag{4.2.48}
$$

This differential equation (4.2.47) is applied to a *Jaffé transformation* having the form

$$
x = \frac{z-1}{z+1},
$$

whose action is

$$
\begin{array}{ccccc}
z: & +\infty & 1 & 0 & -1\\
\downarrow & \downarrow & \downarrow & \downarrow & \downarrow\\
x: & +1 & 0 & -1 & -\infty.
\end{array}
$$

This transformation places the singularity at the origin to $x = -1$ and the infinite singularity to $x = +1$, so that the two singularities are placed symmetrically to the origin. It is important in the following to have such a symmetrical situation.

The application of this transformation along the lines displayed several times above results in the *final differential equation*

$$(x^2 - 1)^2 \frac{d^2 w}{dx^2} + \sum_{i=0}^{3} \Gamma_i \, x^i \frac{dw}{dx} + \sum_{i=0}^{2} \Delta_i \, x^i \, w(x) = 0 \tag{4.2.49}$$

with

$$\begin{aligned}
\Gamma_3 &= -g_{-1} + 2, \\
\Gamma_2 &= 2\, g_2 - 2\, g_1 + 2\, g_0 + 3\, g_{-1} + 2, \\
\Gamma_1 &= 4\, g_2 - 4\, g_{0d} - 3\, g_{-1} - 2, \\
\Gamma_0 &= 2\, g_{2d} + 2\, g_1 + 2\, g_{0d} + g_{-1d} - 2, \\
\Delta_2 &= d_{-2}, \\
\Delta_1 &= -2\, d_{-2} - 2\, d_{-1}, \\
\Delta_0 &= d_{-2d} + 2\, d_{-1} + 4\, d_0.
\end{aligned} \tag{4.2.50}$$

It is important to understand this final differential equation (4.2.49). It has two finite and an infinite singularity; the latter is regular, while the two finite ones are irregular singularities of the differential equation, the s-rank of which is two: $R(-1) = R(+1) = 2$. The relevant interval of the CTCP is $[0, 1]$. Moreover, all particular solutions of this differential equation are holomorphic at the origin $z = 0$. Therefore, these particular solutions may be expanded in power series

$$w(z) = \sum_{n=0}^{\infty} a_n \, z^n.$$

Inserting this ansatz results in an irregular *difference equation* of Poincaré–Perron type having the form

$$\begin{aligned}
&a_0, \, a_1 \text{ arbitrary}, \\
&2\, a_2 + \Gamma_0 \, a_1 + \Delta_0 \, a_0 = 0, \\
&6\, a_3 + 2\, \Gamma_0 \, a_2 + (\Gamma_1 + \Delta_0)\, a_1 + \Delta_1 \, a_0 = 0, \\
&\left(1 + \frac{\alpha_2}{n} + \frac{\beta_2}{n^2}\right) a_{n+2} + \left(\frac{\alpha_1}{n} + \frac{\beta_1}{n^2}\right) a_{n+1} \\
&\quad + \left(-2 + \frac{\alpha_0}{n} + \frac{\beta_0}{n^2}\right) a_n + \left(\frac{\alpha_{-1}}{n} + \frac{\beta_{-1}}{n^2}\right) a_{n-1} \\
&\quad + \left(1 + \frac{\alpha_{-2}}{n} + \frac{\beta_{-2}}{n^2}\right) a_{n-2} = 0, \; n \geq 2
\end{aligned} \tag{4.2.51}$$

with

$$\begin{aligned} \alpha_2 &= 3, \quad \beta_2 = 2, \\ \alpha_1 &= \beta_1 = \Gamma_0, \\ \alpha_0 &= \Gamma_1 + 2, \quad \beta_0 = \Delta_0, \\ \alpha_{-1} &= \Gamma_2, \quad \beta_{-1} = \Delta_1 - \Gamma_2, \\ \alpha_{-2} &= \Gamma_3 - 5, \quad \beta_{-2} = \Delta_2 - 2\,\Gamma_3 + 6. \end{aligned} \tag{4.2.52}$$

Variable-Asymptotic Behaviour of Power Series

It should be mentioned that the variable asymptotics is the same as in the single confluent case, with the exception of the fact that there is a second double root of the characteristic equation (4.2.53), below, of the difference equation (4.2.51), (4.2.52). This means that there are two more Birkhoff solutions with respect to $t = -1$, the asymptotic behaviour of which is the same as the first one.

We start the *index asymptotic investigations of this difference equation* by considering the zeroth-order asymptotics of this difference equation, determined by an algebraic fourth-order equation

$$(t-1)^2\,(t+1)^2 = 0, \tag{4.2.53}$$

the solutions of which are given by the double roots

$$\begin{aligned} t_{1,2} &= 1, \\ t_{3,4} &= -1. \end{aligned}$$

The leading-order asymptotics is determined by the asymptotic solutions of the characteristic equation of (4.2.51), that is, the fourth-order algebraic equation

$$\left(1+\frac{\alpha_2}{n}\right)t^4(n) + \frac{\alpha_1}{n}\,t^3(n) + \left(-2+\frac{\alpha_0}{n}\right)t^2(n) + \frac{\alpha_{-1}}{n}\,t(n) + \left(1+\frac{\alpha_{-2}}{n}\right) = 0,$$

the solutions of which are given by

$$\begin{aligned} t_{1,2} &= 1 \pm \sqrt[4]{\frac{-\sum_{i=-2}^{i=+2}\alpha_i}{n}} + O\left(\frac{1}{n^{\frac{1}{2}}}\right), \\ t_{3,4} &= -1 \pm \sqrt[4]{\frac{-\sum_{i=-2}^{i=+2}(-1)^i\,\alpha_i}{n}} + O\left(\frac{1}{n^{\frac{1}{2}}}\right) \end{aligned}$$

as $n \to \infty$. These solutions are sketched in Figure 4.5.

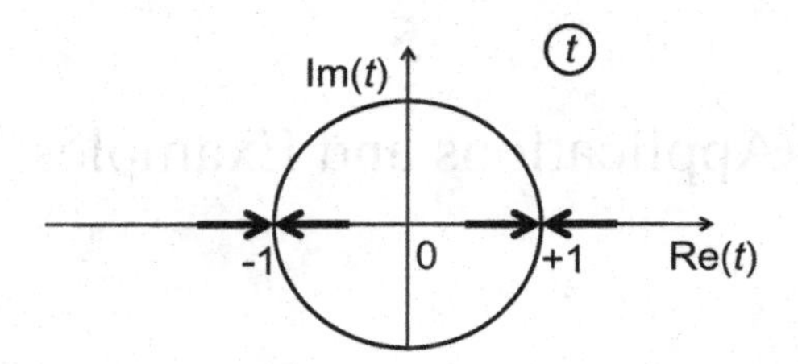

Figure 4.5 First-order asymptotics of irregular difference equations (double confluent).

Eventually, in order to give the full solution, the *Birkhoff set* of the difference equation (4.2.51) is given by

$$s_1(n) = \varrho_1^n \exp\left(\gamma_1 n^{\frac{1}{2}}\right) n^{r_1} \left[1 + \frac{C_{11}}{n^{\frac{1}{2}}} + \frac{C_{12}}{n^{\frac{2}{2}}} + \cdots\right],$$

$$s_2(n) = \varrho_1^n \exp\left(\gamma_2 n^{\frac{1}{2}}\right) n^{r_2} \left[1 + \frac{C_{21}}{n^{\frac{1}{2}}} + \frac{C_{22}}{n^{\frac{2}{2}}} + \cdots\right],$$

$$s_3(n) = \varrho_2^n \exp\left(\gamma_3 n^{\frac{1}{2}}\right) n^{r_3} \left[1 + \frac{C_{31}}{n^{\frac{1}{2}}} + \frac{C_{32}}{n^{\frac{2}{2}}} + \cdots\right],$$

$$s_4(n) = \varrho_2^n \exp\left(\gamma_4 n^{\frac{1}{2}}\right) n^{r_4} \left[1 + \frac{C_{41}}{n^{\frac{1}{2}}} + \frac{C_{42}}{n^{\frac{2}{2}}} + \cdots\right],$$

with

$$\varrho_1 = +1,$$

$$\gamma_1 = \sqrt{-\sum_{i=-2}^{+2} \alpha_i} \neq 0,$$

$$\gamma_2 = -\gamma_1,$$

$$r_1 = r_2 = -\frac{1}{8}\,(2\,\alpha_2 + \alpha_1 - \alpha_{-1} - 2\,\alpha_{-2} - 2),$$

$$\varrho_2 = -1,$$

$$\gamma_3 = \sqrt{-\sum_{i=-2}^{+2} (-1)^i\,\alpha_i} \neq 0,$$

$$\gamma_4 = -\gamma_3,$$

$$r_1 = r_2 = -\frac{1}{8}\,(2\,\alpha_2 - \alpha_1 + \alpha_{-1} - 2\,\alpha_{-2} - 2)\,.$$

For more details and results, the reader is referred to Bay et al. (1997).

5

Applications and Examples

This chapter is, in a sense, the most important one of this book, since it mainly presents examples of *higher special functions*. Since many details may not be contained in the theoretical discussions of the field, it is these examples that may mostly contribute to enabling the reader to treat self-reliantly his or her concrete problem. My aim in writing this book is that the reader will acquire autonomy in order to create new special functions on applying the methods developed here. I am convinced that there are as yet unseen aspects, a variety of hitherto unknown special functions or even mathematically relevant new discoveries lying in the field. It would give me pleasure to know that I have contributed to future developments.

5.1 Oscillations of Dislocations in Crystals: Heun

5.1.1 Linearisation

Linear differential equations of problems of classical mechanics are usually approximations to nonlinear processes, thus are linearisations about certain special solutions of nonlinear equations. Linearisation is one of the most important mathematical operations in the field of classical mechanical problems.

In the following I show how to get a Fuchsian differential equation from linearising a nonlinear partial time-dependent second-order differential equation about a static solution. The resulting differential equation describes one-dimensional linear oscillations $u = u(x,t)$ of an elastic vibrating medium about the static (i.e., time-independent) solution according to the principles of classical mechanics.

The fundamental differential equation describing the classical dynamics is given by

$$\frac{\mathrm{d}^2u}{\mathrm{d}x^2} - \frac{\mathrm{d}^2u}{\mathrm{d}t^2} = \frac{\mathrm{d}U}{\mathrm{d}u}. \tag{5.1.1}$$

This is a nonlinear, second-order partial differential equation. Since the nonlinearity does not occur in the highest derivative, but may arise at most in the term

$$\frac{dU}{du},$$

thus only in the zeroth derivative of $u = u(x, t)$, (5.1.1) is sometimes called *quasilinear*.

Calculation of Static Solutions

The static (viz. time-independent, thus ordinary) nonlinear second-order counterpart of (5.1.1) [the solution of which is denoted $u_0(x)$] is

$$\frac{d^2u_0}{dx^2} = \frac{dU}{du_0}. \tag{5.1.2}$$

A first integration of this static equation leads to

$$\left(\frac{du_0}{dx}\right)^2 = 2\,[U(u_0) + C] \tag{5.1.3}$$

and, finally, to

$$\pm\frac{du_0}{\sqrt{2\,[U(u_0) + C]}} = dx, \tag{5.1.4}$$

where C is a constant of the first integration. (5.1.4) may be solved by means of a further integration:

$$\boxed{x - x_0 = \pm\int\frac{du_0}{\sqrt{2\,[U(u_0) + C]}} = \pm F(U, C),} \tag{5.1.5}$$

where x_0 is a constant of the second integration and $F(U, C)$ is the antiderivative of

$$f[U(u_0), C] = \frac{1}{\sqrt{2\,[U(u_0) + C]}}. \tag{5.1.6}$$

From (5.1.5) follows the solution of the differential equation (5.1.2):

$$\boxed{u_0(x) = F^{-1}\,[\pm(x - x_0), U, C],} \tag{5.1.7}$$

where F^{-1} is the inverse function of the antiderivative of (5.1.7).

Simple Example

Suppose $2\,[U(u_0) + C]$ is a second-order polynomial in u_0:

$$2\,[U(u_0) + C] = A^2\,(u_0 - u_1)^2,$$

where u_1 is a constant. From (5.1.6) follows

$$f(u_0, C) = \frac{1}{\sqrt{2\,[U(u_0) + C]}} = \frac{1}{A\,(u_0 - u_1)}. \tag{5.1.8}$$

Then

$$F(u_0, C) = \frac{1}{A}\ln\,[u_0 - u_1] + K,$$

where K is a constant and thus, according to (5.1.5):

$$x - x_0 = \pm\frac{1}{A}\ \ln\left[u_0 - u_1\right] + K,$$

from which follows

$$u_0(x) = F^{-1}\left\{\pm[x - (x_0 + K)],\, C\right\} = \exp\left\{\pm A\ [x - (x_0 + K)]\right\} + u_1\,. \qquad (5.1.9)$$

It should be mentioned that the highest order of polynomials $2\,[U(u_0) + C]$ for which the integral (5.1.5) is an algebraic function (thus is representable in closed form) is four. Then, the solutions $u_0(x)$ are generally inverse *Jacobian elliptic functions* (Byrd and Friedman, 1971). This case is discussed below.

Linear Oscillations about the Static Solutions

In the following, the eigenfrequencies of the oscillations of these static solutions of the nonlinear wave equation (5.1.1) are calculated.

The linearisation ansatz for the time-dependent solution [describing oscillations about the static solution $u_0(x)$] is

$$u(x,t) = u_0(x) + u_d(x,t).$$

The implementation of the Taylor expansion of the nonlinear function $U(u_0)$ about the static solution $u_0(x)$

$$\frac{\mathrm{d}U}{\mathrm{d}u} = \frac{\mathrm{d}U}{\mathrm{d}u_0} + u_d\,\frac{\mathrm{d}^2U}{\mathrm{d}u_0^2}$$

into the differential equation (5.1.1) yields a linear, ordinary, second-order differential equation

$$\frac{\mathrm{d}^2u_d}{\mathrm{d}x^2} - \frac{\mathrm{d}^2u_d}{\mathrm{d}t^2} = u_d\,\frac{\mathrm{d}^2U}{\mathrm{d}u_0^2}$$

for the linear oscillations $u_d(x,t)$ about the static solution $u_0(x)$ of (5.1.9). This differential equation may be solved by means of a separation ansatz for $u_d(x,t)$

$$u_d(x,t) = y(x)\,T(t)$$

that yields a differential equation for the time variable t:

$$\frac{1}{T}\,\frac{\mathrm{d}^2T}{\mathrm{d}t^2} = -\omega^2 \rightsquigarrow T(t) = \exp\left(\imath\,\omega\,t\right).$$

Thus an oscillation, and an ordinary second-order differential equation for the space variable x, which is actually is of Schrödinger type:

$$\frac{\mathrm{d}^2y}{\mathrm{d}x^2} + \left(\omega^2 - \frac{\mathrm{d}^2U}{\mathrm{d}u_0^2}\right)\,y = 0. \qquad (5.1.10)$$

This differential equation may be written in dependence on u_0. Using the chain rule of derivatives once:

$$\frac{\mathrm{d}y}{\mathrm{d}x} = \frac{\mathrm{d}y}{\mathrm{d}u_0}\frac{\mathrm{d}u_0}{\mathrm{d}x}$$

and once again:

$$\frac{\mathrm{d}^2y}{\mathrm{d}x^2} = \frac{\mathrm{d}^2y}{\mathrm{d}u_0^2}\left(\frac{\mathrm{d}u_0}{\mathrm{d}x}\right)^2 + \frac{\mathrm{d}y}{\mathrm{d}u_0}\frac{\mathrm{d}^2u_0}{\mathrm{d}x^2}. \tag{5.1.11}$$

Taking advantage of (5.1.2) and (5.1.3) eventually leads to

$$\boxed{\frac{\mathrm{d}^2y}{\mathrm{d}u_0^2} + \frac{1}{2}\frac{U'}{U(u_0)+C}\frac{\mathrm{d}y}{\mathrm{d}u_0} + \frac{1}{2}\frac{\omega^2 - U'}{U(u_0)+C}\,y = 0} \tag{5.1.12}$$

with

$$U' = \frac{\mathrm{d}U}{\mathrm{d}u_0}, \quad U'' = \frac{\mathrm{d}^2U}{\mathrm{d}u_0^2}.$$

Static Solutions

As awaited, the crucial item for calculating the static solutions of equation (5.1.7) is the function[1]

$$U = U(z) \tag{5.1.13}$$

determining the type of (5.1.12). In the following discussion this function is supposed to be a (normalised) polynomial of order four, called the *Eshelby potential*:

$$U(z) = \frac{1}{2}z^2(1-z)^2 - \frac{s}{6\sqrt{3}}z = \frac{1}{2}z^4 - z^3 + \frac{1}{2}z^2 - \frac{s}{6\sqrt{3}}z \tag{5.1.14}$$

with

$$0 \le s \le 1.$$

It is physically necessary that $U(z)$ describes a double-well potential (cf. Figure 5.1 and Seeger and Schiller, 1966). The range of definition for the asymmetry parameter s in combination with the prefactor $\frac{1}{6\sqrt{3}}$ in front of the parameter s takes care of this necessity. This need for physical reasons is maintained for the *shifted Eshelby potential* (see Figure 5.10)

$$U(z) + C \tag{5.1.15}$$

with respect to (5.1.5) (cf. Figure 5.1).

The value $s = 0$ of the asymmetry parameter indicates a symmetrical Eshelby as well as a symmetrical shifted Eshelby potential. The value $s = 1$ indicates an unstable

[1] The static solution $u_0 = u_0(x)$ above is denoted $z = z(x)$ in the following.

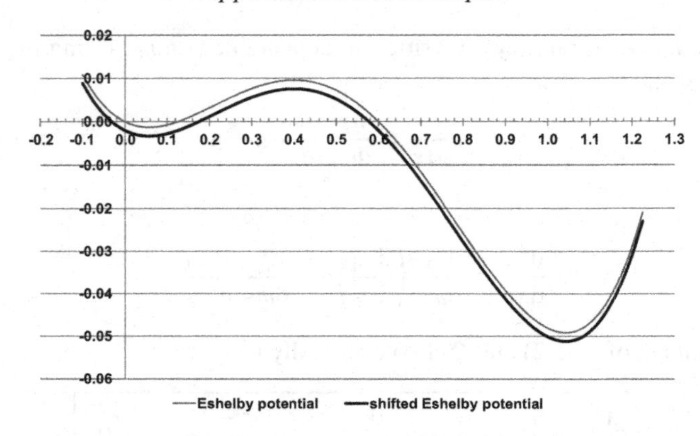

Figure 5.1 The Eshelby and the shifted Eshelby potential.

constant static solution z = const. of the differential equation (5.1.3), such that the derivative at the hump between the two neighbouring wells is zero.

Because of all the zeros being real-valued, the shifted Eshelby potential $U(z) + C$ may be written in the form

$$\begin{aligned} U(z) + C &= \frac{1}{2}\,\left[z^4 + b\,z^3 + c\,z^3 + d\,z^2 + e\right] \\ &= \frac{1}{2}\,(z - z_1)\,(z - z_2)\,(z - z_3)\,(z - z_4), \quad z_1,\, z_2,\, z_3,\, z_4 \in \mathbb{R}, \end{aligned}$$

or

$$\begin{aligned} 2\,[U(z) + C] &= z^4 - 2\,z^3 + z^2 - \frac{s}{3\,\sqrt{3}}\,z + 2\,C \\ &= z^4 + b\,z^3 + c\,z^2 + d\,z + e \\ &= (z - z_1)\,(z - z_2)\,(z - z_3)\,(z - z_4), \quad z_1,\, z_2,\, z_3,\, z_4 \in \mathbb{R}, \end{aligned} \tag{5.1.16}$$

and, without losing generality,

$$z_1 \le z_2 \le z_3 \le z_4 \tag{5.1.17}$$

may be assumed. These four zeros may be calculated by algebraic means (cf. Bronstein et al., 2001, p. 43), yielding the result

$$z_1 = \frac{1}{2}\left\{-\left[\frac{b + A_1}{2}\right] - \sqrt{\left[\frac{b + A_1}{2}\right]^2 - 4\left[y + \frac{b\,y - d}{A_1}\right]}\right\},$$

$$z_2 = \frac{1}{2}\left\{-\left[\frac{b+A_2}{2}\right] - \sqrt{\left[\frac{b+A_2}{2}\right]^2 - 4\left[y + \frac{b\,y-d}{A_2}\right]}\right\},$$

$$z_3 = \frac{1}{2}\left\{-\left[\frac{b+A_1}{2}\right] + \sqrt{\left[\frac{b+A_1}{2}\right]^2 - 4\left[y + \frac{b\,y-d}{A_1}\right]}\right\}, \tag{5.1.18}$$

$$z_4 = \frac{1}{2}\left\{-\left[\frac{b+A_2}{2}\right] + \sqrt{\left[\frac{b+A_2}{2}\right]^2 - 4\left[y + \frac{b\,y-d}{A_2}\right]}\right\},$$

with

$$\begin{aligned}
A_1 &= +\sqrt{8\,y + b^2 - 4\,c},\\
A_2 &= -\sqrt{8\,y + b^2 - 4\,c},\\
y &= -2\,r\,\cos\left(\frac{\varphi}{3}\right),\\
r &= -\sqrt{|p|},\\
\varphi &= \arccos(s),\\
p &= \frac{1}{36}\left(3\,b\,d - c^2 - 12\,e\right),
\end{aligned}$$

where the values of the parameters b, c, d, e are from (5.1.16):

$$\begin{aligned}
b &= -2,\\
c &= 1,\\
d &= -\frac{s}{3\sqrt{3}},\\
e &= 2\,C
\end{aligned}$$

or

$$\begin{aligned}
A_1 &= +2\sqrt{2\,y},\\
A_2 &= -2\sqrt{2\,y},\\
p &= \frac{1}{36}\left(\frac{2\sqrt{3}}{3}\,s - 1 - 24\,C\right).
\end{aligned}$$

Thus, the zeros z_i, $i = 1, 2, 3, 4$, depend on the two parameters s and C and are calculable by algebraic means (see Figure 5.11).

In the following, the derivatives of the Eshelby potential $U(z)$ in (5.1.14) are given for later use:

$$\frac{\mathrm{d}U}{\mathrm{d}z} = z\,(z-1)\,(2\,z-1) - \frac{s}{6\sqrt{3}} = \Pi_{i=1}^{3}(z - z_{ei}),$$

$$
\begin{aligned}
\frac{d^2U}{dz^2} &= 6z^2 - 6z + 1 = 6\left(z - \frac{\sqrt{3}+1}{2\sqrt{3}}\right)\left(z - \frac{\sqrt{3}-1}{2\sqrt{3}}\right), \\
\frac{d^3U}{dz^3} &= 6(2z-1), \\
\frac{d^4U}{dz^4} &= 12.
\end{aligned}
\tag{5.1.19}
$$

Here, $z = z_{ei} \in \mathbb{R}$, $i = 1,2,3$, are the locations of the three turning points of the potential $U = U(z)$. The result is

$$z_{ei} = \frac{1}{2} + \frac{c_i}{\sqrt{3}}, \quad i = 1,2,3, \tag{5.1.20}$$

with

$$c_i = \cos\left(\frac{\arccos(s) + 2\pi i}{3}\right) \tag{5.1.21}$$

(cf. Figure 5.2). It should be mentioned that the z_{ei} do not depend on the constant C.

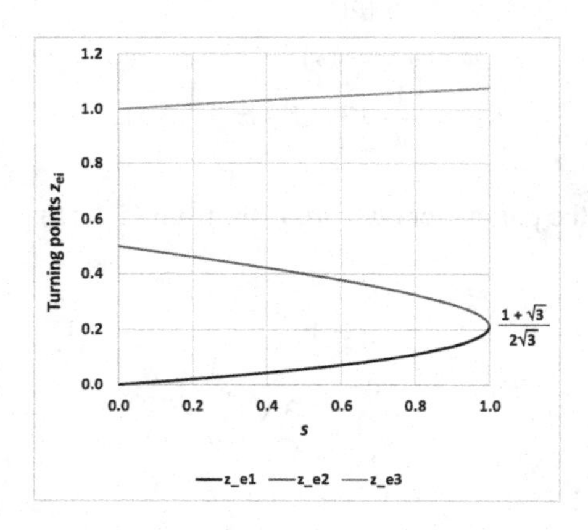

Figure 5.2 The turning points z_{ei}, $i = 1,2,3$ of the Eshelby potential as functions of s.

The form of the static solution $z(x)$ in (5.1.7)

$$z(x) = F^{-1}\left[\pm(x - x_0, s, C)\right] \tag{5.1.22}$$

is completely determined by the asymmetry parameter s and the constant of integration C. These two parameters determine the four real-valued zeros (5.1.17) of the potential (5.1.14) that are given above. The static solutions $z(x)$ of (5.1.1), described by the formula (5.1.5), are completely determined by the positions of these zeros.

In the following, the static solutions $z = z(x)$ of equation (5.1.7) that are possible under the action of the Eshelby potential (5.1.14) are discussed.

In order to calculate the static solutions of the differential equation (5.1.7), it is crucial to consider the integral in (5.1.5):

$$\int \frac{z}{\sqrt{2\,[U(z)+C]}}, \tag{5.1.23}$$

with $U(z)$ from (5.1.14) and C being a constant in z.

It is supposed, in a first step, that all the zeros $z_i, i = 1, 2, 3, 4$ of the shifted Eshelby potential $U + C = U(z) + C$ are simple ones. As a consequence, (5.1.23) is an elliptic integral. This may be seen from the following formulae. It is supposed that the four zeros of the shifted potential (5.1.14), z_1, z_2, z_3, z_4, may be ordered according to

$$z_1 < z_2 \leq z \leq z_3 < z_4 \tag{5.1.24}$$

where z is the independent variable. Thus, z varies between z_2 and z_3. Then

$$\int_{z_2}^{z} \frac{\mathrm{d}t}{\sqrt{(t-z_1)(t-z_2)(z_3-t)(z_4-t)}} = g\,\mathrm{sn}^{-1}\left(\sqrt{\frac{(z_3-z_1)(z-z_2)}{(z_3-z_2)(z-z_1)}},\, k\right) \tag{5.1.25}$$

with

$$k = \sqrt{\frac{(z_3-z_2)(z_4-z_1)}{(z_4-z_2)(z_3-z_1)}}, \tag{5.1.26}$$

$$g = \frac{2}{\sqrt{(z_4-z_2)(z_3-z_1)}} \tag{5.1.27}$$

is an *incomplete elliptic integral*, where $\mathrm{sn}^{-1}(z,\, k)$ is the inverse of the function, being called *sinus amplitudinis* $\mathrm{sn}(z,\, k)$ (cf. Byrd and Friedman, 1971, p. 112, no. 254.00). Here, the parameter k is called the *modulus*, the range of which is

$$0 \leq k \leq 1.$$

It is to be remarked that (5.1.25) is derived from a standardised form, the *normal incomplete elliptic integral of the first kind* (cf. Byrd and Friedman, 1971, p. 8, no. 110.02):

$$\int_0^z \frac{\mathrm{d}t}{\sqrt{(1-t^2)(1-k^2 t^2)}} = \mathrm{sn}^{-1}(z,\, k).$$

The sinus amplitudinis is one of the three Jacobian elliptic functions, being defined according to the trigonometric functions as

$$\mathrm{sn}(z,\, k), \quad \mathrm{cn}(z,\, k) = \sqrt{1-\mathrm{sn}^2(z,\, k)}, \quad \mathrm{dn}(z,\, k) = \sqrt{1-k^2\,\mathrm{sn}^2(z,\, k)}$$

on the complex plane $z \in \mathbb{C}$. These functions are the most general functions having addition theorems; they are double periodic in the complex plane. Jacobian elliptic functions are generalisations of the trigonometric functions and thus have similar mathematical characteristics (see Figure 5.3). In particular, the period p is not a constant but is a function of the modulus k. For $k = 0$ it is the same as the period of the sine, $p = 2\,\pi$; for increasing k the period increases as well, becoming infinite for the maximum value $k = 1$. For more details about these marvellous functions, see Byrd and Friedman (1971) and the second part of Hurwitz et al. (1964, p. 147).

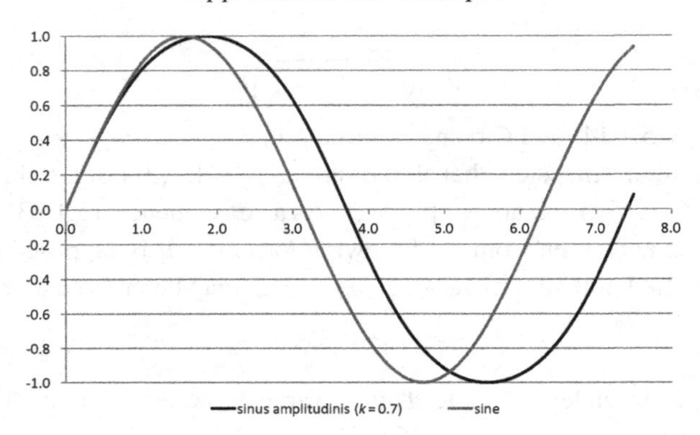

Figure 5.3 The sinus amplitudinis.

It is crucial for understanding the static solutions of the differential equation (5.1.2) that the range of the pair of parameters s, C is a curvy-shaped triangle, given in Figure 5.4. The edges as well as the corners of this triangle characterise specific solutions. Therefore, I call this triangle in the parameter space of the parameters s and C the *characteristic triangle*.

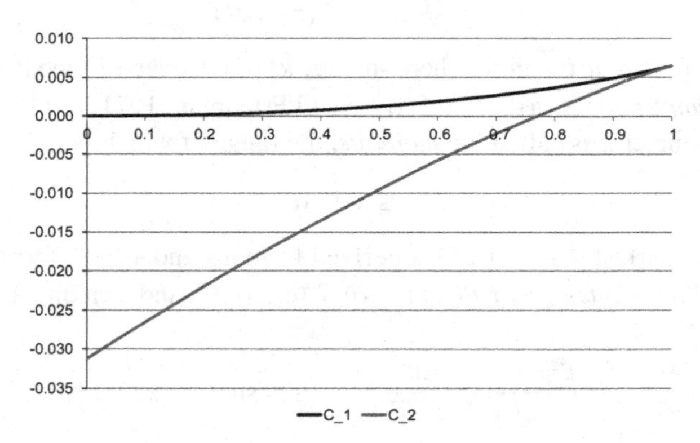

Figure 5.4 The characteristic triangle of static solutions.

The parameter s is normalised such that $0 \leq s \leq 1$, and the parameter C is bounded by an upper and a lower curve $C = C_1(s) = C_{\max}(s)$, $C = C_2(s) = C_{\min}(s)$, respectively, such that

$$C_2(s) \leq C \leq C_1(s)$$

holds. It should be mentioned here that, besides the asymmetry parameter s, it is possible and even convenient to normalise the integration parameter C. This is done by means of

$$C_n = \frac{C - C_{\min}}{C_{\max} - C_{\min}}$$

for each fixed value of s, such that $0 \leq C_n \leq 1$.

It should be recognised that all but one of the static solutions $z(x)$ of the differential equation (5.1.1), represented by the characteristic triangle, are unstable in the sense that any perturbation will cause them to decay into a straight, viz. constant, position at the bottom of the lower well of the Eshelby potential (5.1.14). The exception is the solution that is characterised by $s = C = 0$; this solution is called the *kink solution.*

It is possible to calculate the upper and lower limits $C_{\max}(s) = C_1(s)$, $C_{\min}(s) = C_2(s)$ of the characteristic triangle, as will be done in the following.

The defining criterion of the upper bound $C_{\max}(s) = C_1(s)$ is given by

$$C_1(s) = -U(z_{e1}),$$

where z_{e1} is the location of the lowest turning points of the Eshelby potential $U(z)$ as well as of the shifted Eshelby potential $U(z) + C$.

The lowest turning point z_{e1} is given as [cf. equations (5.1.20), (5.1.21)]

$$z_{e1} = \frac{1}{2} + \frac{1}{\sqrt{3}} \cos\left[\frac{\arccos(s) + 2\pi}{3}\right].$$

As may be calculated by elementary means:

$$U(z_{ei}) = -\frac{1}{2} z_{ei}^2 (1 - z_{ei})(3 z_{ei} - 1), \; i = 1, 2, 3, \tag{5.1.28}$$

thus

$$C_1(s) = \frac{1}{2} z_{e1}^2 \, (1 - z_{e1}) \, (3 z_{e1} - 1) \, .$$

Similarly, the defining criterion of the lower bound $C_{\min}(s) = C_2(s)$ is given by

$$C_2(s) = -U(z_{e2}),$$

where z_{e2} is the location of the central turning point

$$z_{e1} \leq z_{e2} \leq z_{e3}$$

of the Eshelby potential $U(z)$. Thus[2]

$$z_{e2} = \frac{1}{2} + \frac{1}{\sqrt{3}} \cos\left[\frac{\arccos(s) + 4\pi}{3}\right]$$

and [using equation (5.1.28) for $i = 2$]

$$C_2(s) = -U(z_{e2}) = \frac{1}{2} z_{e2}^2 (1 - z_{e2})(3 z_{e2} - 1).$$

The functions $C_1(s)$ and $C_2(s)$ may be seen from Figure 5.4.

In the following, the general solution of the static differential equation (5.1.2) is

[2] For the sake of completeness, it is mentioned that $z_{e3} = \frac{1}{2} + \frac{1}{\sqrt{3}} \cos\left[\frac{\arccos(s)+6\pi}{3}\right]$ holds.

given for arbitrary values of the pairs (s, C) of parameters from within the characteristic triangle; it results from the integral in (5.1.23). Taking four real-valued zeros of $U(z) + C$:

$$U(z) + C = \frac{1}{2}\, z^2\, (z-1)^2 - \frac{s}{6\sqrt{3}}\, z + C(s), \quad 0 \le s \le 1, \quad C_2(s) \le C \le C_1(s)$$

while assuming here

$$z_1 \le z_2 \le z_3 \le z_4,$$

the integral in (5.1.5) becomes an incomplete elliptic integral [Byrd and Friedman, 1971, (5.1.25)]

$$x - x_0 = \pm \int_{z_2}^{z} \frac{dz}{\sqrt{2\,[U(z) + C]}} = \pm g\, \mathrm{sn}^{-1}\left(\sqrt{\frac{z_3 - z_1}{z_3 - z_2}\,\frac{z - z_2}{z - z_1}}, k\right) + K$$

where K is a constant of integration.

Solving this formula for $z = z(x)$ step by step in the following leads to the general static solution of a string in an Eshelby potential, an unstable periodic configuration:

$$\pm\frac{1}{g}\,[x - (x_0 + K)] = \mathrm{sn}^{-1}\left(\sqrt{\frac{z_3 - z_1}{z_3 - z_2}\,\frac{z - z_2}{z - z_1}}, k\right)$$

$$\rightsquigarrow \mathrm{sn}^2\left[\pm\frac{x - (x_0 + K)}{g}, k\right] = \frac{z_3 - z_1}{z_3 - z_2}\,\frac{z - z_2}{z - z_1}$$

$$\rightsquigarrow \frac{z_3 - z_2}{z_3 - z_1}\,\mathrm{sn}^2\left[\pm\frac{x - (x_0 + K)}{g}, k\right] = \frac{z - z_2}{z - z_1}$$

$$\rightsquigarrow \frac{z_3 - z_2}{z_3 - z_1}\,\mathrm{sn}^2\left[\pm\frac{x - (x_0 + K)}{g}, k\right] z - \frac{z_3 - z_2}{z_3 - z_1}\,\mathrm{sn}^2\left[\pm\frac{x - (x_0 + K)}{g}, k\right] z_1 = z - z_2$$

$$\rightsquigarrow \left\{\frac{z_3 - z_2}{z_3 - z_1}\,\mathrm{sn}^2\left[\pm\frac{x - (x_0 + K)}{g}, k\right] - 1\right\} z = -z_2 + z_1\,\frac{z_3 - z_2}{z_3 - z_1}\,\mathrm{sn}^2\left[\pm\frac{x - (x_0 + K)}{g}, k\right].$$

Eventually, the functional form is thus given by

$$\boxed{z = z(x) = \frac{-z_2 + z_1\,\dfrac{z_3 - z_2}{z_3 - z_1}\,\mathrm{sn}^2\left[\pm\dfrac{x - (x_0 + K)}{g}, k\right]}{-1 + \dfrac{z_3 - z_2}{z_3 - z_1}\,\mathrm{sn}^2\left[\pm\dfrac{x - (x_0 + K)}{g}, k\right]}.} \tag{5.1.29}$$

An appropriate parameter for characterising the period p of the solutions is the modulus k from (5.1.26), $0 \le k \le 1$, and the quantity g from (5.1.27), since

$$p = 2\, g\, K(k) \tag{5.1.30}$$

where K is the complete elliptic integral of first order (cf. Byrd and Friedman, 1971, pp. 8, 9, no. 110.02, 110.06) (see Figure 5.12):

$$K(k) = \int_{t=0}^{1} \frac{\mathrm{d}t}{\sqrt{(1-t^2)(1-k^2 t^2)}}.$$

In Figure 5.5 are displayed the solutions (5.1.29) from within the characteristic triangle having constant modulus k.

In Figure 5.6 are displayed the solutions (5.1.29) from within the characteristic triangle having constant period p.

Eventually, in Figure 5.7 are displayed the period p of the solution (5.1.29) along the line $C = C_{\min}(s)$.

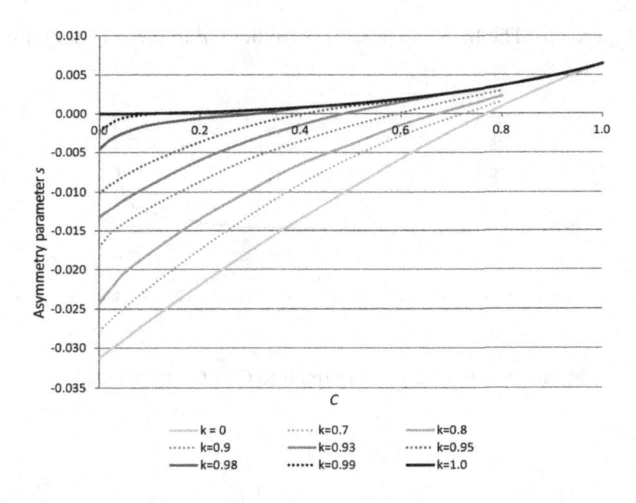

Figure 5.5 The lines of constant modulus k in the s–C diagram.

In the following are discussed all the special cases of the static solutions $z(x)$ that occur with respect to the edges and to the corners of the characteristic triangle.

- For $s = C = 0$ the integral in (5.1.5) is of the type (see Bronstein et al., 2001, p. 1050)

$$x - \tilde{x}_0 = \pm \int \frac{\mathrm{d}z}{\sqrt{2\,[U(z) + C]}} = \pm \int \frac{\mathrm{d}z}{\sqrt{2\,[U(z)]}} = \pm \int \frac{\mathrm{d}z}{\sqrt{[z^2\,(z-1)^2]}}$$

$$= \pm \int \frac{\mathrm{d}z}{\sqrt{[z^2\,(1-z)^2]}} = \pm \int \frac{\mathrm{d}z}{z\,(1-z)}$$

$$= \mp \ln \left[\frac{1-z}{z}\right] + K.$$

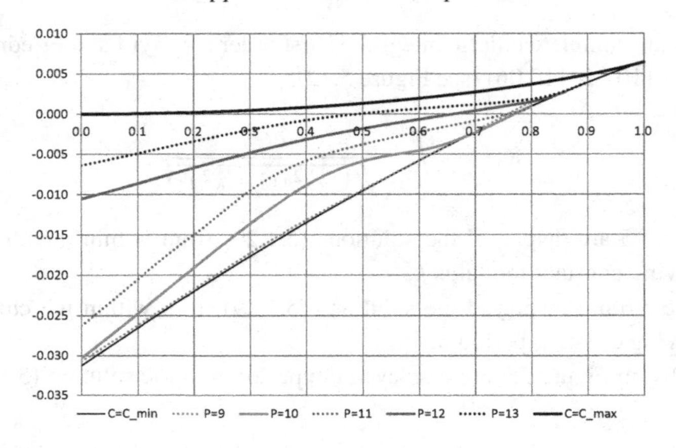

Figure 5.6 The lines of constant periodicity p in the s–C diagram.

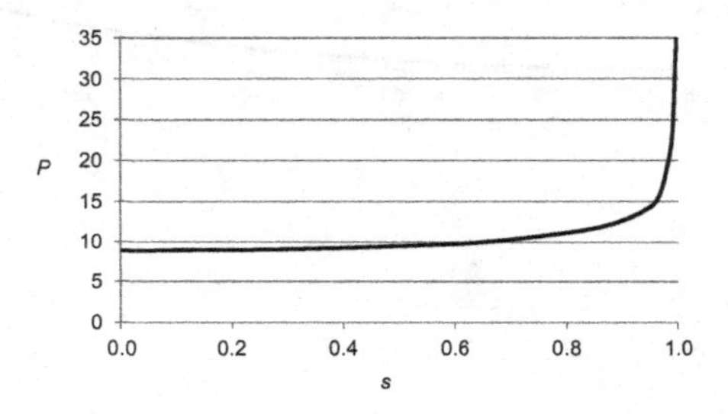

Figure 5.7 Period p along the line $C = C_2(s) = C_{\min}(s)$.

Thus

$$\pm (x - x_0) = \ln \left[\frac{1 - z}{z} \right]$$

with

$$x_0 = \tilde{x}_0 - K$$

or

$$\exp [\pm (x - x_0)] = \frac{1}{z} - 1,$$

and thus the solution $z(x)$ of (5.1.5) becomes a single transition from one well of (5.1.14) to the neigbouring one of the Eshelby potential (cf. Figure 5.8):

$$\boxed{z(x) = \frac{1}{1 + \exp [\pm (x - x_0)]}.} \tag{5.1.31}$$

Here, $k = 1$ and the period p of the solution $z(x)$ is infinite [because $K(k = 1)$ is infinite]. This is the above-mentioned kink solution shown in Figure 5.8.

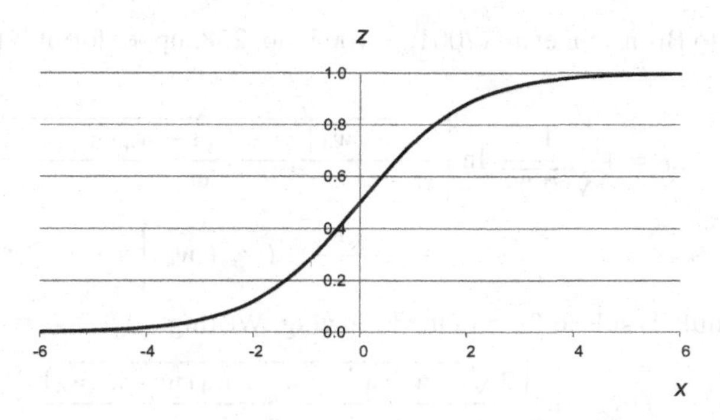

Figure 5.8 The single transition solution (kink solution) of the differential equation (5.1.7).

- The upper bound of C, thus $C = C_1(s) = C_{\max}(s)$, represents a bounced transition called a *breather solution*. The crucial point here is that the Eshelby potential is adjusted by $C = C_1(s)$ such that the lowest zero of the shifted Eshelby potential first is a double one and second coincides with the lowest turning point $z_1 = z_2 = z_{e1}$.

The integral in (5.1.5) is of the type (see Bronstein et al., 2001, pp. 1051, 1051)

$$x - x_0 = \pm \int \frac{\mathrm{d}z}{\sqrt{2\,[U(z) + C]}} = \pm \int \frac{\mathrm{d}z}{\sqrt{2\,[U(z) + C_1(s)]}}.$$

It is the defining characteristic of the function $U(z) + C_1(s)$ that the shifted Eshelby potential

$$U(z) + C_1(s)$$

does have a double root, because z_1 and z_2 coincide: $z_1 = z_2 = z_{e1}$. Thus

$$\begin{aligned} x - x_0 &= \pm \int \frac{\mathrm{d}z}{\sqrt{(z - z_{e1})^2\,(z - z_3)\,(z - z_4)}} \\ &= \pm \int \frac{\mathrm{d}z}{(z - z_{e1})\,\sqrt{(z - z_3)\,(z - z_4)}}. \end{aligned}$$

Defining

$$\begin{aligned} w &= z - z_{e1}, \\ w_3 &= z_3 - z_{e1}, \\ w_4 &= z_4 - z_{e1}, \end{aligned}$$

this integral yields

$$\begin{aligned} x - x_0 &= \pm \int \frac{\mathrm{d}w}{w\,\sqrt{(w - w_3)\,(w - w_4)}} \\ &= \pm \int \frac{\mathrm{d}w}{w\,\sqrt{w^2 - (w_3 + w_4)\,w + w_3\,w_4}}. \end{aligned}$$

According to Bronstein et al. (2001, p. 1065, no. 258, upper formula), this integral is given by

$$x - x_0 = \mp \frac{1}{\sqrt{w_3\, w_4}} \ln \left[\frac{2\sqrt{w_3\, w_4\,[w^2 - (w_3 + w_4)\, w + w_3\, w_4]}}{w} + \frac{2\, w_3\, w_4}{w} - (w_3 + w_4) \right] + C.$$

The formula is solved for z in the following. Writing

$$f(x) = \ln \left[\frac{2\sqrt{w_3\, w_4\,[w^2 - (w_3 + w_4)\, w + w_3\, w_4]}}{w} + \frac{2\, w_3\, w_4}{w} - (w_3 + w_4) \right]$$

with

$$f(x) = \pm\sqrt{w_3\, w_4}\; [x - (x_0 + C)]$$

yields

$$\exp[f(x)] = \frac{2\sqrt{w_3\, w_4\,[w^2 - (w_3 + w_4)\, w + w_3\, w_4]}}{w} + \frac{2\, w_3\, w_4}{w} - (w_3 + w_4)$$

or

$$\frac{1}{2}\, \{\exp[f(x)] + w_3 + w_4\}\; w - w_3\, w_4 = \sqrt{w_3\, w_4\,[w^2 - (w_3 + w_4)\, w + w_3\, w_4]}$$

or

$$\left(\frac{1}{2}\, \{\exp[f(x)] + w_3 + w_4\}\; w - w_3\, w_4 \right)^2 = w_3\, w_4\,[w^2 - (w_3 + w_4)\, w + w_3\, w_4]$$

or

$$(A\, w - w_3\, w_4)^2 = w_3\, w_4\,[w^2 - (w_3 + w_4)\, w + w_3\, w_4] \qquad (5.1.32)$$

with

$$A = \frac{1}{2}\, \{\exp[f(x)] + w_3 + w_4\}.$$

Extracting (5.1.32) yields

$$A^2\, w^2 - 2\, A\, w_3\, w_4\, w + w_3^2\, w_4^2 = w_3\, w_4\,[w - (w_3 + w_4)]\, w + w_3^2\, w_4^2$$

or

$$A^2\, w^2 - 2\, A\, w_3\, w_4\, w = w_3\, w_4\,[w - (w_3 + w_4)]\, w$$

or

$$A^2\, w - 2\, A\, w_3\, w_4 = w_3\, w_4\,[w - (w_3 + w_4)]$$

or

$$(A^2 - w_3\, w_4)\, w = 2\, A\, w_3\, w_4 - w_3\, w_4\,(w_3 + w_4).$$

As a result, the breather solution is given by

$$
\begin{aligned}
w(x) = z(x) - z_{e1} &= \frac{2\,A\,w_3\,w_4 - w_3\,w_4\,(w_3 + w_4)}{A^2 - w_3\,w_4} \\
&= \frac{w_3\,w_4\,\exp[f(x)] + w_3\,w_4\,(w_3 + w_4) - w_3\,w_4\,(w_3 + w_4)}{\frac{1}{4}\exp[2\,f(x)] + \frac{1}{2}(w_3 + w_4)\exp[f(x)] + \frac{1}{4}(w_3 + w_4)^2 - w_3\,w_4} \\
&= \frac{w_3\,w_4\,\exp[f(x)]}{\frac{1}{4}\exp[2\,f(x)] + \frac{1}{2}(w_3 + w_4)\exp[f(x)] + \frac{1}{4}(w_3 - w_4)^2}.
\end{aligned}
$$

Thus, the solution $z(x)$ of (5.1.5) becomes a pair of transitions (cf. Figure 5.9)

$$
\boxed{z(x) = z_{e1} + \frac{2\,w_3\,w_4}{w_3 + w_4 + \dfrac{1}{2}\exp[f(x)] + \dfrac{1}{2}(w_3 - w_4)^2\exp[-f(x)]}} \tag{5.1.33}
$$

sometimes, as mentioned above, called a breather solution because of its vibrational character as a whole.

It should be mentioned that in a coordinate system in which $w_3 - w_4$ is normalised, thus

$$
w_3 - w_4 = 1
$$

holds, the expression

$$
\frac{1}{2}\exp[f(x)] + \frac{1}{2}(w_3 - w_4)^2\exp[-f(x)]
$$

in (5.1.33) may be written as

$$
\cosh[f(x)].
$$

The modulus k along this edge $C = C_1(s)$ of the characteristic triangle is $k = 1$ and thus, because of (5.1.30), the period p is infinite (because $K(k = 1)$ is infinite).

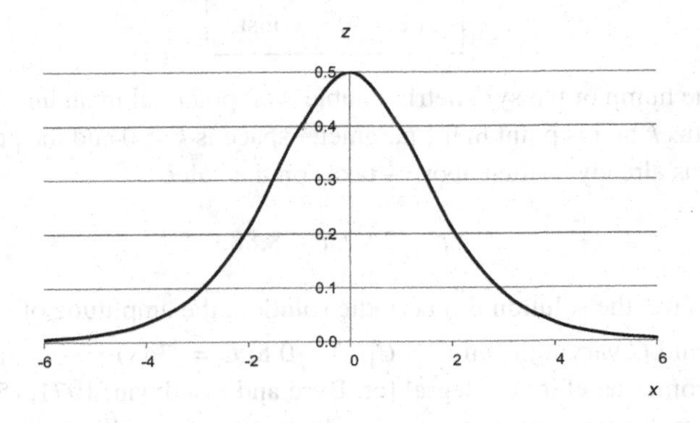

Figure 5.9 The breather solution of the differential equation (5.1.2).

- For $s = 1$ and

$$C = C_1(1) = C_2(1) = C_{\max} = C_{\min} = \frac{(\sqrt{3}-1)^2}{48\sqrt{3}} \approx 0.00645,$$

the breather solution degenerates into a constant

$$\boxed{z(x) = z_{e1}(s=1) = z_{e2}(s=1) = \tfrac{1+\sqrt{3}}{2\sqrt{3}} = \text{const.}} \tag{5.1.34}$$

lying in an unstable straight configuration on the edge of the single well of the Eshelby potential. The modulus k at this point in the parameter space is discontinuous, thus $k = 1$ and $k = 0$ at the same point.

- The static solution indicated by the lower bound of C, $C = C_2(s) = C_{\min}(s)$ of the characteristic triangle is a constant in z, thus a straight configuration:

$$\boxed{z(x) = z_{e2}(s) = \frac{1}{2} + \frac{\cos\left[\dfrac{\arccos(s) + 4\pi}{3}\right]}{\sqrt{3}} = \text{const.}|_s\,.}$$

The modulus k along this edge is $k = 0$ and the period p varies between infinity and the value

$$p = 2\sqrt{2}\,\pi.$$

This is remarkable, since for all values of s the configuration – as written above – is a straight one!

- For $s = 0$ and

$$C = C_2(0) = C_{\min} = -\frac{1}{32} = -0.03125,$$

the solution is a constant

$$\boxed{z(x) = 0 = \text{const.}}$$

lying on the hump of the symmetric double-well potential in an unstable position. The modulus k at this point in the parameter space is $k = 0$ and the period p of the solution – as already written above – takes on the value

$$p = 2\sqrt{2}\,\pi \approx 8.88.$$

Thus, as above, the solution is a periodic solution, the amplitude of which is zero.

- For $s = 0$ and C varying from $C = C_1(s) = 0$ to $C = C_2(s) = -\frac{1}{32}$, the integral in (5.1.5) becomes an elliptic integral [cf. Byrd and Friedman, 1971, (5.1.25)]. As a consequence, the static solution $z(x)$ results in a Jacobian elliptic function. This is shown in the following.

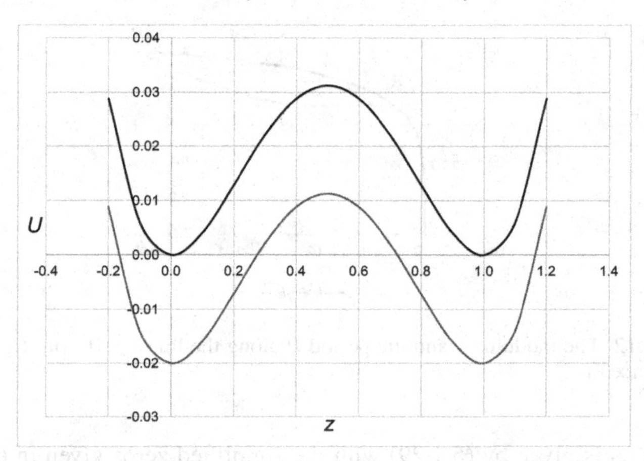

Figure 5.10 Eshelby potential $U(z)$ as well as shifted Eshelby potential $U(z) + C$ with $C = -0.02$.

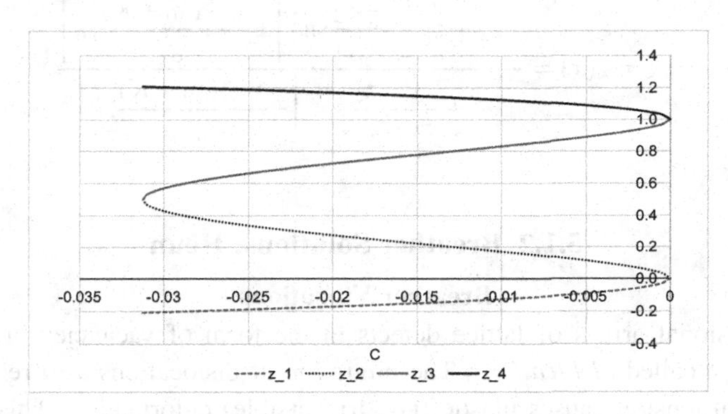

Figure 5.11 Zeros of the shifted Eshelby potential for $s = 0$ in dependence on C.

Since the term $-\frac{s}{6\sqrt{6}}\, z$ in the Eshelby potential (5.1.14) is an asymmetry parameter, the Eshelby potential becomes symmetric for $s = 0$ with respect to the line $z = \frac{1}{2}$. The zeros are given by

$$\begin{aligned}
z_1 &= \frac{1 - \sqrt{1 + \sqrt{-32\, C}}}{2}, \\
z_2 &= \frac{1 - \sqrt{1 - \sqrt{-32\, C}}}{2}, \\
z_3 &= \frac{1 + \sqrt{1 - \sqrt{-32\, C}}}{2}, \\
z_4 &= \frac{1 + \sqrt{1 + \sqrt{-32\, C}}}{2}.
\end{aligned} \tag{5.1.35}$$

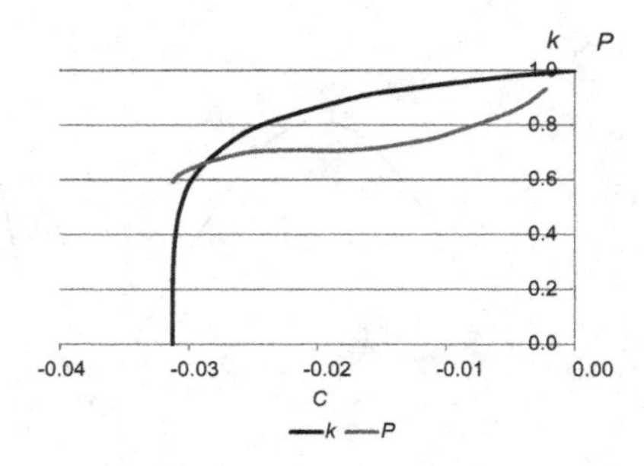

Figure 5.12 The modulus k and the period P along the line $s = 0$ from C_1 to C_2 in dependence on C.

The solution is given by (5.1.29) with the simplified zeros given in (5.1.35) (see Esslinger, 1990, p. 62):

$$z = z(x) = \frac{-z_2 + z_1 \dfrac{z_3 - z_2}{z_3 - z_1} \mathrm{sn}^2 \left[\pm \dfrac{x - (x_0 + K)}{g}, k \right]}{-1 + \dfrac{z_3 - z_2}{z_3 - z_1} \mathrm{sn}^2 \left[\pm \dfrac{x - (x_0 + K)}{g}, k \right]}.$$

5.1.2 Breather Solutions: Heun

Breather Vibrations

One-dimensional arrays of lattice defects in the form of vacancies in crystalline materials are called *dislocations*. The migration of dislocations as a result of heat and external tension causes plastic (i.e., irreversible) deformation. This migration as the basic physical process of plastic deformation under external stress in turn is determined by the lowest eigenfrequency of the dislocations moving in periodic potentials.

A mathematical model of such a situation is given by the oscillation of the breather solutions moving in a sloped double-well potential, i.e., the oscillations of the breather solution (5.1.33) of equation (5.1.2) (cf. Seeger and Schiller, 1966). This is the physical motivation of the interest in calculating the low-lying eigenfrequencies of a breather solution (5.1.33) of equation (5.1.2).

The corresponding differential equation describing these oscillations is (5.1.10)

$$\frac{\mathrm{d}^2 y}{\mathrm{d}x^2} + \left(\omega^2 - \frac{\mathrm{d}^2 U}{\mathrm{d}z^2} \right) y = 0. \tag{5.1.36}$$

As was shown in §5.1.1, this differential equation may be written in dependence on the dependent variable z of the static solution, yielding (5.1.12)

$$\frac{\mathrm{d}^2 y}{\mathrm{d}z^2} + \frac{1}{2} \frac{U'}{U(z) + C} \frac{\mathrm{d}y}{\mathrm{d}z} + \frac{1}{2} \frac{\omega^2 - U''}{U(z) + C} y = 0. \tag{5.1.37}$$

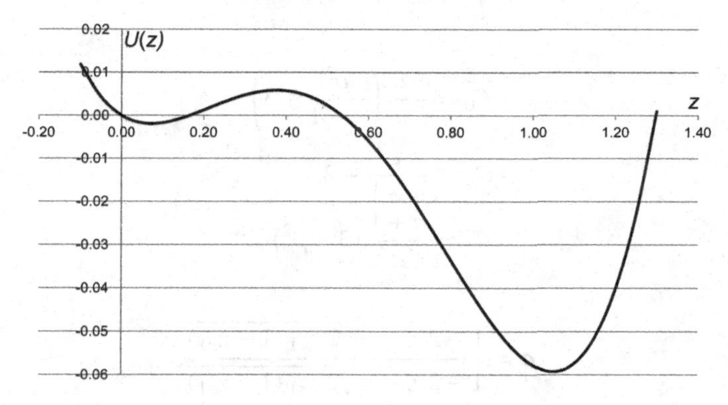

Figure 5.13 The Eshelby potential.

The potential $U = U(z)$ – as the outcome of the surrounding atoms – in which the dislocation dynamics takes place is given by a fourth-order polynomial called the *Eshelby potential* [cf. equation (5.1.14) and Figure 5.13]

$$U(z) = \frac{1}{2}\,z^2\,(z-1)^2 - \frac{s}{6\sqrt{3}}\,z = \frac{1}{2}\,z^4 - z^3 + \frac{1}{2}\,z^2 - \frac{s}{6\sqrt{3}}\,z. \tag{5.1.38}$$

Thus, the periodic potential is, for mathematical reasons, substituted by a double-well potential that is sufficient for investigating an elementary step from one well to a neighbouring one.

As may be seen below, from §5.1.1 the calculation of the oscillations of dislocations in crystals leads to a Fuchsian differential equation (5.1.12) having four singularities.

Deriving the function of the Eshelby potential (5.1.38) once [cf. equations (5.1.19)]

$$U' = \frac{\mathrm{d}U}{\mathrm{d}z} = 2\,z^3 - 3\,z^2 + z - \frac{s}{6\sqrt{3}} = z\,(z-1)\,(2\,z-1) - \frac{s}{6\sqrt{3}}$$

and deriving it once again

$$U'' = \frac{\mathrm{d}^2U}{\mathrm{d}z^2} = 6\,z^2 - 6\,z + 1$$

results in the coefficient

$$\frac{1}{2}\,\frac{U'}{U+C} = \frac{1}{z} + \frac{\frac{1}{2}}{z-a} + \frac{\frac{1}{2}}{z-1}$$

as well as the coefficient

$$\frac{1}{2}\,\frac{\omega^2 - U''}{U+C} = -\frac{\rho^2}{z^2} + \frac{C_0}{z} + \frac{C_a}{z-a} + \frac{C_1}{z-1}$$

with[3]

$$C_0 = -\frac{1}{a}\left(\frac{\rho^2}{a-1}+3\right),$$
$$C_a = \frac{a\,\rho^2}{a-1} - 3,$$
$$C_1 = \frac{a+1}{a}(3-\rho^2)$$

and

$$a = \frac{1 - 2\,z_{e1} - \sqrt{2\,z_{e1}\,(1-z_{e1})}}{1 - 2\,z_{e1} + \sqrt{2\,z_{e1}\,(1-z_{e1})}},$$
$$\rho = \sqrt{1 - \frac{\omega^2}{\beta^2}}, \tag{5.1.39}$$
$$\beta^2 = 6\,z_{e1}^2 - 6\,z_{e1} + 1$$

with [cf. second equation of (5.1.19)]

$$\beta^2 = \left.\frac{\mathrm{d}^2 U}{\mathrm{d}z^2}\right|_{z_{e1}}$$

for the differential equation (5.1.12). Because

$$\left.\frac{\mathrm{d}^2 U}{\mathrm{d}z^2}\right|_{z_{ei}} = 2\,c_i^2 - \frac{1}{2}, \quad i = 1, 2, 3,$$

with c_i from equation (5.1.21), this may also be written in the form

$$\beta^2 = 2\,c_1^2 - \frac{1}{2} = 2\,\cos^2\left(\frac{\arccos(s) + 2\,\pi}{3}\right) - \frac{1}{2}.$$

The value z_{e1} is the lowest-lying turning point of the Eshelby potential $U(z)$, as well as of the shifted Eshelby potential $U(z) + C$. Thus

$$0 \le \varrho \le 1, \quad \text{real-valued},$$
$$0 \le a \le 1.$$

The resulting differential equation (5.1.12) is

$$\boxed{\frac{\mathrm{d}^2 y}{\mathrm{d}z^2} + \left\{\frac{1}{z} + \frac{1/2}{z-a} + \frac{1/2}{z-1}\right\}\frac{\mathrm{d}y}{\mathrm{d}z} + \left\{-\frac{\rho^2}{z^2} + \frac{C_0}{z} + \frac{C_a}{z-a} + \frac{C_1}{z-1}\right\} y = 0} \tag{5.1.40}$$

or

$$\frac{\mathrm{d}^2 y}{\mathrm{d}z^2} + \left\{\frac{1}{z} + \frac{1/2}{z-a} + \frac{1/2}{z-1}\right\}\frac{\mathrm{d}y}{\mathrm{d}z} + \left\{-\frac{\rho^2\,a}{z} + 3\,(1+a) - 6\,z\right\}\frac{y}{z\,(z-a)\,(z-1)} = 0. \tag{5.1.41}$$

[3] As may be verified, $\Sigma_{i=1}^{3} C_i = 0$.

The differential equation (5.1.40), (5.1.41) is a *Fuchsian differential* equation having four singularities, the conventional Riemann symbol of which is given by

$$P\begin{pmatrix} 1 & 1 & 1 & 1 & \\ 0 & a & 1 & \infty & z,\, 3(1+a) \\ \rho & 0 & 0 & 3 & \\ -\rho & \frac{1}{2} & \frac{1}{2} & -2 & \end{pmatrix}.$$

The s-rank symbol of (5.1.40), (5.1.41) is thus

$$\{1, 1, 1; 1\}$$

and the relevant interval on which this boundary eigenvalue problem is defined is

$$[0, a].$$

The physical problem determines the boundary eigenvalue problem. In order to calculate particular solutions $y(z)$ of (5.1.40), (5.1.41) that describe vibrations of the static solution (5.1.33) (viz. the so-called breather solution), they have to behave like z^ρ at the origin $z = 0$ and, simultaneously, either like $z - a$ at $z = a$ or like $(z - a)^{\frac{1}{2}}$ at $z = a$. The former describe even vibrations, the latter describe odd vibrations in the physical coordinate. Thus, these boundary eigenvalue conditions split the problem into two parts: the one yields the even, the other the odd eigensolutions with respect to the physical coordinate system.

Even Eigensolutions
From the boundary eigenvalue condition, formulated above, it follows that the asymptotic behaviour of the solutions $y(z)$ has to be

$$z^\rho$$

as $z \to 0$ and, simultaneously, it has to be holomorphic at $z = a$.
The transformation

$$y(z) = z^\rho\, w(z)$$

yields the differential equation

$$\frac{d^2w}{dz^2} + \left\{\frac{1+2\rho}{z} + \frac{1/2}{z-a} + \frac{1/2}{z-1}\right\}\frac{dw}{dz} + \frac{(\rho+3)(\rho-2)\,z - q}{z\,(z-a)(z-1)}\,w = 0 \tag{5.1.42}$$

with

$$q = \frac{1}{2}\,(1+a)\,(2\rho-3)\,(\rho+2)$$

or

$$\frac{d^2w}{dz^2} + \left\{\frac{1+2\rho}{z} + \frac{1/2}{z-a} + \frac{1/2}{z-1}\right\}\frac{dw}{dz} + \left(\frac{C_0}{z} + \frac{C_a}{z-a} + \frac{C_1}{z-1}\right) w = 0$$

with

$$C_0 = -\frac{q}{a},$$
$$C_a = \frac{a\,(\rho-2)\,(\rho+3) - q}{a\,(a-1)},$$
$$C_1 = -\frac{(\rho-2)\,(\rho+3) + q}{a-1},$$

the conventional Riemann symbol of which is

$$P\begin{pmatrix} 1 & 1 & 1 & 1 & \\ 0 & a & 1 & \infty & z,\, q \\ 0 & 0 & 0 & \rho+3 & \\ -2\rho & \frac{1}{2} & \frac{1}{2} & \rho-2 & \end{pmatrix}.$$

Then, a power series ansatz

$$w(z) = \sum_{n=0}^{\infty} a_n\, z^n \tag{5.1.43}$$

yields solutions $w(z)$ that are holomorphic at $z = 0$ and, simultaneously, at $z = a$ if its radius of convergence r is larger than $r = a$. This, in turn, means it is unity (since there is no other singularity of the underlying differential equation in between):

$$r = 1.$$

Thus, for a given value of a, it looks for values $\rho = \rho(a)$ such that the series (5.1.43) converges for all complex values of z for which holds

$$|z| < 1.$$

Odd Eigensolutions

The transformation

$$y(z) = z^{\rho}\,(z-a)^{\frac{1}{2}}\,(z-1)^{\frac{1}{2}}\,w(z)$$

yields the differential equation

$$\frac{d^2w}{dz^2} + \left(\frac{1+2\rho}{z} + \frac{3/2}{z-a} + \frac{3/2}{z-1}\right)\frac{dw}{dz} + \frac{(\rho+4)\,(\rho-1)\,z - q}{z\,(z-a)\,(z-1)}\,y = 0 \tag{5.1.44}$$

with

$$q = \frac{1}{2}\,(1+a)\,(2\rho+5)\,(\rho-1)$$

or

$$\frac{d^2w}{dz^2} + \left(\frac{1+2\rho}{z} + \frac{3/2}{z-a} + \frac{3/2}{z-1}\right)\frac{dw}{dz} + \left(\frac{C_0}{z} + \frac{C_a}{z-a} + \frac{C_1}{z-1}\right) w = 0$$

with

$$C_0 = -\frac{q}{a},$$
$$C_a = \frac{a(\rho-1)(\rho+4) - q}{a(a-1)},$$
$$C_1 = -\frac{(\rho-1)(\rho+4) + q}{a-1},$$

the conventional Riemann symbol of which is

$$P\begin{pmatrix} 1 & 1 & 1 & 1 & \\ 0 & a & 1 & \infty & z, q \\ 0 & 0 & 0 & \rho+4 & \\ -2\rho & -\frac{1}{2} & -\frac{1}{2} & \rho-1 & \end{pmatrix}.$$

Then, a power series ansatz

$$w(z) = \sum_{n=0}^{\infty} a_n z^n \tag{5.1.45}$$

yields solutions $w(z)$ that are holomorphic at $z = 0$ and, simultaneously, at $z = a$ if its radius of convergence r is larger than $r = a$. This, in turn, means it is unity (since there is no other singularity of the underlying differential equation in between):

$$r = 1.$$

Thus, for a given value of a, it looks for values $\rho = \rho(a)$ such that the series (5.1.45) converges for all complex values of z for which holds

$$|z| < 1.$$

In both of these cases, the even and the odd ones, the differential equations (5.1.42) and (5.1.44), respectively, have particular solutions $w(z)$ that are holomorphic at the origin $z = 0$ as well as at $z = a$. For the even eigensolutions, the series (5.1.45) is denoted

$$w_e(z) = \sum_{n=0}^{\infty} a_{n,e} z^n \tag{5.1.46}$$

and for the odd ones it is denoted

$$w_o(z) = \sum_{n=0}^{\infty} a_{n,o} z^n. \tag{5.1.47}$$

As discussed already, the radius of convergence r of these power series ansatzes (5.1.42) and (5.1.44) about the origin for the eigensolutions is generally $r = a$. For specific values of $\rho = \rho(a)$ (viz. the eigenvalues) it is enlarged to $r = 1$, viz. to the singularity of the differential equation (5.1.42) and (5.1.44), respectively, that is next to the one located at $z = a$. These solutions have been called *Heun functions* in §3.2.3.

A Jaffé ansatz of the form (3.2.34) (i.e., a power series in this case) in both cases yields a second-order regular difference equation (3.2.36) of Poincaré–Perron type including the initial condition, and thus may be solved recursively:

$$\begin{aligned} \frac{2a}{1+a}(1+2\rho)\,a_1 - (2\rho-3)(\rho+2)\,a_0 = 0, \\ P_n\,a_{n+1} - Q_n\,a_n + R_n\,a_{n+1} = 0, \quad n \geq 1, \end{aligned} \tag{5.1.48}$$

with

$$\begin{aligned} P_n &= 2a(n+1)(n+1+2\rho) \\ &= 2a + \frac{4a(\rho+1)}{n} + \frac{2a(2\rho+1)}{n^2}, \\ Q_n &= (1+a)(2n+2\rho-3)(n+\rho+2) \\ &= 2(a+1) + \frac{(4\rho+1)(a+1)}{n} + \frac{(a+1)(\rho+2)(2\rho-3)}{n^2}, \\ R_n &= 2(n+\rho-3)(n+\rho+2) \\ &= 2 + \frac{2(2\rho-1)}{n} + \frac{2(\rho+2)(\rho-3)}{n^2} \end{aligned}$$

in the even case and

$$\begin{aligned} 2a(1+2\varrho)\,a_1/(1+a) - (2\varrho+5)(\varrho-1)\,a_0 \\ = 0, P_n\,a_(n-1) - Q_n\,a_n + R_n\,a_(n+1) = 0,\ n \leq 1 \end{aligned} \tag{5.1.49}$$

with

$$\begin{aligned} P_n &= 2a(n+1)(n+1+2\rho) \\ &= 2a + \frac{4a(\rho+1)}{n} + \frac{2a(2\rho+1)}{n^2}, \\ Q_n &= (1+a)(2n+2\rho+5)(n+\rho-1) \\ &= 2(a+1) + \frac{(a+1)(4\rho+3)}{n} + \frac{(a+1)(\rho-1)(2\rho+5)}{n^2}, \\ R_n &= 2(n+\rho+3)(n+\rho-2) \\ &= 2 + \frac{2(2\rho+1)}{n} + \frac{2(\rho-2)(\rho+3)}{n^2} \end{aligned}$$

in the odd case.

As appears from the theory, the eigenvalue curves $\rho = \rho(a)$ are given by the condition (4.2.22), which is an infinite continued fraction. A numerical evaluation results in four eigenvalue curves $\rho = \rho_i(a)$, $i = 1, 2, 3, 4$, that are given in Figure 5.14. Here, 'EV' stands for 'eigenvalue', meaning ρ and the relation to the frequency ω is given by the second equation of (5.1.39).

While the eigenvalue curve $\rho = \rho_1(a)$ indicates the mode of decay of the unstable static solution (5.1.22) and the trivial eigenvalue curve $\rho = \rho_2(a) \equiv 0$ the sliding of

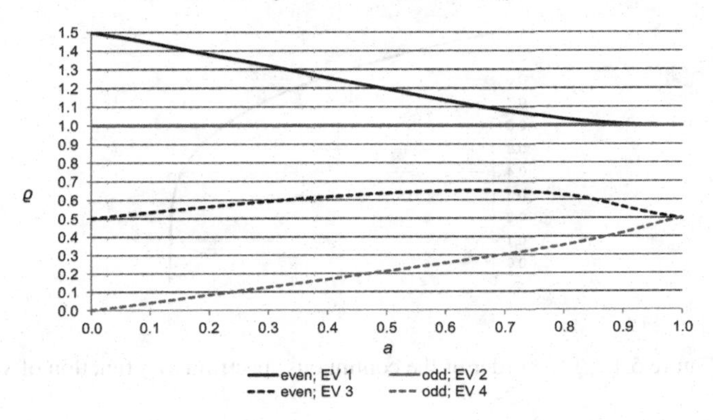

Figure 5.14 Discrete boundary eigenvalue spectrum of a vibrating string.

this solution along the Eshelby potential V from (5.1.38), viz. a transition mode, there are two localised vibrations of the static solution, namely $\rho = \rho_3(a)$ and $\rho = \rho_4(a)$, the former related to a symmetrical vibration with respect to the physical coordinate $x = 0$, the latter related to an asymmetrical one.

It is mentioned here that these four discrete eigenvalue curves may be expanded in series about $a = 0$:

$$\begin{aligned}
\rho_1(a) &= \frac{4}{35}\,\xi + \frac{4\,5471}{9\,(35)^2}\,\xi^2 + \frac{295157028}{9^2\,(35)^5}\,\xi^3 + \cdots,\\
\rho_2(a) &= 1,\\
\rho_3(a) &= \frac{1}{2} + \frac{3}{35}\,\xi + \frac{1429}{5^3\,7^3}\,\xi^2 + \frac{70715377}{2^3\,5^5\,7^5\,11}\,\xi^3 + \cdots,\\
\rho_4(a) &= \frac{3}{2} - \frac{1}{7}\,\xi - \frac{347}{2^2\,3^2\,7^3}\,\xi^2 - \frac{425871}{2^4\,3\,7^5\,11}\,\xi^3 + \cdots,
\end{aligned}$$

with $\xi = \frac{4\,a}{(1+a)^2}$.

Eventually, there is a continuous spectrum for frequencies ω larger than β, thus above the curve (cf. Figure 5.15)

$$\beta = \beta(s) = \sqrt{6\,z_{e1}^2 - 6\,z_{e1} + 1},$$

with z_{e1} from (5.1.20), (5.1.21). The condition for this curve is given by [cf. second equation of (5.1.39)]

$$\varrho = \sqrt{1 - \frac{\omega^2}{\beta^2}} = 0.$$

The Case of Maximum Stress

Carrying out the confluence process $a \to 0$ in equation (5.1.40), viz. the asymmetry parameter s tends to the maximum $s = 1$ (i.e., the maximum value of the stress), results in the static solution (5.1.34) that is a constant $u_0(x) = z = \frac{1}{2}$, meaning that

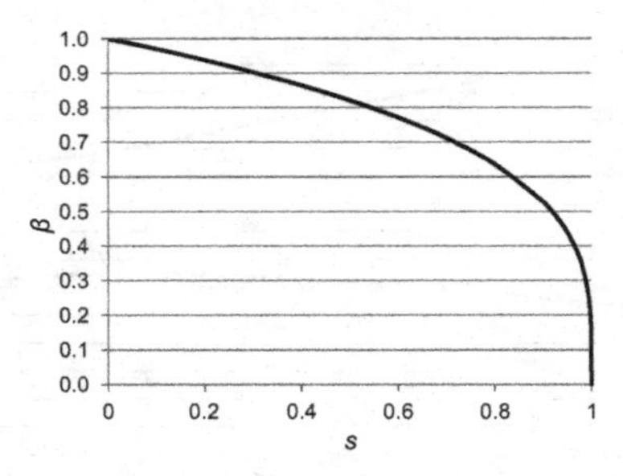

Figure 5.15 The border of the continuous spectrum as a function of s.

the string is positioned in an unstable straight configuration. The differential equation (5.1.40), being of Fuchsian type, then degenerates to having three singularities:

$$\frac{d^2y}{dz^2} + \left\{\frac{A_{z1}}{z} + \frac{A_{z2}}{z-1}\right\}\frac{dy}{dz} + \left(\frac{B_{z1}}{z^2} + \frac{C}{z} - \frac{C}{z-1}\right) y = 0 \tag{5.1.50}$$

with

$$\begin{aligned} A_{z1} &= \frac{3}{2}, \\ A_{z2} &= \frac{1}{2}, \\ B_{z_1} &= -3, \\ C &= 3. \end{aligned}$$

The Riemann symbol of this differential equation is

$$P\begin{pmatrix} 1 & 1 & 1 & \\ 0 & 1 & \infty & z \\ \frac{3}{2} & 0 & +3 & \\ -2 & \frac{1}{2} & -2 & \end{pmatrix}$$

and its s-rank symbol is thus

$$\{1, 1; 1\}.$$

It is easily seen that not only the parameter a disappears, but also the eigenvalue parameter ρ under the confluence process, so that the resulting differential equation (5.1.50) does not have parameters any more.

In this special case, in order to get a perspicuous representation, it is necessary to resort to the normal form (5.1.10) of (5.1.12):

$$\frac{d^2y}{dx^2} + \left(\omega^2 - \frac{d^2U}{dz^2}\right) y = 0 \tag{5.1.51}$$

with (5.1.24)

$$\left.\frac{d^2U}{dz^2}\right|_{z=\frac{1+\sqrt{3}}{2\sqrt{3}}} = 6\,z^2 - 6\,z + 1\Big|_{z=\frac{1+\sqrt{3}}{2\sqrt{3}}} = 0. \tag{5.1.52}$$

The general solution of (5.1.51) with (5.1.52) is given by

$$y(x) = a\,\sin(\omega\,z) + b\,\cos(\omega\,z), \tag{5.1.53}$$

where a and b are any constants in z.

The Case of Vanishing Stress

Carrying out the confluence process $a \to 1$ in equation (5.1.37), viz. the asymmetry parameter s tends to the minimum $s = 0$ (i.e., the value of vanishing stress), results in the static solution (5.1.31) that is the kink solution, meaning that the string is positioned in a stable transition configuration from one valley of the Eshelby potential to a neighbouring one. The differential equation (5.1.37), being of Fuchsian type, then degenerates to having three singularities:

$$\frac{d^2y}{dz^2} + \left[\frac{1}{z} + \frac{1}{z-1}\right]\frac{dy}{dz} + \frac{1}{z\,(z-1)}\left[\frac{\rho^2}{z} - \frac{\rho^2}{z-1} - 6\right]y = 0$$

or

$$\frac{d^2y}{dz^2} + \left[\frac{1}{z} + \frac{1}{z-1}\right]\frac{dy}{dz} + \left[-\frac{\rho^2}{z^2} - \frac{\rho^2}{(z-1)^2} + \frac{2\,(3-\rho^2)}{z} + \frac{2\,(\rho^2-3)}{z-1}\right]y = 0,$$

the conventional Riemann symbol of which is

$$P\begin{pmatrix} 1 & 1 & 1 & \\ 0 & 1 & \infty & z \\ \rho & \rho & +3 & \\ -\rho & -\rho & -2 & \end{pmatrix}.$$

This differential equation describes the oscillations of the kink solution (5.1.31) in the case of vanishing stress. The result is two discrete eigenvalues, namely

$$\varrho_1 = 1,$$
$$\varrho_2 = \frac{1}{2}$$

or

$$\omega_1^2 = 0,$$
$$\omega_2^2 = \frac{3}{4}.$$

5.1.3 Periodic Solutions: Arscott

Consider the general static solutions (5.1.29) of equation (5.1.2). These are represented by the inner points of the characteristic triangle as sketched in Figure 5.4. In the following, the vibrations of these static solutions are discussed. According to §5.1.1, these oscillations are described by the differential equation (5.1.12)

$$\frac{\mathrm{d}^2 y}{\mathrm{d}z^2} + P(z)\,\frac{\mathrm{d}y}{\mathrm{d}z} + Q(z)\,y = 0, \tag{5.1.54}$$

the coefficients $P(z)$ and $Q(z)$ of which are

$$P(z) = \frac{U'(z)}{2\,[U(z) + C]} \tag{5.1.55}$$

and

$$Q(z) = \frac{\omega^2 - U''(z)}{2\,[U(z) + C]}. \tag{5.1.56}$$

The function $U(z) + C$ in the denominators is given by the shifted Eshelby potential (5.1.14)

$$\begin{aligned} U(z) + C &= \frac{1}{2}\,z^2\,(1 - z)^2 - \frac{s}{6\sqrt{3}}\,z + C \\ &= \frac{1}{2}\,(z - z_1)\,(z - z_2)\,(z - z_3)\,(z - z_4) \end{aligned}$$

having exclusively simple zeros, which may thus be ordered according to

$$z_1 < z_2 < z_3 < z_4.$$

As a result,

$$\begin{aligned} 2\,[U(z) + C] &= z^2\,(1 - z)^2 - a\,z + K \\ &= z^4 - 2\,z^3 + z^2 - a\,z + 2\,C \\ &= (z - z_1)\,(z - z_2)\,(z - z_3)\,(z - z_4) \end{aligned}$$

with

$$a = \frac{s}{3\sqrt{3}}, \qquad K = 2\,C.$$

The first derivative of the Eshelby potential is given by

$$\begin{aligned} U'(z) &= \frac{\mathrm{d}U}{\mathrm{d}z} = z\,(z - 1)\,(2\,z - 1) - 2\,a \\ &= 2\,z^3 - 3\,z^2 + z - 2\,a \\ &= 2\,(z - z_{e1})\,(z - z_{e2})\,(z - z_{e3}) \end{aligned}$$

where all the extreme values z_{e1}, z_{e2}, z_{e3} are real and may thus be ordered according to

$$z_{e1} < z_{e2} < z_{e3}.$$

The second derivative is

$$U''(z) = \frac{dU}{dz} = 6\,z^2 - 6\,z + 1 = 6\,z\,(z-1) + 1.$$

Thus, the coefficients $P(z)$ and $Q(z)$ of the differential equation (5.1.12)

$$\frac{d^2y}{dz^2} + P(z)\,\frac{dy}{dz} + Q(z)\,y = 0 \tag{5.1.57}$$

may be written in the form

$$\begin{aligned} P(z) &= \frac{U'(z)}{2[U(z)+C]} \\ &= \frac{2\,(z-z_{e1})\,(z-z_{e2})\,(z-z_{e3})}{(z-z_1)\,(z-z_2)\,(z-z_3)\,(z-z_4)}, \\ Q(z) &= \frac{\omega^2 - U''(z)}{2[U(z)+C]} \\ &= \frac{\omega^2 - 1 - 6\,z\,(z-1)}{(z-z_1)\,(z-z_2)\,(z-z_3)\,(z-z_4)}, \end{aligned} \tag{5.1.58}$$

where $U(z)$ is the Eshelby potential (5.1.14).

Generally, such a differential equation has the following Riemann P-symbol:

$$P\begin{pmatrix} 1 & 1 & 1 & 1 & 1 & & & \\ z_1 & z_2 & z_3 & z_4 & \infty & & & \\ 0 & 0 & 0 & 0 & \alpha_{\infty_1}; & z; & A; & B \\ \alpha_{z_1} & \alpha_{z_2} & \alpha_{z_3} & \alpha_{z_4} & \alpha_{\infty_2} & & & \end{pmatrix},$$

where A and B are accessory parameters. This is a Fuchsian differential equation having five singularities (and two accessory parameters).

Before these expressions are broken into partial fractions, a Moebius transformation (1.2.26) is applied which puts z_1 to $x = -a$, z_2 to $x = -1$, z_3 to $x = 0$, z_4 to $x = b$ and leaves the point at infinity there. This means that the zeros of the shifted Eshelby potential are shifted according to the following tabular:

$$\begin{array}{ccccccc} z: & z_1 & z_2 & z_{e2} & z_3 & z_4 & \infty \\ \downarrow & \downarrow & \downarrow & \downarrow & \downarrow & \downarrow & \downarrow \\ x: & -a & -b & 0 & 1 & c & \infty \end{array}$$

with

$$\begin{aligned} a &= \frac{z_1 - z_{e2}}{z_3 - z_{e2}} \geq 1, \\ b &= \frac{z_2 - z_{e2}}{z_3 - z_{e2}} \leq 1, \\ c &= \frac{z_4 - z_{e2}}{z_3 - z_{e2}} \geq 1. \end{aligned}$$

It should be recognised that $a \geq 1$, $0 \leq b \leq 1$ and $c \geq 1$ hold.

In order to carry out this shifting, apply the Moebius transformation (1.2.26) having the concrete form

$$x(z) = \frac{z - z_{e2}}{z_3 - z_{e2}} \rightsquigarrow z(x) = x\,(z_3 - z_{e2}) + z_{e2}. \tag{5.1.59}$$

From (5.1.59) follow the quantities that are needed to transform the differential equation (5.1.54)–(5.1.56) from the independent variable z to x:

$$\frac{\mathrm{d}x}{\mathrm{d}z} = \frac{1}{z_3 - z_{e2}},$$

$$\frac{\mathrm{d}^2 x}{\mathrm{d}z^2} = 0,$$

$$\frac{1}{\dfrac{\mathrm{d}x}{\mathrm{d}z}} = z_3 - z_{e2},$$

$$\frac{\dfrac{\mathrm{d}^2 x}{\mathrm{d}z^2}}{\left(\dfrac{\mathrm{d}x}{\mathrm{d}z}\right)^2} = 0.$$

The transformed differential equation is of the same form as (5.1.57), (5.1.58):

$$\frac{\mathrm{d}^2 y}{\mathrm{d}x^2} + \tilde{P}(x)\,\frac{\mathrm{d}y}{\mathrm{d}x} + \tilde{Q}(x)\,y = 0. \tag{5.1.60}$$

Its coefficients $\tilde{P}(z)$ and $\tilde{Q}(z)$, however, are given by

$$\tilde{P} = \frac{P}{\dfrac{\mathrm{d}x}{\mathrm{d}z}} = P\,(z_3 - z_{e2}) \tag{5.1.61}$$

and

$$\tilde{Q} = \frac{Q}{\left(\dfrac{\mathrm{d}x}{\mathrm{d}z}\right)^2} = Q\,(z_3 - z_{e2})^2. \tag{5.1.62}$$

With the help of a computer algebra program, the transformation (5.1.59) turns out to be

$$\tilde{P}(z(x)) = \tilde{P}(x) = \frac{A_a}{x + a} + \frac{A_b}{x + b} + \frac{A_1}{x - 1} + \frac{A_c}{x - c}$$

with

$$A_a = 2\,\frac{z_1^3 - z_1^2\,(z_{e1} + z_{e2} + z_{e3}) + z_1\,[z_{e1}\,(z_{e2} + z_{e3}) + z_{e2}\,z_{e3}] - z_{e1}\,z_{e2}\,z_{e3}}{z_1^3 - z_1^2\,(z_1 + z_2 + z_3) + z_1\,[z_2\,(z_3 + z_4) + z_3\,z_4] - z_{e1}\,z_{e2}\,z_{e3}},$$

$$A_b = 2\,\frac{z_2^3 - z_2^2\,(z_{e1} + z_{e2} + z_{e3}) + z_2\,[z_{e1}\,(z_{e2} + z_{e3}) + z_{e2}\,z_{e3}] - z_{e1}\,z_{e2}\,z_{e3}}{(z_2 - z_1)\,[z_2^2 - z_2\,(z_3 + z_4) + z_3\,z_4]},$$

$$A_1 = 2\,\frac{z_3^3 - z_3^2\,(z_{e1} + z_{e2} + z_{e3}) + z_3\,[z_{e1}\,(z_{e2} + z_{e3}) + z_{e2}\,z_{e3}] - z_{e1}\,z_{e2}\,z_{e3}}{(z_1 - z_3)\,(z_2 - z_3)\,(z_3 - z_4)},$$

$$A_c = 2\,\frac{(z_{e3} - z_4)\,[z_4^2 - z_4\,(z_{e1} + z_{e2}) + z_{e1}\,z_{e2}]}{(z_1 - z_4)\,(z_2 - z_4)\,(z_3 - z_4)},$$

and (5.1.62) turns out to be

$$\tilde{Q}(z(x)) = \tilde{Q}(x) = \frac{C_a}{x + a} + \frac{C_b}{x + b} + \frac{C_1}{x - 1} + \frac{C_c}{x - c}$$

where the coefficients C_a, C_b, C_1, C_c are

$$C_a = \frac{(z_3 - z_{e2})\,\left[\omega^2 - 1 - 6\,z_1\,(z_1 - 1)\right]}{z_1^3 - z_1^2\,[z_2 + z_3 + z_4] + z_1\,[z_2\,(z_3 + z_4) + z_3\,z_4] + z_2\,z_3\,z_4)},$$

$$C_b = \frac{(z_{e2} - z_3)\,[\omega^2 - 1 - 6\,z_2\,(z_2 + 1)]}{(z_1 - z_2)\,[z_2^2 - z_2\,(z_3 + z_4) + z_3\,z_4]},$$

$$C_1 = \frac{(z_3 - z_{e2})\,[\omega^2 - 1 - 6\,z_3\,(z_3 + 1)]}{(z_1 - z_3)\,(z_2 - z_3)\,(z_3 - z_4)},$$

$$C_c = \frac{(z_{e2} - z_3)\,[\omega^2 - 1 - 6\,z_4\,(z_4 - 1)]}{(z_1 - z_4)\,(z_2 - z_4)\,(z_3 - z_4)}.$$

As may also be shown with the help of a computer algebra program,

$$C_a + C_b + C_1 + C_c = 0 \tag{5.1.63}$$

holds, meaning that the singularity at infinity stays regular under the transformation (5.1.59). Moreover

$$A_a + A_{-1} + A_0 + A_b = 2.$$

The Riemann P-symbol is

$$P\begin{pmatrix} 1 & 1 & 1 & 1 & 1 & & & \\ a & b & 1 & c & \infty & & & \\ 0 & 0 & 0 & 0 & \alpha_{\infty_1}; & z; & A; & B \\ \frac{1}{2} & \frac{1}{2} & \frac{1}{2} & \frac{1}{2} & \alpha_{\infty_2} & & & \end{pmatrix},$$

where A and B are the accessory parameters

$$\begin{aligned} A &= a\,[b\,(C_1 + C_c) - c\,(C_1 + C_b) - C_b - C_c] - b\,[c\,(C_1 + C_a) + C_a + C_c] \\ &\quad + c\,(C_a + C_b), \\ B &= a\,[c\,C_b - b\,(c\,C_1 + C_c)] + b\,c\,C_a. \end{aligned}$$

The reason for carrying out the Moebius transformation (5.1.59) is that the eigensolutions of the problems may be written in terms of a Taylor series

$$y(x) = \sum_{n=0}^{\infty} a_n\, x^n.$$

Inserting this ansatz into the differential equation (5.1.60)–(5.1.62) allows us to convert it into a regular difference equation of Poincaré–Perron type of order three. This is done in the following.

First, the whole differential equation (5.1.60)–(5.1.62) is multiplied by

$$(x+a)\,(x+b)\,(x-1)\,(x+c)$$

to give

$$(x+a)\,(x+b)\,(x-1)\,(x+c) = x^4 + E_3\,x^3 + E_2\,x^2 + E_1\,x + E_0$$

with

$$\begin{aligned}
E_3 &= a+b-c-1,\\
E_2 &= a\,(b-c-1) - b\,(c+1) + c,\\
E_1 &= b\,c - a\,b\,(c+1) - c,\\
E_0 &= a\,b\,c.
\end{aligned}$$

The coefficient in front of the first derivative is

$$\left(\frac{A_a}{x+a} + \frac{A_b}{x+b} + \frac{A_1}{x-1} + \frac{A_c}{x-c}\right)(x+a)\,(x+b)\,(x-1)\,(x+c)$$
$$= D_3\,x^3 + D_2\,x^2 + D_1\,x + D_0$$

with

$$\begin{aligned}
D_3 &= A_1 + A_a + A_b + A_c = 2,\\
D_2 &= a\,[A_1 + A_b + A_c] + A_1\,(b-c) + A_a\,[b-c-1] - A_b\,(c+1) + A_c\,(b-1),\\
D_1 &= a\,[A_1\,(b-c) - A_b\,(c+1) + A_c\,(b-1)] - A_1\,b\,c + A_a\,[c - b\,(c+1)]\\
&\quad + A_b\,c - A_c\,b,\\
D_0 &= -a\,(A_1\,b\,c - A_b\,c + A_c b) + A_a\,b\,c.
\end{aligned}$$

The coefficient in front of $y(x)$ is

$$\left(\frac{C_a}{x+a} + \frac{C_b}{x+b} + \frac{C_1}{x-1} + \frac{C_c}{x-c}\right)(x+a)\,(x+b)\,(x-1)\,(x+c)$$
$$= F_3\,x^3 + F_2\,x^2 + F_1\,x + F_0$$

with

$$
\begin{aligned}
F_3 &= C_1 + C_a + C_b + C_c = 0, \\
F_2 &= a\,(C_1 + C_b + C_c) + b\,(C_1 + C_a + C_c) - c\,(C_1 + C_a + C_b) - (C_a + C_b + C_c), \\
F_1 &= a\,[b\,(C_1 + C_c) - c\,(C_1 + C_b) - C_b - C_c] - b\,[c\,(C_1 + C_a) + C_a + C_c] \\
&\quad + c\,(C_a + C_b), \\
F_0 &= a\,[C_b\,c - b\,(C_1\,c + C_c) + C_c\,b] + C_a\,b\,c.
\end{aligned}
$$

Inserting these formulae into the differential equation results in

$$
\begin{aligned}
&\left[x^4 + E_3\,x^3 + E_2\,x^2 + E_1\,x + E_0\right] \sum_{n=0}^{\infty} a_n\,n\,(n-1)\,x^{n-2} \\
&\left[D_3\,x^3 + D_2\,x^2 + D_1\,x + D_0\right] \sum_{n=0}^{\infty} a_n\,n\,x^{n-1} \\
&\left[F_2\,x^2 + F_1\,x + F_0\right] \sum_{n=0}^{\infty} a_n\,x^n = 0
\end{aligned}
$$

or

$$
\begin{aligned}
&\sum_{n=0}^{\infty} a_n\,n\,(n-1)\,x^{n+2} + E_3 \sum_{n=0}^{\infty} a_n\,n\,(n-1)\,x^{n+1} + E_2 \sum_{n=0}^{\infty} a_n\,n\,(n-1)\,x^{n} \\
&\qquad + E_1 \sum_{n=0}^{\infty} a_n\,n\,(n-1)\,x^{n-1} + E_0 \sum_{n=0}^{\infty} a_n\,n\,(n-1)\,x^{n-2} \\
&2 \sum_{n=0}^{\infty} a_n\,n\,x^{n+2} + D_2 \sum_{n=0}^{\infty} a_n\,n\,x^{n+1} + D_1 \sum_{n=0}^{\infty} a_n\,n\,x^{n} + D_0 \sum_{n=0}^{\infty} a_n\,n\,x^{n-1} \\
&+ F_2 \sum_{n=0}^{\infty} a_n\,x^{n+1} + F_1 \sum_{n=0}^{\infty} a_n\,x^{n} + F_0 \sum_{n=0}^{\infty} a_n\,x^{n-1} = 0, \quad n = 0, 1, 2, 3, \ldots,
\end{aligned}
$$

or

$$
\begin{aligned}
&a_2\,2\cdot 1\,x^4 + a_3\,3\cdot 2\,x^5 + \cdots \\
\\
&+ E_3\,a_2\,2\cdot 1\,x^3 + E_3\,a_3\,3\cdot 2\,x^4 + \cdots \\
&+ E_2\,a_2\,2\cdot 1\,x^2 + E_2\,a_3\,3\cdot 2\,x^3 + E_2\,a_4\,4\cdot 3\,x^4 + \cdots \\
&+ E_1\,a_2\,2\cdot 1\,x^1 + E_1\,a_3\,3\cdot 2\,x^2 + E_1\,a_4\,4\cdot 3\,x^3 + E_1\,a_5\,5\cdot 4\,x^4 + \cdots \\
&+ E_0\,a_2\,2\cdot 1\,x^0 + E_0\,a_3\,3\cdot 2\,x^1 + E_0\,a_4\,4\cdot 3\,x^2 + E_0\,a_5\,5\cdot 4\,x^3 + E_0\,a_6\,6\cdot 5\,x^4 + \cdots \\
\\
&+ 2\,a_1\,1\,x^3 + 2\,a_2\,2\,x^4 + 2\,a_3\,3\,x^5 + \cdots \\
&+ D_2\,a_1\,1\,x^2 + D_2\,a_2\,2\,x^3 + D_2\,a_3\,3\,x^4 + \cdots \\
&+ D_1\,a_1\,1\,x^1 + D_1\,a_2\,2\,x^2 + D_1\,a_3\,3\,x^3 + D_1\,a_4\,4\,x^4 + \cdots \\
&+ D_0\,a_1\,1\,x^0 + D_0\,a_2\,2\,x^1 + D_0\,a_3\,3\,x^2 + D_0\,a_4\,4\,x^3 + D_0\,a_5\,5\,x^4 + \cdots
\end{aligned}
$$

$$
\begin{aligned}
&+ F_2\, a_0\, x^2 + F_2\, a_1\, x^3 + F_2\, a_2\, x^4 + \cdots \\
&+ F_1\, a_0\, x^1 + F_1\, a_1\, x^2 + F_1\, a_2\, x^3 + F_1\, a_3\, x^4 + \cdots \\
&+ F_0\, a_0\, x^0 + F_0\, a_1\, x^1 + F_0\, a_2\, x^2 + F_0\, a_3\, x^3 + F_0\, a_4\, x^4 \cdots \\
&\qquad = 0.
\end{aligned}
$$

A decomposition into powers of x yields

$$
\begin{aligned}
x^0 &: F_0\, a_0 + D_0\, 1\, a_1 + E_0\, a_2\, 2 \cdot 1 = 0, \\
x^1 &: F_1\, a_0 + (F_0 + D_1\, 1)\, a_1 + (D_0\, 2 + E_1\, 2 \cdot 1)\, a_2 + E_0\, 3 \cdot 2\, a_3 = 0, \\
x^2 &: F_2\, a_0 + (F_1 + D_2\, 1)\, a_1 + (F_0 + D_1\, 2 + E_2\, 2 \cdot 1)\, a_2 \\
&\qquad + (D_0\, 3 + E_1\, 3 \cdot 2)\, a_3 + E_0\, 4 \cdot 3\, a_4 = 0, \\
x^3 &: (F_2 + 2 \cdot 1)\, a_1 + (F_1 + D_2\, 2 + E_3\, 2 \cdot 1)\, a_2 + (F_0 + D_1\, 3 + E_2\, 3 \cdot 2)\, a_3 \\
&\qquad + (4D_0 + E_0\, 5 \cdot 4)\, a_4 + (4D_0 + E_0\, 5 \cdot 4)\, a_5 = 0, \\
x^n &: [F_2 + 2\,(n-2) + (n-2)\,(n-3)]\, a_{n-2} \\
&\qquad + [F_1 + (n-1)\, D_2 + (n-1)\,(n-2)\, E_3]\, a_{n-1} + [F_0 + n\, D_1 + n\,(n-1)\, E_2]\, a_n \\
&\qquad + (n+1)\,(D_0 + n\, E_1)\, a_{n+1} + (n+2)\,(n+1)\, E_0\, a_{n+2} = 0, \quad n = 4, 5, 6 \ldots,
\end{aligned}
$$

or

$$
\begin{aligned}
&F_0\, a_0 + D_0\, a_1 + 2\, E_0\, a_2 = 0, \\
&F_1\, a_0 + (F_0 + D_1)\, a_1 + 2\,(D_0 + E_1)\, a_2 + 6\, E_0\, a_3 = 0, \\
&F_2\, a_0 + (F_1 + D_2)\, a_1 + (F_0 + 2\, D_1 + 2\, E_2)\, a_2 + (D_0\, 3 + 6\, E_1)\, a_3 + 12\, E_0\, a_4 = 0, \\
&(F_2 + 2)\, a_1 + (F_1 + 2\, D_2 + 2\, E_3)\, a_2 + (F_0 + 3\, D_1 + 6\, E_2)\, a_3 + 20\, E_0\, a_4 = 0,
\end{aligned}
$$

$$
\begin{aligned}
&[F_2 + 2\,(n-2) + (n-2)\,(n-3)]\, a_{n-2} + [F_1 + (n-1)\, D_2 + (n-1)\,(n-2)\, E_3]\, a_{n-1} \\
&\qquad + [F_0 + n\, D_1 + n\,(n-1)\, E_2]\, a_n + (n+1)\,(D_0 + n\, E_1)\, a_{n+1} \\
&\qquad + (n+2)\,(n+1)\, E_0\, a_{n+2} = 0, \quad n = 4, 5, 6 \ldots.
\end{aligned}
$$

Thus, the result is the difference equation

$$
\begin{aligned}
&\left(\kappa_2 + \frac{\alpha_2}{n} + \frac{\beta_2}{n^2}\right) a_{n+2} + \left(\kappa_1 + \frac{\alpha_1}{n} + \frac{\beta_1}{n^2}\right) a_{n+1} + \\
&\left(\kappa_0 + \frac{\alpha_0}{n} + \frac{\beta_0}{n^2}\right) a_n + \left(\kappa_{-1} + \frac{\alpha_{-1}}{n} + \frac{\beta_{-1}}{n^2}\right) a_{n-1} + \left(\kappa_{-2} + \frac{\alpha_{-2}}{n} + \frac{\beta_{-2}}{n^2}\right) a_{n-2} = 0
\end{aligned}
\tag{5.1.64}
$$

with

$$
\begin{aligned}
\kappa_2 &= E_0, \quad \alpha_2 = 3\, E_0, \quad \beta_2 = 2\, E_0, \\
\kappa_1 &= E_1, \quad \alpha_1 = D_0 + E_1, \quad \beta_1 = D_0, \\
\kappa_0 &= 0, \quad \alpha_0 = D_1 + E_2, \quad \beta_0 = F_0 - E_2,
\end{aligned}
$$

$$\kappa_{-1} = E_3, \quad \alpha_{-1} = D_2 - 3\,E_3, \quad \beta_{-1} = -(D_2 - 2\,E_3 - F_1),$$
$$\kappa_{-2} = 1, \quad \alpha_{-2} = -3, \quad \beta_{-2} = F_2 + 2.$$

Because z_{e2} is mapped to $x = 0$ by means of (5.1.59), the totality of the eigenfunctions decomposes into even and odd solutions. These are calculated via (5.1.64) by means of the initial conditions $a_0 = 0,\ a_1 = 1$ and $a_0 = 1,\ a_1 = 0$, respectively. The eigenvalue condition in both cases is

$$\lim_{n\to\infty} \frac{a_n}{a_n^{(l)}} = 0,$$

where $\{a_n^{(l)}\}$ is a linearly independent solution of (5.1.64) with respect to $\{a_n\}$. This independent solution may be obtained by means of a linearly independent initial condition, for example,

$$a_0 = 0,\ a_1 = 1;\ a_0^{(l)} = 1,\ a_1^{(l)} = 0.$$

It is possible – instead of (5.1.59) – to apply a Moebius transformation, such that the relevant interval $[z_2, z_3]$ is put onto the x-interval $[0, 1]$. The disadvantage, however, of such a procedure is that the central two-point connection problem cannot be solved in the whole parameter space, i.e., within the whole characteristic triangle as shown in Figure 5.4.

The numerical calculation indicates that the problem has an infinite discrete spectrum of eigenvalue curves. Figures 5.16–5.19 show the four lowest eigenvalue curves of the vibrating string in the Eshelby potential for several values of s ('S' stands for symmetric and 'A' stands for antisymmetric and the number indicates the energy level, i.e., the number of knots of the eigensolutions).

Note that the equation (5.1.63) may not serve as an eigenvalue condition, since the coefficients C_a, C_0, C_1, C_b, although dependent on the parameters s and C as

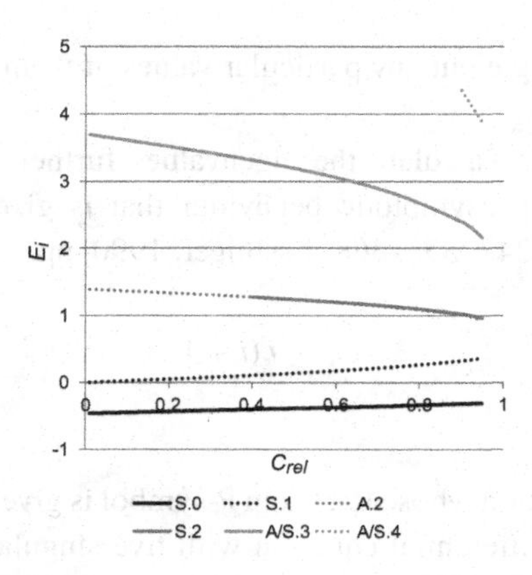

Figure 5.16 The eigenvalue curves of the vibrating string in a quartic potential for $s = 0.4$.

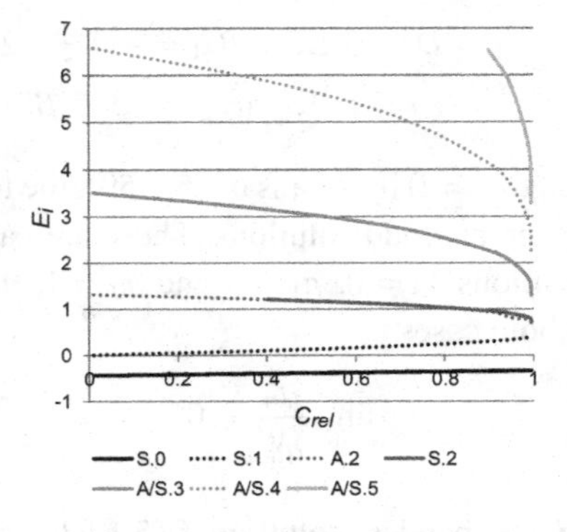

Figure 5.17 The eigenvalue curves of the vibrating string in a quartic potential for $s = 0.5$.

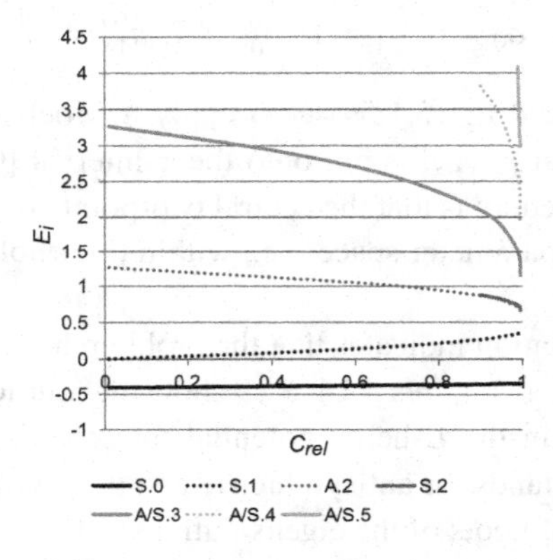

Figure 5.18 The eigenvalue curves of the vibrating string in a quartic potential for $s = 0.6$.

well as on ω, do not single out any particular values of them, since it holds for all of them.

It is not necessary to calculate the eigenvalues further than $i = 4$, since their behaviours flow into an asymptotic behaviour that is given by (cf. Courant and Hilbert, 1968, pp. 111, 247, 353–368; Esslinger, 1990, pp. 97, 98)

$$\omega_i^2 \sim i\,(i + 1)$$

as $i \to \infty$.

The differential equation whose Riemann P-symbol is given by (5.1.61) is a special form of the Fuchsian differential equation with five singularities. As already mentioned above, I refer to the Fuchsian differential equation having five singularities as the *Arscott differential equation*, since it was Felix Medland Arscott who seems to

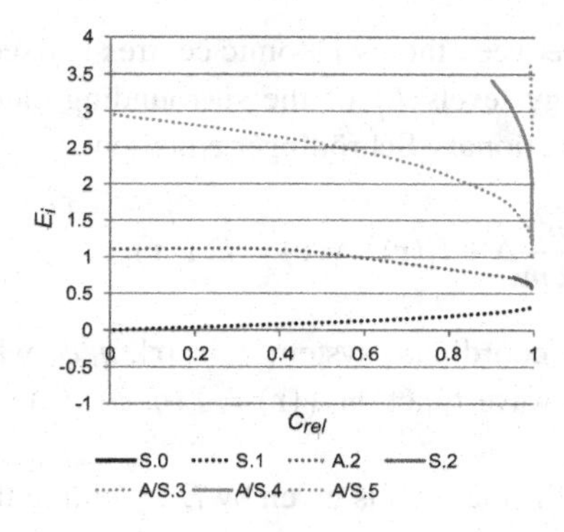

Figure 5.19 The eigenvalue curves of the vibrating string in a quartic potential for $s = 0.7$.

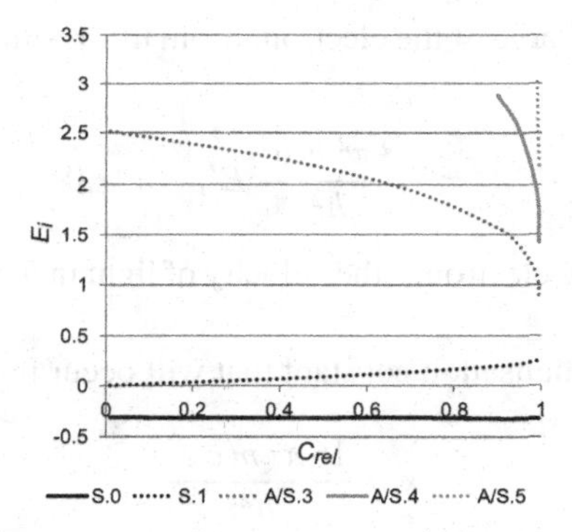

Figure 5.20 The eigenvalue curves of the vibrating string in a quartic potential for $s = 0.8$.

have treated this differential equation, as well as some of its confluent cases, for the first time, for instance in Arscott et al. (1983).

Moreover, I refer to all the confluent and reduced cases of the Fuchsian differential equation with five singularities as the *class of Arscott differential equations.*

5.2 Hydrogen Molecule Ion: Single Confluent Heun

The hydrogen molecule ion consists of two hydrogen atomic kernels and one electron that brings them together, forming a two-atomic molecule. In the early twentieth century the scientific questions were virulent: whether such a constitution can be stable and what the energy levels of such a molecule are. In the following calculation an affirmative answer to these questions is given, as shown by Jaffé (1933) by means of a sophisticated quantum calculation.

Suppose that the distance between the two atomic centres is fixed at r_{12} (and thus is not dependent on the energy levels E_i' of the surrounding electron). Then, the described three-dimensional stationary Schrödinger equation

$$\left[-\frac{\hbar^2}{2m}\Delta + U(\mathbf{r})\right] \chi(\mathbf{r}) = E\,\chi(\mathbf{r})$$

is separable in an ellipsoidal coordinate system $z = |\mathbf{r}|, \eta, \phi$, where the angular-dependent part $Y(\eta, \phi)$ of the wave function $\chi(\mathbf{r}) = Y(\eta, \phi)\, y(z)$ is not focussed on here, but the radial $y(z)$.

Suppose that the energy of the electron is given by E', yielding the total energy of the molecule as

$$E = E' - \frac{e^2}{r_{12}},$$

where e is the electric charge of the electron. From this results the normalised energy parameter

$$\lambda^2 = -\frac{8\,\pi^2\, m\, c^2}{h^2}\, E', \quad \lambda \geq 0,$$

with m as the mass of the electron, c the velocity of light in the vacuum and h Planck's constant.

There is another normalisation constant that will occur in the quantum equations:

$$\kappa = \frac{16\,\pi^2\, m\, c\, e^2}{h^2}.$$

Since this is a separable three-dimensional problem, there occurs a separation constant μ and an (integer) quantum number n_3 of the angular-dependent dimensions.

From these parameters are derived other that are used in the calculation:

$$p = \frac{\kappa}{2\,\lambda},$$
$$\nu = \mu + \lambda^2 + (1 - p)\,(n_3 + 1 + 2\,\lambda) + 2\,\lambda\, n_3.$$

Having determined these quantities, the energy levels $\lambda = \lambda_i$, $i = 0, 1, 2, 3, \ldots$, of the hydrogen molecule ion may be calculated in dependence on κ, n_3 and μ by means of a quantum theoretical calculation as done in the following, giving a non-trivial example of an application of the CTCP of the *single confluent case* of the Fuchsian differential equation having four singularities.

The radial component of the quantum mechanical calculation of the energy levels of a hydrogen molecule ion leads to the *single confluent case* of the Heun differential equation:

$$\frac{\mathrm{d}^2 y}{\mathrm{d}z^2} + P(z)\,\frac{\mathrm{d}y}{\mathrm{d}z} + Q(z)\, y = 0 \qquad (5.2.1)$$

with

$$
\begin{aligned}
P(z) &= \frac{A_{+1}}{z-1} + \frac{A_{-1}}{z+1} + G_0, \\
Q(z) &= \frac{B_{+1}}{(z-1)^2} + \frac{B_{-1}}{(z+1)^2} + \frac{C_{+1}}{z-1} + \frac{C_{-1}}{z+1} + D_0,
\end{aligned}
\tag{5.2.2}
$$

whereby

$$
\begin{aligned}
A_{+1} &= 1, \\
A_{-1} &= 1, \\
G_0 &= 0, \\
B_{+1} &= \frac{n_3^2}{4}, \\
B_{-1} &= \frac{n_3^2}{4}, \\
C_{+1} &= \frac{-2\,\lambda^2 + 2\,\kappa - 2\,\mu - n_3^2}{4}, \\
C_{-1} &= \frac{2\,\lambda^2 + 2\,\kappa + 2\,\mu + n_3^2}{4}, \\
D_0 &= -\lambda^2.
\end{aligned}
$$

As radial component of the three-dimensional problem, the relevant interval of the independent variable z is defined as $1 \leq z < \infty$ and the s-rank symbol of equation (5.2.1), (5.2.2) is

$$\{1, 1; 2\}.$$

The CTCP consists of looking for the values $\lambda = \Lambda_i$ such that the particular solution $y(z)$ of (5.2.1), (5.2.2) becomes square integrable on $+1 \leq z \leq +\infty$ along the real axis.

First, consider the local solutions about the two finite singularities, both of which are regular. The two pairs of characteristic exponents at the two finite regular singularities at $z = +1$ and at $z = -1$ are

$$
\begin{aligned}
\alpha_{0r+1,1} &= \alpha_{0r-1,1} = -\frac{n_3}{2}, \\
\alpha_{0r+1,2} &= \alpha_{0r-1,2} = +\frac{n_3}{2}.
\end{aligned}
$$

According to the formulation of the CTCP, the solution at the infinite singularity that is irregular is considered with respect to its asymptotic behaviour as $z \to \infty$ along the positive real axis. Therefore, it is appropriate to consider its Thomé solutions. In terms of these Thomé solutions, the general solution $y^{(g)}(z)$ of equation (5.2.1), (5.2.2) is given by [cf. (2.3.25), (1.2.53)]

$$y^{(g)}(z) = C_1 \exp(\alpha_{11} z)\, z^{\alpha_{0i1}} \left(1 + \frac{C_{11}}{z} + \frac{C_{12}}{z^2} + \cdots\right)$$
$$+ C_2 \exp(\alpha_{12} z)\, z^{\alpha_{0i2}} \left(1 + \frac{C_{21}}{z} + \frac{C_{22}}{z^2} + \ldots\right)$$

as $z \to +\infty$ with arbitrary complex-valued constants C_1 and C_2 and with

$$\alpha_{11} = \frac{1}{2}\left(-G_0 + \sqrt{G_0^2 - 4\,D_0}\right) = +\sqrt{-D_0} = +\lambda,$$
$$\alpha_{12} = \frac{1}{2}\left(-G_0 - \sqrt{G_0^2 - 4\,D_0}\right) = -\sqrt{-D_0} = -\lambda$$

and

$$\alpha_{0i1} = -\frac{(A_{z_1} + A_{z_2})\,\alpha_{11} + C_{z_1} + C_{z_2}}{G_0 + 2\,\alpha_{11}} = -1 - p,\ p = \frac{\kappa}{2\,\lambda},$$
$$\alpha_{0i2} = -\frac{(A_{z_1} + A_{z_2})\,\alpha_{12} + C_{z_1} + C_{z_2}}{G_0 + 2\,\alpha_{12}} = -1 + p,$$

whereby

$$A_{z_1} + A_{z_2} = 2,$$
$$C_{z_1} + C_{z_2} = \kappa.$$

From the knowledge of the characteristic exponents, the requirement of square integrability on $+1 \le z < +\infty$ can be made concrete. It means that $y(z)$ has to behave finitely at $z = +1$ (as indicated by the characteristic exponent $\alpha_{0r+1,2}$) and simultaneously has to tend to zero as $z \to +\infty$ (as indicated by the characteristic exponent α_{12} of the second kind).

The Jaffé ansatz for this CTCP of the single confluent Heun equation is given by

$$y(z) = \exp(-\lambda\, z)\, (z-1)^{\alpha_{0r+1,2}}\, (z+1)^{\alpha_{0r-1,2}}\, w(x) \tag{5.2.3}$$

with

$$w(x) = (1-x)^{\alpha_{0r+1,2}+\alpha_{0r-1,2}-\alpha_{0i2}}\, v(x) = (1-x)^{n_3+1-p}\, v(x)$$

and

$$x = \frac{z-1}{z+1} = \frac{1-\frac{1}{z}}{1+\frac{1}{z}}. \tag{5.2.4}$$

The significant points of the real axis by this Moebius transformation (5.2.4) are mapped according to

$$z = +\infty \to x = +1,$$
$$z = +1 \to x = 0,$$
$$z = 0 \to x = -1,$$
$$z = -1 \to x = -\infty.$$

The power terms in the ansatz (5.2.3) guarantee that the correct asymptotic behaviour results as $z \to +1$ as well as $z \to +\infty$ along the real axis:

$$(z-1)^{\alpha_{0r+1,2}} = (z-1)^{\frac{n_3}{2}} \text{ as } z \to +1,$$
$$z^{\alpha_{0i2}} \text{ as } z \to +\infty.$$

For the latter statement, recognise that by using

$$1 - x = 1 - \frac{z-1}{z+1} = \frac{2}{z+1}$$

the power terms of the Jaffé ansatz are given by

$$\begin{aligned}
&(z-1)^{\frac{n_3}{2}} (z+1)^{\frac{n_3}{2}} (1-x)^{n_3+1-p} \\
&\quad = (z-1)^{\frac{n_3}{2}} (z+1)^{\frac{n_3}{2}} \left(\frac{2}{z+1}\right)^{n_3+1-p} \\
&\quad = 2^{n_3+1-p} (z-1)^{\frac{n_3}{2}} (z+1)^{-\frac{n_3}{2}-1+p} \\
&\quad = 2^{n_3+1-p}\, z^{\frac{n_3}{2}-\frac{n_3}{2}-1+p} \underbrace{\left(1-\frac{1}{z}\right)^{\frac{n_3}{2}}}_{\to 1 \text{ as } z\to\infty} \underbrace{\left(1+\frac{1}{z}\right)^{-\frac{n_3}{2}-1+p}}_{\to 1 \text{ as } z\to\infty} \\
&\quad \sim z^{-1+p} = z^{\alpha_{0i2}} \text{ as } z \to \infty.
\end{aligned}$$

The differential equation for $v(x)$ results in

$$\frac{\mathrm{d}^2 v}{\mathrm{d}x^2} + \tilde{\tilde{P}}(x) \frac{\mathrm{d}v}{\mathrm{d}x} + \tilde{\tilde{Q}}(x)\, v = 0$$

with

$$\tilde{\tilde{P}}(x) = -\frac{a\,x^2 + b\,x + c}{x\,(x-1)\,(x+1)} = \frac{c}{x} - \frac{a+b+c}{2\,(x-1)} - \frac{a-b+c}{2\,(x+1)}$$

and

$$\tilde{\tilde{Q}}(x) = -\frac{d\,x+e}{x\,(x-1)\,(x+1)} = -\frac{d\,x+e}{x\,(x-1)\,(x+1)} = \frac{e}{x} - \frac{d+e}{2\,(x-1)} + \frac{d-e}{2\,(x+1)},$$

whereby

$$\begin{aligned}
a &= n_3 - 2\varrho - 1, \\
b &= 2\,(2\,\lambda + \varrho + 1), \\
c &= -(n_3 + 1), \\
d &= \varrho\,(n_3 - \varrho), \\
e &= \nu.
\end{aligned}$$

Control Calculation

$$
\begin{aligned}
P(x) &= \frac{A\,x}{1-x^2} + \frac{B}{1-x^2} + \frac{C}{x\,(1-x^2)} \\
&= \frac{C}{x} + \frac{1}{2}\,\frac{A+B+C}{1-x} - \frac{1}{2}\,\frac{A-B+C}{x\,(1+x)}, \\
Q(x) &= \frac{D}{1-x^2} + \frac{E}{x\,(1-x^2)} \\
&= \frac{E}{x} + \frac{1}{2}\,\frac{D+E}{1-x} + \frac{1}{2}\,\frac{D-E}{x\,(1+x)}.
\end{aligned}
$$

Partial decompositions:

$$
\begin{aligned}
\frac{x}{1-x^2} &= \frac{\frac{1}{2}}{1-x} - \frac{\frac{1}{2}}{1+x}, \\
\frac{1}{1-x^2} &= \frac{\frac{1}{2}}{1-x} + \frac{\frac{1}{2}}{1+x}, \\
\frac{1}{x\,(1-x^2)} &= \frac{1}{x} + \frac{\frac{1}{2}}{1-x} - \frac{\frac{1}{2}}{1+x}.
\end{aligned}
$$

The solution $v(x)$ is holomorphic at $x = 0$ and thus may be represented in a convergent power series

$$v(x) = \sum_{n=0}^{\infty} a_n\, x^n,$$

the radius of convergence of which is the disc $|x| = 1$. The resulting difference equation of this power series as an ansatz may be converted into a three-term recurrence relation by means of an initial condition:

$$(-x^3 + x)\,\frac{\mathrm{d}^2 v}{\mathrm{d}x^2} + \left(A\,x^2 + B\,x + C\right)\frac{\mathrm{d}v}{\mathrm{d}x} + (D\,x + E)\,v = 0$$

with

$$
\begin{aligned}
A &= -(n_3 - 2\,\varrho - 1), \\
B &= -2\,(2\,\lambda + \varrho + 1), \\
C &= n_3 + 1, \\
D &= \varrho\,(\varrho - n_3), \\
E &= -\nu
\end{aligned}
$$

or

$$(-x^3 + x)\sum_{n=0}^{\infty} a_n\, n\,(n-1)\,x^{n-2}$$

$$+A\,x^2 \sum_{n=0}^{\infty} a_n\, n\, x^{n-1} + B\,x \sum_{n=0}^{\infty} a_n\, n\, x^{n-1} + C \sum_{n=0}^{\infty} a_n\, n\, x^{n-1}$$
$$+D\,x \sum_{n=0}^{\infty} a_n\, x^n + E \sum_{n=0}^{\infty} a_n\, x^n = 0$$

or

$$-\sum_{n=0}^{\infty} a_n\, n\,(n-1)\, x^{n+1} + \sum_{n=0}^{\infty} a_n\, n\,(n-1)\, x^{n-1}$$
$$+A \sum_{n=0}^{\infty} a_n\, n\, x^{n+1} + B \sum_{n=0}^{\infty} a_n\, n\, x^n + C \sum_{n=0}^{\infty} a_n\, n\, x^{n-1}$$
$$+D \sum_{n=0}^{\infty} a_n\, x^{n+1} + E \sum_{n=0}^{\infty} a_n\, x^n = 0$$

yielding

$$a_1 + p_0\, a_0 = 0, \quad a_0 \neq 0,$$
$$a_{n+1} + p_n\, a_n - q_{n-1}\, a_{n-1} = 0,\ n = 1, 2, 3, \ldots,$$

with

$$p_n = -\frac{2\,n^2 + 2\,n\,(2\,\lambda + n_3 + 1 - p) + \nu}{(n+1)(n+n_3+1)},$$
$$q_n = -\frac{(n-p)(n+n_3-p)}{(n+1)(n+n_3+1)} \tag{5.2.5}$$

with $n = 0, 1, 2, 3, \ldots$, or

$$a_1 + p_0\, a_0 = 0, \quad a_0 \neq 0,$$
$$\left[1 + \frac{\alpha_1}{n} + \frac{\beta_1}{n^2}\right] a_{n+1} + \left[-2 + \frac{\alpha_0}{n} + \frac{\beta_0}{n^2}\right] a_n + \left[-1 + \frac{\alpha_{-1}}{n} + \frac{\beta_{-1}}{n^2}\right] a_{n-1} = 0,$$
$$n = 1, 2, 3, \ldots, \tag{5.2.6}$$

whereby

$$\alpha_1 = n_3 + 2, \quad \beta_1 = n_3 + 1,$$
$$\alpha_0 = -2\,(2\,\lambda + n_3 + 1 - p), \quad \beta_0 = -\nu,$$
$$\alpha_{-1} = 2\,p - n_3, \quad \beta_{-1} = p\,(n_3 - p).$$

The linear, ordinary, homogeneous difference equation (5.2.6) is of second order, therefore, its general solution consists of a two-dimensional fundamental system. This pair of linear independent particular solutions may be characterised by means of their different asymptotic behaviour as $n \to \infty$:

$$\frac{a_{n+1}}{a_n} = 1 \pm \sqrt{\frac{4\lambda}{n}}$$

Thus, the boundary eigenvalue condition may be formulated by means of an infinite continued fraction according to (4.2.22):

$$F(\lambda = \lambda_i) = 0$$

with

$$F(\lambda) = p_0(\lambda) + q(\lambda)$$

and with $q(\lambda)$ being the infinite continued fraction [cf. §§1.3.1 and 4.2.1]

$$q(\lambda) = \cfrac{q_0}{p_1 + \cfrac{q_1}{p_2 + \cfrac{q_2}{p_3 + \cfrac{q_3}{\cdots}}}}$$

where the p_n and the q_n, $n = 0, 1, 2, \ldots$, are from (5.2.5). The result of the calculation is shown in Figure 5.21.

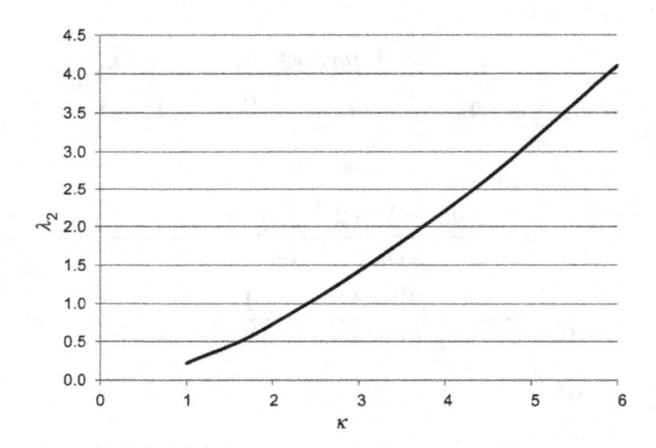

Figure 5.21 Lowest ground-state energies of the ionised hydrogen molecule ion dependent on the distance of the hydrogen atoms.

Thus, the exact quantum calculation of the energy levels of the hydrogen molecule ion may be carried out in a quite satisfactory manner.

5.3 Versiera d'Agnesi: Single Confluent Heun

In all the above-given examples it is characteristic that the singularities of the underlying differential equations are located on the real axis. The theory, however, is more general, covering the situation of complex-valued locations of singularities. Such an example is given in the following.

In §4.2.2 I gave what I call the *Jaffé transformation*, a specific *Moebius transformation* with the purpose of contracting the relevant interval of a central two-point connection problem such that it becomes less than the distance to the neighbouring singularity of the underlying differential equation with respect to the origin. This allows a series ansatz of the solution about the most regular singularity of the underlying differential equation to be convergent over the whole relevant interval of the CTCP, i.e., the positive real axis.

In the following is discussed a famous example from the starting times of modern function theory: the versiera of the Milanese Maria Gaetana Agnesi (1718–1799).[4]

The versiera is a third-order algebraic curve, the functional form of which is given by

$$V(z) = \frac{a^3}{a^2 + z^2}, \qquad z, a \in \mathbb{R}. \tag{5.3.7}$$

It was investigated already in 1703 by Pierre de Fermat and Guido Grandi and was published by Maria Agnesi in 1748.

In the following, it is crucial that the independent variable z is taken to be complex-valued: $z \in \mathbb{C}$. The versiera may be constructed according to Figure 5.22.

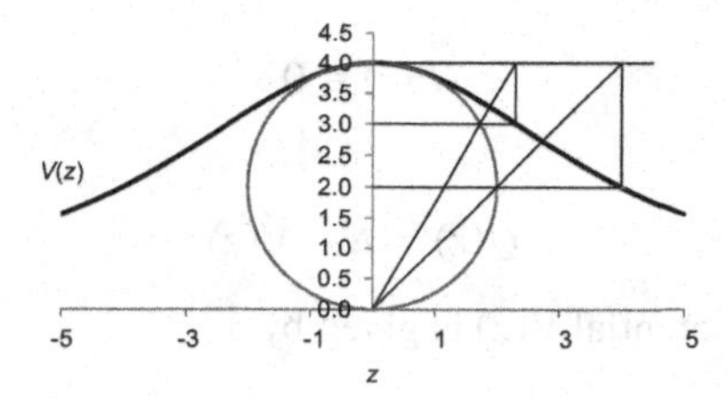

Figure 5.22 The construction of the versiera of Maria Gaetana Agnesi.

Incidentally, the versiera is the first derivative of the arcus tangens function

$$a^2 \frac{\mathrm{d}\left[\arctan(\frac{z}{a})\right]}{\mathrm{d}z} = V(z)$$

such that

$$\int V(z)\,\mathrm{d}z = \int \frac{a^3}{a^2 + z^2}\,\mathrm{d}z = a^2 \arctan\left(\frac{z}{a}\right).$$

Quantum mechanically interpreted, for negative values of a, thus $a \in \mathbb{R}^-$, the versiera is a single well (cf. Figure 5.23).

Thus, consider the one-dimensional Schrödinger equation

$$\frac{\mathrm{d}^2 y}{\mathrm{d}z^2} + Q(z)\,y = 0, \tag{5.3.8}$$

[4] 'Versiera' is the Latin for a sheet rope, a rope used for positioning the sails on sailing ships. Maria Agnesi was an Italian mathematician, philosopher, theologian and humanitarian of the eighteenth century. She was the first woman to write a mathematics handbook and the first woman appointed as a mathematics professor at a university. The Italian notion 'la versiera di Agnesi' is after the Latin 'versiera' and the 'sinus versus'. This was misinterpreted by John Colson, a professor at Cambridge University, as 'l'avversiera di Agnesi', that is, a woman playing against God, i.e., a witch. Therefore, the curve is sometimes spuriously called the 'Witch of Agnesi'.

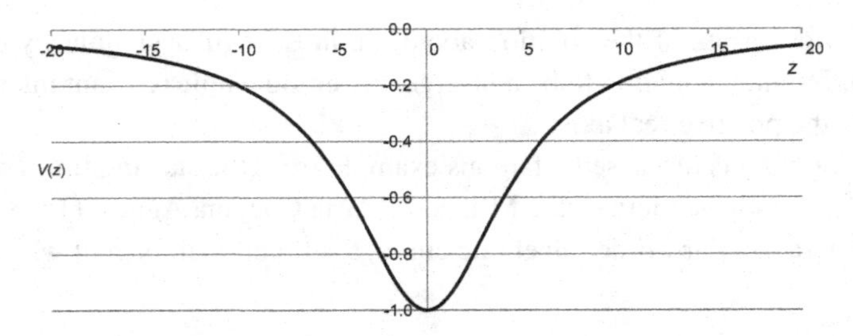

Figure 5.23 The versiera of Maria Gaetana Agnesi as quantum mechanical potential.

a normal form of (2.1.2)

$$\frac{\mathrm{d}^2 y}{\mathrm{d}z^2} + P(z)\,\frac{\mathrm{d}y}{\mathrm{d}z} + Q(z)\,y = 0$$

with

$$P(z) \equiv 0$$

and

$$Q(z) = E - V(z) \tag{5.3.9}$$

whereby the quantum potential $V(z)$ is given by

$$V(z) = \frac{a^3}{a^2 + z^2},\ a \in \mathbb{R}^-. \tag{5.3.10}$$

Thus, the coefficient $Q(z)$ in this quantum situation is given by

$$Q(z) = E - V(z) = E - \frac{a^3}{a^2 + z^2} = E - \frac{a^3}{(z + \imath a)(z - \imath a)} = E - \left(\frac{\frac{a^2}{2}\,\imath}{z + \imath a} - \frac{\frac{a^2}{2}\,\imath}{z - \imath a} \right) \tag{5.3.11}$$

whereby[5] $\imath = \sqrt{-1}$ is the imaginary unit. In the terminology of singular differential equations used in this book, equation (5.3.8), (5.3.9) with (5.3.10) represents a single confluent case (3.4.1) of the Fuchsian differential equation with four singularities (3.2.32), thus has s-rank symbol

$$\{1, 1; 2\}\,.$$

The quantum mechanical problem consists of looking for values E_j of the parameter E for which the differential equation (5.3.8), (5.3.9) with (5.3.10) has square integrable particular solutions $y = y(z)$ on the real axis $z \in \mathbb{R}$: $y \in L_2(\mathbb{R})$. It is clear from the physical situation that $a < E_i < 0$ holds. Because of the symmetrical form of the versiera, this is a CTCP of equation (5.3.8), (5.3.9) with (5.3.10) on the positive real axis $z \in \mathbb{R}_0^+$, meaning that the eigensolutions separate into even and odd ones with respect to the origin $z = 0$. The finite singularities of (5.3.8), (5.3.9) with (5.3.10) are regular ones, not lying on the relevant interval of the CTCP but

[5] The imaginary unit $\imath$ is not to be confused with the alphabetic character i in the following.

on the imaginary axis of the complex plane at $z = \pm\imath\, a$. The left end of the relevant interval, the origin $z = 0$, is an ordinary point of the differential equation, thus all the particular solutions of the differential equation are holomorphic there.

The Thomé solutions at the point at infinity are given by [cf. (3.4.10), (3.4.11)]

$$y(z) = \exp(\alpha_1\, z)\, z^{\alpha_{0i}} \sum_{n=0}^{\infty} C_n\, z^{-n}. \tag{5.3.12}$$

The identifications

$$\begin{aligned}
z_1 &= +\imath\, a,\\
z_2 &= -\imath\, a = -z_1,\\
G_0 &= 0,\\
A_{z_1} &= A_{z_2} = 0,\\
D_0 &= E,\\
B_{z_1} &= B_{z_2} = 0,\\
C_{z_1} &= \imath\,\frac{a^2}{2} = C_{\imath\, a},\\
C_{z_2} &= -\imath\,\frac{a^2}{2} = C_{-\imath\, a} = -C_{\imath\, a}
\end{aligned}$$

thus yield [cf. (3.4.12)]

$$C_{z_1} + C_{z_2} = 0$$

with characteristic exponents

$$\begin{aligned}
\alpha_{11} &= \frac{1}{2}\left(-G_0 + \sqrt{G_0^2 - 4\, D_0}\right) = \sqrt{-E},\\
\alpha_{12} &= \frac{1}{2}\left(-G_0 - \sqrt{G_0^2 - 4\, D_0}\right) = -\sqrt{-E}
\end{aligned}$$

and [cf. (3.4.12)]

$$\begin{aligned}
\alpha_{0i1} &= -\frac{(A_{z_1} + A_{z_2})\,\alpha_{11} + C_{z_1} + C_{z_2}}{G_0 + 2\,\alpha_{11}} = 0,\\
\alpha_{0i2} &= -\frac{(A_{z_1} + A_{z_2})\,\alpha_{12} + C_{z_1} + C_{z_2}}{G_0 + 2\,\alpha_{12}} = 0.
\end{aligned}$$

The characteristic exponents at the finite singularities at $z = \pm\imath\, a$ are given by

$$\begin{aligned}
\alpha_{0\imath a1} &= 0, \quad \alpha_{0\imath a2} = 1,\\
\alpha_{0-\imath a1} &= 0, \quad \alpha_{0-\imath a2} = 1.
\end{aligned}$$

Thus, these two finite singularities, lying on the imaginary axis, are apparent singularities (cf. §2.3.1). The generalised Riemann symbol of equation (5.3.8), (5.3.9) with (5.3.10) is given by

$$P\begin{pmatrix} 1 & 1 & 2 & \\ a\imath & -a\imath & \infty & z \\ \alpha_{0\imath a1} & \alpha_{0-\imath a1} & \alpha_{0i1} & E \\ \alpha_{0\imath a2} & \alpha_{0-\imath a2} & \alpha_{0i2} & \\ & & \alpha_{11} & \\ & & \alpha_{12} & \end{pmatrix}.$$

5.3.1 Jaffé Ansatz

I consider the differential equation (5.3.8) having the coefficient

$$\begin{aligned} Q(z) &= E - \frac{a^3}{a^2+z^2} = E - \frac{a^3}{(z+\imath a)(z-\imath a)} \\ &= E - \left(\frac{\frac{a^2}{2}\imath}{z+a\imath} - \frac{\frac{a^2}{2}\imath}{z-a\imath} \right). \end{aligned}$$

Writing

$$(z^2+a^2)\,Q(z) = E\,z^2 + a^2\,(E-a) \tag{5.3.13}$$

yields

$$(z^2+a^2)\,\frac{\mathrm{d}^2y}{\mathrm{d}z^2} + \left[E\,z^2 + a^2\,(E-a)\right]\,y = 0. \tag{5.3.14}$$

The coefficients of this differential equation are both holomorphic at the origin $z = 0$. Therefore, particular solutions may be expanded into Taylor series about the origin:

$$y(z) = \sum_{n=0}^{\infty} a_n\,z^n, \tag{5.3.15}$$

the derivatives of which are given by

$$\begin{aligned} \frac{\mathrm{d}y}{\mathrm{d}z} &= \sum_{n=0}^{\infty} a_n\,n\,z^{n-1}, \\ \frac{\mathrm{d}^2y}{\mathrm{d}z^2} &= \sum_{n=0}^{\infty} a_n\,n\,(n-1)\,z^{n-2}. \end{aligned}$$

Inserting these expressions into the differential equation (5.3.14) results in the three-term recurrence relation

$$\boxed{\begin{aligned} &2\,a_2 + (E-a)\,a_0 = 0, \\ &6\,a_3 + (E-a)\,a_1 = 0, \\ &a^2\left(1+\frac{3}{n}+\frac{2}{n^2}\right)a_{n+2} + \left[1-\frac{1}{n}+\frac{a^2\,(E-a)}{n^2}\right]a_n \\ &\qquad\qquad + \frac{E}{n^2}\,a_{n-2} = 0\ ,\ n = 2,3,\ldots. \end{aligned}}$$

The characteristic equation of (5.3.16) is the second-order algebraic equation

$$a^2 t^2 + t = t(a^2 t + 1) = 0, \tag{5.3.16}$$

the two solutions of which are

$$t_1 = 0, \tag{5.3.17}$$

$$t_2 = -\frac{1}{a^2}. \tag{5.3.18}$$

Thus, as may be seen, the radius of convergence r of the series in the ansatz (5.3.15) is either

$$r_1 = a \tag{5.3.19}$$

or infinite. The particular solution $y(z)$ of (5.3.14) in this latter case is thus an entire function.

Since both the differential equation and the difference equation are of second order, we can make a one-to-one assignment of their respective fundamental solutions. The exponentially increasing solutions of the differential equation as $z \to +\infty$ are represented by the Taylor series (5.3.15) whose radius of convergence takes the value $r = r_1 = a$, and are therefore singular at the finite singularities of the differential equation at $z = \pm a\,\imath$. And the exponentially decreasing solutions of the differential equation (5.3.13) as $z \to +\infty$ are represented by the Taylor series (5.3.15) whose radius of convergence is infinite, i.e., they are holomorphic at the finite singularities of the differential equation at $z = \pm a\,\imath$ itself. These are exactly the eigenfunctions of the singular boundary eigenvalue problem. This is a quite noteworthy feature of this problem.

As in §4.2.1, the eigenvalues $E = E_\imath$ may be calculated by means of an infinite continued fraction. This is carried out in the next section.

5.3.2 Asymptotics

By means of a perturbation theory of the Schrödinger equation, it is possible to establish an approximation of the eigenvalues of the CTCP. This, in turn, is based on an approximation to the Versiera d'Agnesi as a quantum potential of the according Schrödinger equation (5.3.8) with (5.3.2). This idea is outlined in the following.

So, we start with the Schrödinger equation

$$\frac{d^2 y}{dz^2} + [E - V(z)]\, y = 0$$

having the potential (Versiera d'Agnesi)

$$V(z) = \frac{a^3}{a^2 + z^2}, \quad a \in \mathbb{R}^-.$$

As this potential is holomorphic over the real axis, it may be expanded into a Taylor series on each point, thus also about the origin $z = 0$:

$$V(z) = \frac{a^3}{a^2 + z^2} = V(0) + \left.\frac{d^2V}{dz^2}\right|_{z=0} \frac{z^2}{2!} + \left.\frac{d^4V}{dz^4}\right|_{z=0} \frac{z^4}{4!} + \left.\frac{d^6V}{dz^6}\right|_{z=0} \frac{z^6}{6!} + \cdots .$$

The derivatives of the versiera may be given in closed form:

$$\begin{aligned}
\frac{dV}{dz} &= -\frac{2\,a^3\,z}{(a^2 + z^2)^2}, \\
\frac{d^2V}{dz^2} &= \frac{2\,a^3\,(3\,z^2 - a^2)}{(a^2 + z^2)^3}, \\
\frac{d^3V}{dz^3} &= -\frac{24\,a^3\,z\,(z^2 - a^2)}{(a^2 + z^2)^4}, \\
\frac{d^4V}{dz^4} &= \frac{24\,a^3\,(5\,z^4 - 10\,a^2\,z^2 + a^4)}{(a^2 + z^2)^5}, \\
\cdots &\quad \cdots
\end{aligned}$$

This is also the case for the derivatives of the Versiera d'Agnesi at the origin:

$$\begin{aligned}
V(0) &= a, \\
\left.\frac{dV}{dz}\right|_{z=0} &= 0, \\
\left.\frac{d^2V}{dz^2}\right|_{z=0} &= -\frac{2}{a}, \\
\left.\frac{d^3V}{dz^3}\right|_{z=0} &= 0, \\
\left.\frac{d^4V}{dz^4}\right|_{z=0} &= \frac{24}{a^3}, \\
\left.\frac{d^5V}{dz^5}\right|_{z=0} &= 0, \\
\left.\frac{d^6V}{dz^6}\right|_{z=0} &= -\frac{720}{a^5}, \\
\cdots &\quad \cdots
\end{aligned}$$

In general, the derivatives of the Versiera d'Agnesi at the origin of arbitrary order are given, for odd orders, as

$$\left.\frac{d^sV}{dz^s}\right|_{z=0} = 0, \quad s = 2\,k + 1 = 1, 3, 5, \ldots,$$

and for even orders as

$$\left.\frac{d^sV}{dz^s}\right|_{z=0} = (-1)^k\,\frac{s!\,a^{s+3}}{a^{2\,(s+1)}} = (-1)^k\,\frac{s!}{a^{s-1}}, \quad s = 2\,k = 0, 2, 4, 6, \ldots,$$

thus with $k = 0, 1, 2, 3, \ldots$. As a result,

$$\frac{\left.\frac{d^s V}{dz^s}\right|_{z=0}}{s!} = (-1)^k \frac{1}{a^{s-1}}, \quad s = 2\,k = 0, 2, 4, 6, \ldots.$$

Therefore, the Taylor expansion of the Versiera d'Agnesi about the origin is given by the geometric series

$$\boxed{V(z) = \frac{a^3}{a^2 + z^2} = a - \frac{z^2}{a} + \frac{z^4}{a^3} - \frac{z^6}{a^5} + - \cdots .} \tag{5.3.20}$$

Basic Solution

The approximation theory, presented here, consists of two parts: the basic solution and the approximated solution. Since the CTCP for a quadratic potential may be solved in closed form, the basic solution consists of the first two terms in the series (5.3.20). Thus, the basic potential is given by

$$V_0(z) = V_0(0) + \left.\frac{d^2 V}{dz^2}\right|_{z=0} \frac{z^2}{2!} = a - \frac{z^2}{a}, \quad a \in \mathbb{R}^-. \tag{5.3.21}$$

Thus, the basic differential equation is

$$\frac{d^2 y_0}{dz^2} + \left[\tilde{E}_0 - V_0(z)\right] y_0(z) = 0$$

or

$$\frac{d^2 y_0}{dz^2} + \left[\tilde{E}_0 - \left(a - \frac{z^2}{a}\right)\right] y_0(z) = 0.$$

In order to make vanish the coefficient $1/a$ in front of the term z^2, a Moebius transformation (cf. §§1.2.2 and 4.2.2) is carried out with respect to this equation:

$$x = \frac{1}{\sqrt[4]{-a}}\, z,$$

yielding

$$\frac{d^2 y_0}{dx^2} + \left[E_0 + x^2\right] y_0(x) = 0.$$

The eigenvalues of this basic CTCP are given by [cf. Schubert and Weber, 1980, p. 153, eq. (5.95)]

$$E_{0n} = \sqrt{-a}\,\left(\tilde{E}_{0n} - a\right) = 2\,n + 1, \quad n = 0, 1, 2, 3, \ldots$$

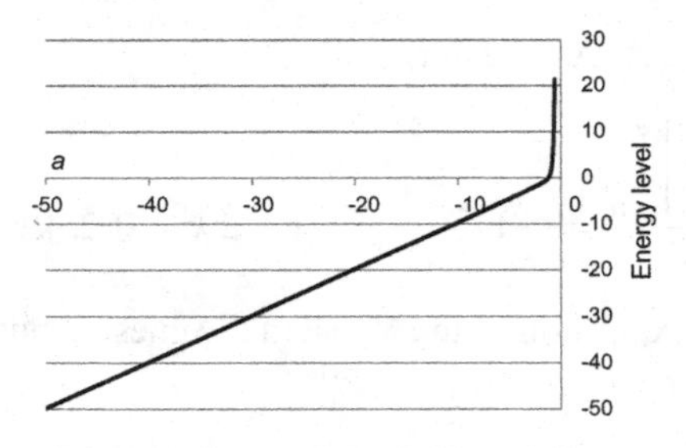

Figure 5.24 The lowest eigenvalue of the basic problem.

thus

$$\tilde{E}_{0n} = a + \frac{2n+1}{\sqrt{-a}}, \quad n = 0, 1, 2, 3, \ldots, a \in \mathbb{R}^-.$$

The eigenfunctions of the basic problem are given by [cf. Schubert and Weber, 1980, p. 153, eq. (5.96)]

$$y_{0n}(x) = \exp\left(-\frac{x^2}{2}\right) H_n(x) \tag{5.3.22}$$

with $H_n(x)$ being the Hermite polynomials of order n.

Specifying the calculation to the lowest eigenvalue, with numbering $n = 0$, yields

$$E_{00} = 1$$

or (cf. Figure 5.24)

$$\tilde{E}_{00} = a + \frac{1}{\sqrt{-a}}, \quad a \in \mathbb{R}^-. \tag{5.3.23}$$

It should be recognised that

$$\lim_{a \to -\infty} \left(\tilde{E}_{00} - a\right) = 0.$$

This means that the lowest eigenvalue $\bar{E}_{00}(a)$ of the Versiera d'Agnesi comes to lie between a and $a + \frac{1}{\sqrt{-a}}$ with $a \in \mathbb{R}^-$, and it holds that

$$\lim_{a \to -\infty} \left[\bar{E}_{00}(a) - \tilde{E}_{00}(a)\right] = 0.$$

Some values of $\tilde{E}_{00}(a)$ for negative natural values of a are tabulated below:

a	-1	-2	-3
$\tilde{E}_{00}(a)$	$-1+1=0$	$-2+\frac{1}{\sqrt{2}}\approx -1.29$	$-3+\frac{1}{\sqrt{3}}\approx -2.42$

a	-4	-5	-10
$\tilde{E}_{00}(a)$	$-4+\frac{1}{\sqrt{4}}\approx -3.50$	$-5+\frac{1}{\sqrt{5}}\approx -4.55$	$-10+\frac{1}{\sqrt{10}}\approx -9.68$

a	-20	-30	-50
$\tilde{E}_{00}(a)$	$-20+\frac{1}{\sqrt{20}}\approx -19.78$	$-30+\frac{1}{\sqrt{30}}\approx -29.82$	$-50+\frac{1}{\sqrt{50}}\approx -49.86$

The eigenfunction of this lowest energy level is the lowest Hermitian polynomial $H_0(x)$:

$$H_0(x) = 1.$$

Thus, the basic eigenfunction of the lowest energy level is the asymptotic factor in (5.3.22):

$$y_{00}(x) = \exp\left(-x^2\right).$$

Approximation

The approximation to the Versiera d'Agnesi is a truncated part $V_1(z)$ of the Taylor series (5.3.21) starting with order four:

$$\begin{aligned} V_1(z) &= \sum_{k=2}^{N} c_{2k}\, z^{2k} \\ &= c_4\, z^4 - c_6\, z^6 + - \cdots + (-1)^k\, c_{2N}\, z^{2N} \\ &= \frac{z^4}{a^3} - \frac{z^6}{a^5} + - \cdots + (-1)^k\, \frac{z^{2N}}{a^{2N-1}} \end{aligned}$$

with

$$c_{2k} = (-1)^k\, \frac{1}{a^{s-1}}, \quad s = 2\,k, \quad k = 2, 3, 4, 5, 6, \ldots, 2\,N$$

thus

$$\begin{aligned} V(z) = V_0(z) + V_1(z) &= a - \frac{z^2}{a} + \sum_{k=2}^{N} c_{2k}\, z^{2k} \\ &= a - \frac{z^2}{a} + \frac{z^4}{a^3} - \frac{z^6}{a^5} + - \cdots + (-1)^k\, \frac{z^{2N}}{a^{2N-1}}. \end{aligned} \tag{5.3.24}$$

The series (5.3.24) exclusively consists of terms of even order and is truncated after N terms.

Applying the formula (cf. Courant and Hilbert, 1968, Chapter V, §13, p. 297)

$$E_{1n,N} = \frac{2}{2^n\, n!\, \sqrt{\pi}} \sum_{k=2}^{N} c_{2\,k} \int_{x=0}^{+\infty} x^{\nu}\, H_n^2(x) \exp\left(-x^2\right) \mathrm{d}x$$

and specifying $n = 0$ yields

$$E_{10,N} = 2 \sum_{k=2}^{N} \frac{1 \cdot 3 \cdots (2\,k-1)}{2^{k+1}}\, c_{2\,k}$$

and thus the eigenvalues of the approximation part of the CTCP are

$$\begin{aligned} E_{10,N}(a) &= 2 \sum_{k=2}^{N} (-1)^k \frac{1 \cdot 3 \cdots (2\,k-1)}{2^{k+1}} \frac{1}{a^{2\,k-1}} \\ &= \frac{3}{4\,a^3} - \frac{15}{8\,a^5} + - \cdots + (-1)^k \frac{1 \cdot 3 \cdots (2\,N-1)}{2^N\, a^{2\,N-1}}. \end{aligned} \tag{5.3.25}$$

As may be seen, these series diverge, thus are asymptotic, meaning they are approximately valid only for large absolute values of a, or, in other words, as $a \longrightarrow -\infty$.

As an example of this statement, take $N = 2$. Then, the series (5.3.25) is valid all the more as the value of a becomes large. This may be seen by considering the first two terms:

$$\frac{3}{4\,a^3} \quad \text{and} \quad -\frac{15}{8\,a^5}.$$

The smaller the ratio $C(a)$ of the latter term to the former one, the more accurate is the term itself:

$$C(a) = \left| \frac{\frac{15}{8\,a^5}}{\frac{3}{4\,a^3}} \right| = \frac{5}{2\,a^2} \ll 1.$$

Then, the approximated eigenvalue $\bar{E}_2(a)$ of the quantum problem is given by the sum of the basic term (5.3.23) and the correction term (5.3.25), yielding

$$\bar{E}_N(a) \approx \tilde{E}_{00}(a) + E_{10,2}(a) = a + \frac{1}{\sqrt{-a}} + \frac{3}{4\,a^3} - \frac{15}{8\,a^5}. \tag{5.3.26}$$

In the following, some numerical values are tabulated:

a	−1	−2	−3	−4	−5	−10	−20
$C(a)$	2.500	0.625	0.278	0.156	0.100	0.025	0.00625
$\tilde{E}_{00}(a)$	0	−1.29	−2.42	−3.50	−4.55	−9.68	−19.78
$E_{10,2}(a)$	1.125	-0.035	-0.020	-0.010	-0.005	-0.001	-0.0001
$\bar{E}_2(a)$	1.13	−1.33	−2.44	−3.51	−4.56	−9.68	−19.78

Regarding the fact that the exact value of the lowest eigenvalue $\bar{E}(a)$ of the quantum problem even for general values of N lies in the interval $a < \bar{E}(a) < \bar{E}_N(a)$ (determined by a lowest possible value that is just the bottom of the well a as a floor and the value of $\bar{E}_N(a)$ as a cap), the approximated value $\bar{E}_2(a)$ even from $a = -2$ on is quite remarkable.

Figure 5.25 shows the approximation of the versiera of Maria Gaetana Agnesi as a quantum potential.

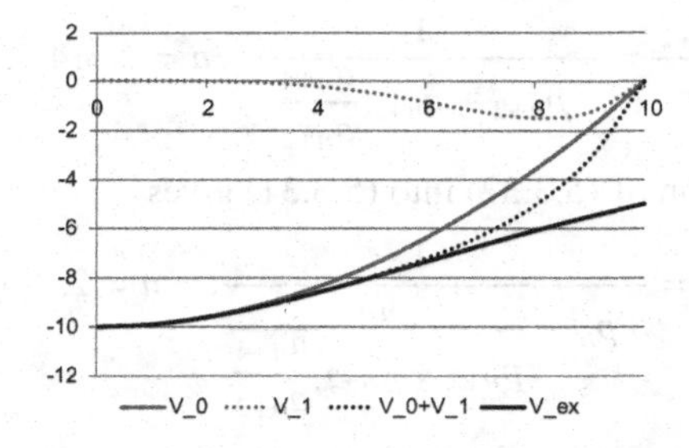

Figure 5.25 The approximation of the versiera of Maria Gaetana Agnesi.

Numerics

We consider the difference equation (5.3.16) including its initial condition having the general form

$$p_0\, a_0 + q_0\, a_2 = 0, \tag{5.3.27}$$

$$p_1\, a_1 + q_1\, a_3 = 0, \tag{5.3.28}$$

$$a_{n-2} + p_n\, a_n + q_n\, a_{n+2} = 0, \quad n = 2, 3, 4, \ldots, \tag{5.3.29}$$

with

$$\boxed{\begin{aligned} p_n &= \frac{n(n-1) + a^2(E-a)}{E}, \quad n = 0, 1, 2, 3, 4, \ldots, \\ q_n &= \frac{a^2(n+2)(n+1)}{E}, \quad n = 0, 1, 2, 3, 4, \ldots. \end{aligned}} \tag{5.3.30}$$

From (5.3.30) we have explicitly

$$p_0 = \frac{a^2(E-a)}{E}, \quad q_0 = \frac{2a^2}{E},$$

$$p_1 = \frac{a^2(E-a)}{E}, \quad q_1 = \frac{6a^2}{E}.$$

After dividing (5.3.29) by a_n, it becomes

$$\frac{a_{n-2}}{a_n} = -p_n - q_n \frac{a_{n+2}}{a_n}, \quad n = 2, 3, 4, \ldots.$$

By inversion

$$\frac{a_n}{a_{n-2}} = -\frac{1}{p_n + q_n \dfrac{a_{n+2}}{a_n}}, \quad n = 2, 3, 4, \ldots. \tag{5.3.31}$$

Replacing n by $n + 2$ in (5.3.31) yields

$$\frac{a_{n+2}}{a_n} = -\frac{1}{p_{n+2} + q_{n+2} \dfrac{a_{n+4}}{a_{n+2}}}, \quad n = 2, 3, 4, \ldots. \tag{5.3.32}$$

Inserting the expression of (5.3.32) into (5.3.31) gives

$$\frac{a_n}{a_{n-2}} = -\frac{1}{p_n - \dfrac{q_n}{p_{n+2} + q_{n+2} \dfrac{a_{n+4}}{a_{n+2}}}}, \quad n = 2, 3, 4, \ldots.$$

Continuing in this way indefinitely results in an infinite continued fraction

$$\frac{a_n}{a_{n-2}} = -\frac{1}{p_n - \dfrac{q_n}{p_{n+2} - \dfrac{q_{n+2}}{p_{n+4} - \ldots}}}, \quad n = 2, 3, 4, \ldots. \tag{5.3.33}$$

Turning to the even eigenfunctions means specifying this formula (5.3.33) to $n = 2$:

$$\frac{a_2}{a_0} = -\frac{1}{p_2 - \dfrac{q_2}{p_4 - \dfrac{q_4}{p_6 - \ldots}}}. \tag{5.3.34}$$

The infinite continued fraction on the right-hand side of equation (5.3.34) is just the ratio

$$\frac{a_2}{a_0}$$

of the first two terms of the difference equation (5.3.29). Moreover, as we know from §§1.3.1 and 4.2.1, this is the ratio of the recessive solution of the difference equation (5.3.29).

On the other hand, there is the initial condition (5.3.27), which prescribes this ratio to be

$$\frac{a_2}{a_0} = -\frac{p_0}{q_0} = -\frac{E-a}{2}.$$

From these two prescriptions, requiring the same value, the condition for the eigenvalues of the even eigenfunctions is derived to be

$$\left.\frac{a_2}{a_0}\right|_{\mathrm{ic}} = \left.\frac{a_2}{a_0}\right|_{\mathrm{cf}}, \tag{5.3.35}$$

thus

$$F_{a,even}(E) = \left.\frac{a_2}{a_0}\right|_{\mathrm{ic}} - \left.\frac{a_2}{a_0}\right|_{\mathrm{cf}} = 0,$$

where the index ic means 'initial condition' and cf means 'continued fraction'.

As a final result, the (transcendental) eigenvalue condition of the singular eigenvalue problem of the quantum potential of the Versiera d'Agnesi in the case of even eigensolutions is given by

$$\boxed{F_{a,even}(E) = -\frac{E-a}{2} + \cfrac{1}{p_2 - \cfrac{q_2}{p_4 - \cfrac{q_4}{p_6 - \dots}}} = 0.}$$

Turning to the odd eigenfunctions means specifying the formula (5.3.33) to $n = 3$:

$$\frac{a_3}{a_1} = -\cfrac{1}{p_3 - \cfrac{q_3}{p_5 - \cfrac{q_5}{p_7 - \dots}}}.$$

On the other hand, there is the initial condition (5.3.28), which prescribes this ratio to be

$$\frac{a_3}{a_1} = -\frac{p_1}{q_1} = -\frac{E-a}{6}.$$

From these two prescriptions, requiring the same value, the condition for the eigenvalues of the odd eigenfunctions is derived to be

$$\left.\frac{a_3}{a_1}\right|_{\mathrm{ic}} = \left.\frac{a_3}{a_1}\right|_{\mathrm{cf}},$$

thus

$$F_{a,odd}(E) = \left.\frac{a_3}{a_1}\right|_{\mathrm{ic}} - \left.\frac{a_3}{a_1}\right|_{\mathrm{cf}} = 0,$$

where again the index ic means 'initial condition' and cf means 'continued fraction'.

As a final result, the (transcendental) eigenvalue condition of the singular eigenvalue problem of the quantum potential of the Versiera d'Agnesi in the case of odd eigensolutions is given by

$$\boxed{F_{a,odd}(E) = -\frac{E-a}{6} - \cfrac{1}{p_3 - \cfrac{q_3}{p_5 - \cfrac{q_5}{p_7 - \ldots}}} = 0.}$$

Thus, the eigenvalue conditions, in both cases, require the calculation of the infinite continued fraction (5.3.33). This may be done, e.g., by means of the method given by Salvattore Pincherle (1894). Then, the zeros of the functions $F_{a,.}(E)$ are the eigenvalues of the singular eigenvalue problem that solves the quantum mechanical interpretation of the Versiera d'Agnesi in an exact manner. In the following, a step-by-step solution of this numerical question is given.

Start with a general infinite continued fraction having the form

$$C = b_0 + \cfrac{a_1}{b_1 + \cfrac{a_2}{b_2 + \cfrac{a_3}{b_3 + \cfrac{a_4}{b_4 + \ldots}}}}. \tag{5.3.36}$$

The 'fundamental recurrence relation' of Pincherle to calculate this infinite continued fraction is

$$\begin{aligned} A_{n+1} &= b_{n+1}\, A_n + a_{n+1}\, A_{n-1},\ n = 0,1,2,3,\ldots,\ A_0 = b_0,\ A_{-1} = 1, \\ B_{n+1} &= b_{n+1}\, B_n + a_{n+1}\, B_{n-1},\ n = 0,1,2,3,\ldots,\ B_0 = 1,\ B_{-1} = 0, \end{aligned}$$

from which results the partial ratio of the infinite continued fraction according to

$$C_n = \frac{A_n}{C_n}, \quad n = 0,1,2,3,\ldots.$$

A limiting process eventually leads to the value of the infinite continued fraction (5.3.36):

$$C = \lim_{n\to\infty} C_n.$$

Turning now to the concrete case (5.3.35) for even eigenvalues:

$$F_{a,even}(E) = -\frac{E-a}{2} - \cfrac{1}{p_2 - \cfrac{q_2}{p_4 - \cfrac{q_4}{p_6 - \ldots}}} = 0.$$

A comparison between (5.3.33) and (5.3.2) in the even case yields

$$
\begin{aligned}
& b_0 = -\frac{E-a}{2}, \\
a_1 = -1, \quad & b_1 = p_2, \\
a_2 = -q_2, \quad & b_2 = p_4, \\
a_3 = -q_4, \quad & b_3 = p_6, \\
a_4 = -q_6, \quad & b_4 = p_8, \\
& \ldots
\end{aligned}
$$

Thus

$$
\left.\begin{array}{lll} A_0 & = & b_0 = -\frac{E-a}{2} \\ B_0 & = & 1 \end{array}\right\} \rightsquigarrow C_0 = \frac{A_0}{B_0} = b_0 = -\frac{E-a}{2},
$$

$$
\left.\begin{array}{lll} A_1 = b_1 A_0 + a_1 A_{-1} & = & -\frac{E-a}{2} p_2 - q_2 \\ B_1 = b_1 B_0 + a_1 B_{-1} & = & p_2 \end{array}\right\} \rightsquigarrow C_1 = \frac{A_1}{B_1} = -\frac{E-a}{2} - \frac{1}{p_2},
$$

$$
\left.\begin{array}{lll} A_2 & = & b_2 A_1 + a_2 A_0 = p_4 A_1 - q_4 A_0 \\ B_2 & = & b_2 B_1 + a_2 B_0 = p_4 B_1 - q_4 B_0 \end{array}\right\} \rightsquigarrow C_2 = \frac{A_2}{B_2} = b_0 + \cfrac{a_1}{b_1 + \cfrac{a_2}{b_2}},
$$

$$
\left.\begin{array}{lll} A_3 & = & b_3 A_2 + a_3 A_1 = p_6 A_2 - q_6 A_1 \\ B_3 & = & b_3 B_2 + a_3 B_1 = p_6 B_2 - q_6 B_1 \end{array}\right\} \rightsquigarrow C_3 = \frac{A_3}{B_3} = b_0 + \cfrac{a_1}{b_1 + \cfrac{a_2}{b_2 + \cfrac{a_3}{b_3}}},
$$

$$
\left.\begin{array}{lll} A_4 & = & b_4 A_3 + a_4 A_2 = p_8 A_3 - q_8 A_2 \\ B_4 & = & b_4 B_3 + a_4 B_2 = p_8 B_3 - q_8 B_2 \end{array}\right\} \rightsquigarrow C_4 = \frac{A_4}{B_4} = b_0 + \cfrac{a_1}{b_1 + \cfrac{a_2}{b_2 + \cfrac{a_3}{b_3 + \cfrac{a_4}{b_4}}}}.
$$

$$\ldots$$

Turning now to the case (5.3.35) for eigenvalues of odd eigenfunctions:

$$
F_{a,odd}(E) = -\frac{E-a}{6} - \cfrac{1}{p_3 - \cfrac{q_3}{p_5 - \cfrac{q_5}{p_7 - \ldots}}}.
$$

A comparison between (5.3.33) and (5.3.2) in the odd case yields

$$
\begin{aligned}
& b_0 = -\frac{E-a}{6}, \\
a_1 = -1, \quad & b_1 = p_3,
\end{aligned}
$$

$$a_2 = -q_3, \quad b_2 = p_5,$$
$$a_3 = -q_5, \quad b_3 = p_7,$$
$$a_4 = -q_7, \quad b_4 = p_9,$$
$$\dots$$

Thus

$$\left.\begin{array}{lll} A_0 & = & b_1 = -\frac{E-a}{6} \\ B_0 & = & 1 \end{array}\right\} \rightsquigarrow C_0 = \frac{A_0}{B_0} = b_1 = -\frac{E-a}{6},$$

$$\left.\begin{array}{lll} A_1 = b_1\,A_0 + a_1\,A_{-1} & = & -\frac{E-a}{6}\,p_3 - q_3 \\ B_1 = b_1\,B_0 + a_1\,B_{-1} & = & p_3 \end{array}\right\} \rightsquigarrow C_1 = \frac{A_1}{B_1} = -\frac{E-a}{6} - \frac{1}{p_3},$$

$$\left.\begin{array}{lll} A_2 & = & b_2\,A_1 + a_2\,A_0 = p_5\,A_1 - q_5\,A_0 \\ B_2 & = & b_2\,B_1 + a_2\,B_0 = p_5\,B_1 - q_5\,B_0 \end{array}\right\} \rightsquigarrow C_2 = \frac{A_2}{B_2} = -\frac{E-a}{6} - \cfrac{q_3}{p_3 - \cfrac{q_4}{p_4}},$$

$$\left.\begin{array}{lll} A_3 & = & b_3\,A_2 + a_3\,A_1 = p_7\,A_2 - q_7\,A_1 \\ B_3 & = & b_3\,B_2 + a_3\,B_1 = p_7\,B_2 - q_7\,B_1 \end{array}\right\} \rightsquigarrow C_3 = \frac{A_3}{B_3} = -\frac{E-a}{6} - \cfrac{q_3}{p_3 - \cfrac{q_4}{p_4 - \cfrac{q_5}{p_5}}},$$

$$\left.\begin{array}{lll} A_4 & = & b_4\,A_3 + a_4\,A_2 = p_9\,A_3 - q_9\,A_2 \\ B_4 & = & b_4\,B_3 + a_4\,B_2 = p_9\,B_3 - q_9\,B_2 \end{array}\right\} \rightsquigarrow C_4 = \frac{A_4}{B_4} = -\frac{E-a}{6} - \cfrac{q_3}{p_3 - \cfrac{q_4}{p_4 - \cfrac{q_5}{p_5 - \cfrac{p_6}{q_6}}}}$$

$$\dots$$

Figure 5.26 displays the calculation of the eigenvalue curves $E = E_i(a)$. The numerical evaluation shows one even and one odd eigenvalue curve. This is in accordance with the fundamental postulate that the point $E = 0$ cannot be a cumulation point of eigenvalues, but the potential can have only a finite number of discrete energy levels,

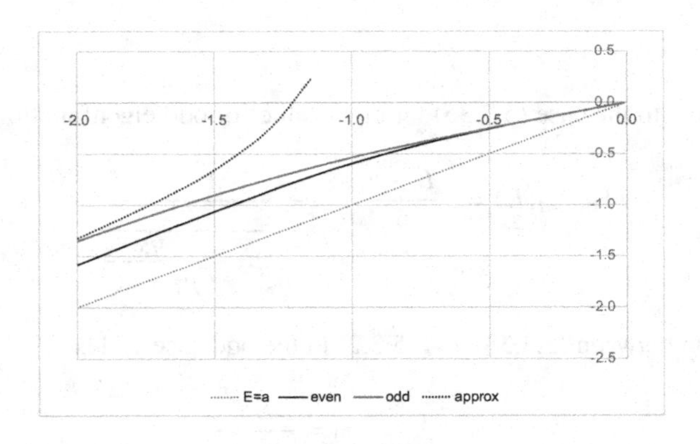

Figure 5.26 The exact eigenvalue curves and an approximated one.

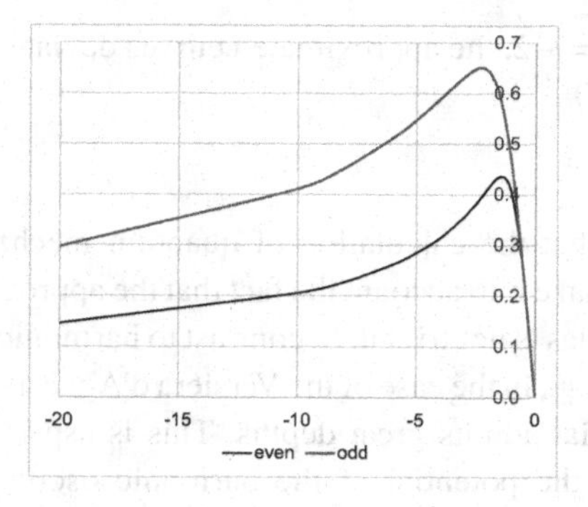

Figure 5.27 The difference of the eigenvalue curves to the asymptote.

since the quantum potential $V(z)$ in (5.3.7) tends to zero as $z \to \pm\infty$ in a quadratic manner. As may also be seen, both of the eigenvalue curves tend to the asymptote $E = a$ as $a \to \infty, a \in \mathbb{R}^-$. For the sake of comparison, this asymptote $E = a$ is drawn as well.

Figure 5.27 shows the distance between the two eigenvalue curves and their asymptote and Figure 5.28 shows the ratio between the two eigenvalue curves and their asymptote.

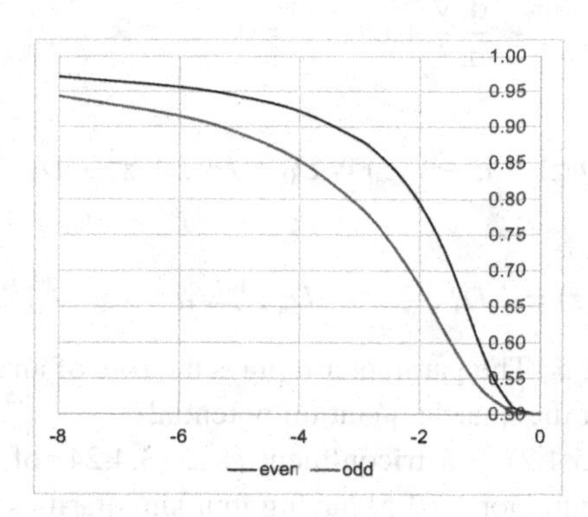

Figure 5.28 The ratio of exact and approximated eigenvalue curves.

Some remarks on the numerical result seem to be in order:

- It can be seen that the exact curve for decreasing values of a, i.e., for $a \to -\infty$, approaches the curve $E = a$, i.e., the eigenvalues approach more and more the potential floor. It should be noted that in contrast to here, this distance remains constant, namely unity, with the linear oscillator (i.e., a quantum particle in a square parabolic potential).
- It is seen, moreover, that the approximate solution for no value of a approaches the exact curve; this is plausible since an infinite series [cf. (5.3.25), (5.3.26)] has been truncated, which naturally results in an offset.

- Around the value $a = -2$, the approximated curves deviates substantially from the exact ones, as $a \to 0$.

Conclusion

This example shows that the calculation of quantum mechanical ground states in potential wells should take into account the fact that the approximation of the potential wells by square parabolas is not trivial. In contrast to harmonic oscillators, the ground states in some cases – e.g., in the case of the Versiera d'Agnesi – approach the potential well, when the potential admits great depths. This is especially true for potentials which, in contrast to the potential of the harmonic oscillator, have only a finite depth.

5.4 The Quantum Quartic Oscillator: Triconfluent Heun

The one-dimensional quartic oscillator is one of the most famous problems in quantum mechanics. It is described by means of the triconfluent case (3.4.24) of the Fuchsian differential equation (3.2.5) having four singularities in the form of a Schrödinger equation

$$\frac{d^2y}{dz^2} + Q(z)\,y = 0, \quad z \in \mathbb{R}, \tag{5.4.1}$$

with

$$Q(z) = E - V(z) = D_0 + D_1\,z + z^2 + D_4\,z^4 \tag{5.4.2}$$

and thus

$$V(z) = -D_1\,z - z^2 - D_4\,z^4 = a\,z + z^2 + \lambda^2\,z^4$$

where $a \in \mathbb{R}^+$ and $\lambda \in \mathbb{R}$. The parameter a plays the role of an asymmetry parameter and λ is the strengh of the quartic quantum potential.

Equation (5.4.1), (5.4.2) is a triconfluent case (3.4.24) of the Heun differential Fuchsian differential equation (3.2.5) having four singularities with

$$D_3 = G_0 = G_1 = G_2 = 0,$$

thus in its normal form. The parameter D_1 may always be transformed to unity $D_1 = 1$ without loss of generality.

The calculation of the spectrum of the quartic oscillator consists of looking for solutions $y(z)$ for which

$$\int_{z=-\infty}^{z=+\infty} y^2(z)\,dz < \infty$$

and may be achieved by means of two coupled central two-point connection problems, as discussed in §4.2.2, matching them at the end. This is carried out in the following.

The CTCP on the Positive Real Axis
Take a Jaffé ansatz having the form

$$y(z) = \exp\left(\frac{\eta}{3}\,z^3 + \frac{\kappa}{2}\,z^2 + \nu\,z\right)\,(z + z_0)^{\alpha}\,w(z) \tag{5.4.3}$$

whereby the characteristic exponent of the second kind and of third order η takes on two values

$$\eta_1 = -\lambda, \quad \eta_2 = +\lambda.$$

The other characteristic exponents are then given by

$$\kappa = -\frac{D_3}{2\,\eta} = 0, \quad \nu = -\frac{D_2}{2} = -\frac{1}{2}, \quad \alpha = \frac{a - 2\,\kappa\,\nu - 2\,\eta}{2\,\eta}.$$

The result of the Jaffé ansatz (5.4.3) is the differential equation for $w(z)$:

$$\frac{\mathrm{d}^2 w}{\mathrm{d}z^2} + \left[g_2\,z^2 + g_1\,z + g_0 + \frac{g_{-1}}{z + z_{+0}}\right]\frac{\mathrm{d}w}{\mathrm{d}z} + \left[d_0 + \frac{d_{-1}}{z + z_0} + \frac{d_{-2}}{(z + z_0)^2}\right]w = 0. \tag{5.4.4}$$

According to the general theory, a Jaffé transformation of the form

$$x = \frac{z}{z + z_0}$$

has to be carried out, transforming the differential equation (5.4.4) to

$$(x - 1)^4\,\frac{\mathrm{d}^2 w}{\mathrm{d}z^2} + \sum_{i=0}^{3}\Gamma_i\,x^i\,\frac{\mathrm{d}w}{\mathrm{d}z} + \sum_{i=0}^{2}\Delta_i\,x^i\,w = 0.$$

The CTCP on the Negative Real Axis
Here, the Jaffé ansatz is

$$y(z) = \exp\left(\frac{\eta}{3}\,z^3 + \frac{\kappa}{2}\,z^2 + \nu\,z\right)\,(-z + z_0)^{\alpha}\,w(z),$$

leading to equation (5.4.3) and a Jaffé transformation of the form

$$x = \frac{z}{z - z_0},$$

eventually resulting in (5.4.4).

The Difference Equation
Inserting a power series for $w(x)$:

$$w(x) = \sum_{n=0}^{\infty} a_n\,x^n$$

convergent for

$$|x| < 1$$

results in an irregular fourth-order difference equation of Poincaré–Perron type with initial condition

$$\begin{aligned}
&a_0,\ a_1 \text{ are arbitrary},\\
&2\,a_2 + \Gamma_0\,a_1 + \Delta_0\,a_0 = 0,\\
&6\,a_3 + (2\,\Gamma_0 - 8)\,a_2 + (\Gamma_1 + \Delta_0)\,a_1 + \Delta_1\,a_0 = 0,
\end{aligned}$$

$$\left(1 + \frac{\alpha_2}{n} + \frac{\beta_2}{n^2}\right) a_{n+2} + \left(1 + \frac{\alpha_2}{n} + \frac{\beta_2}{n^2}\right) a_{n+1} + \left(1 + \frac{\alpha_2}{n} + \frac{\beta_2}{n^2}\right) a_n$$
$$+ \left(1 + \frac{\alpha_2}{n} + \frac{\beta_2}{n^2}\right) a_{n-1} + \left(1 + \frac{\alpha_2}{n} + \frac{\beta_2}{n^2}\right) a_{n-2} = 0, \quad n \geq 2.$$

The parameters occurring in this calculation are given in the following:

$$\begin{aligned}
g_2 = 2\,\eta, \quad g_1 = 2\,\kappa = 0, \quad g_0 = 2\,\nu = -1, \quad g_{-1} = 2\,\alpha,\\
d_0 = D_0 + \nu^2 + \kappa - 2\,\alpha\,(\eta\,z_0 - \kappa),\\
d_{-1} = 2\,\alpha\,(\eta\,z_0^2 - \kappa\,z_0 + \nu), \quad d_{-2} = \alpha\,(\alpha - 1),
\end{aligned}$$

$$\begin{aligned}
\Gamma_3 = -g_{-1} + 2, \quad \Gamma_2 = g_2\,z_0^3 - g_1\,z_0^2 + g_0\,z_0 + 3\,g_{-1} - 6,\\
\Gamma_1 = g_1\,z_0^2 - 2\,g_0\,z_0 - 3\,g_{-1} - 6, \quad \Gamma_0 = -g_0\,z_0 + g_{-1} - 2,\\
\Delta_2 = d_{-2}, \quad \Delta_1 = d_{-1}\,z_0 - 2\,d_{-2}, \quad \Delta_0 = d_0\,z_0^2 - d_{-1}\,z_0 + d_{-2},
\end{aligned}$$

$$d_0 = D_0 + \nu^2 + \kappa + 2\,\alpha\,(\eta\,z_0 - \kappa),$$

$$\begin{aligned}
\alpha_2 = 3, \quad \beta_2 = 2, \quad \alpha_1 = \Gamma_0 - 4, \quad \beta_1 = \Gamma_0, \quad \alpha_0 = \Gamma_1 - 6, \quad \beta_0 = \Delta_0,\\
\alpha_{-1} = \Gamma_2 + 12, \quad \beta_{-1} = -\Gamma_2 + \Delta_1 - 8, \quad \alpha_{-2} = \Gamma_3 - 5, \quad \beta_{-2} = -2\,\Gamma_3 + \Delta_2 + 6.
\end{aligned}$$

The Matching

The matching condition of the two CTCPs at $z = x = 0$ is

$$y_+(z = 0) = y_-(z = 0), \qquad \left.\frac{\mathrm{d}y_+}{\mathrm{d}z}\right|_{z=0} = \left.\frac{\mathrm{d}y_-}{\mathrm{d}z}\right|_{z=0}$$

because under these conditions, the two eigenfunctions on the positive and on the negative real axis become smooth in all derivatives.

The Result

The numerical result is a countable infinite number of discrete positive real-valued eigenvalues $E = E_i$, $i = 0, 1, 2, 3 \ldots$, (cf. Bay et al., 1997; Lay, 1997). Some of the lowest eigenvalues are displayed in Figures 5.29–5.31 in dependence on a as well as λ.

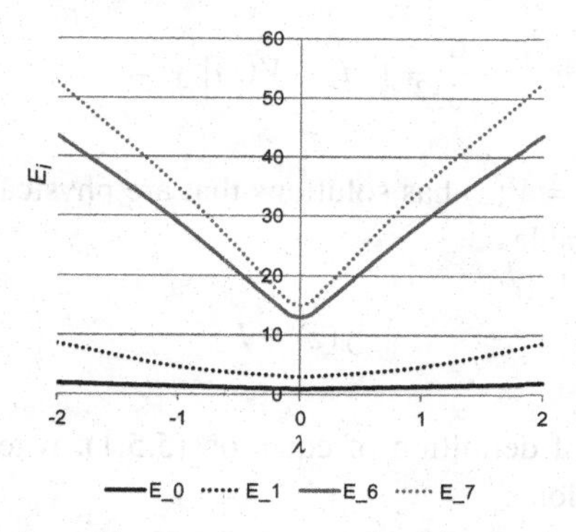

Figure 5.29 The discrete spectrum of the quantum quartic oscillator I for $a = 0$.

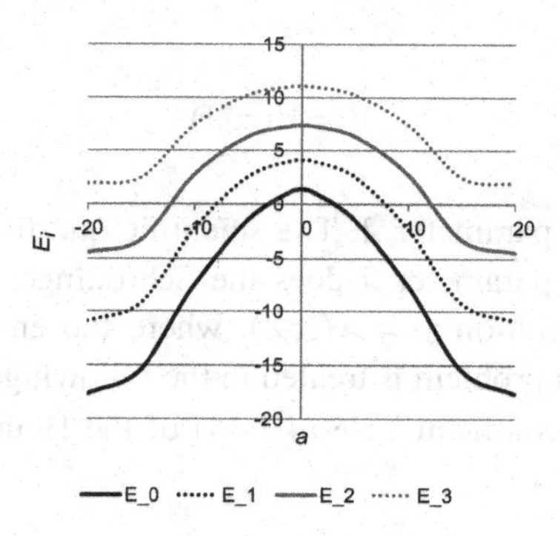

Figure 5.30 The discrete spectrum of the quantum quartic oscillator II for $\lambda = 0.7$.

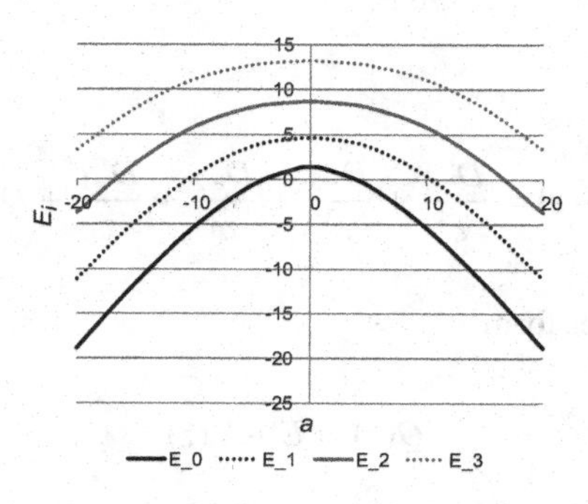

Figure 5.31 The discrete spectrum of the quantum quartic oscillator III for $\lambda = 1.0$.

5.5 The Phenomenon of Avoided Crossings: Double Confluent Heun

A standard quantum mechanical problem is the question for energy values $E = E_i$, $i = 0, 1, 2, \ldots$, for which the Schrödinger equation

$$\frac{\mathrm{d}^2 y}{\mathrm{d}z^2} + [E - V(z)]\, y = 0 \tag{5.5.1}$$

for a given potential $V = V(z)$ has solutions that are physically reasonable, viz. that are quadratically integrable

$$y(z) \in L_2$$

on the relevant range of definition of equation (5.5.1), where L_2 means the set of square integrable functions.

However, there are non-standard but still physically reasonable questions, one of which is to be dealt with in the following. Suppose there is a set of quantum mechanical potentials

$$V = V(z; \lambda)$$

that are dependent on a parameter λ. The scientific question here is the following: for which values of the parameter λ does the Schrödinger equation (5.5.1) have a physically reasonable solution $y = y(z; \lambda)$, where the energy parameter E has a certain fixed value? This problem is treated in the following (cf. Bay et al., 1998).

Consider the double confluent case (4.2.44) of the Heun equation (3.2.5) in its normal form, thus

$$\frac{\mathrm{d}^2 y}{\mathrm{d}z^2} + Q(z)\, y = 0 \tag{5.5.2}$$

with

$$Q(z) = \frac{D_{-4}}{z^4} + \frac{D_{-3}}{z^3} + \frac{D_{-2}}{z^2} + \frac{D_{-1}}{z} + D_0 \tag{5.5.3}$$

being in the Schrödinger form

$$Q(z) = E - V(z)$$

with

$$E = D_0,$$
$$V(z) = -\left(\frac{D_{-4}}{z^4} + \frac{D_{-3}}{z^3} + \frac{D_{-2}}{z^2} + \frac{D_{-1}}{z}\right).$$

Suppose now that there are three real-valued parameters $t > 0$, $a > 0$, $c > 0$ and a further one[6] $\lambda \in \mathbb{R}$ such that

$$\begin{aligned} D_0 &= E = -\frac{t^2}{4}, \\ D_{-4} &= \frac{t^2}{4} = -E, \\ D_{-3} &= -\frac{t}{2}(2-c), \\ D_{-2} &= -\frac{t^2}{2}(2\lambda+1), \\ D_{-1} &= -\frac{t}{2}(c-2a). \end{aligned}$$

In this case there is a quantum well, however, of a different shape as in §5.1 [cf. equation (5.1.10)]. Moreover, it should be recognised that there is no quantum tunnelling between the regions of positive and negative values of z because of the strong increase in the quantum potential $V(z)$ as $z \to 0$. This is called the *phenomenon of avoided crossing*, since there occurs an apparent eigenvalue curve crossing that is actually an avoiding one.

The CTCP on the positive real axis between $z = 0$ and infinity is to be solved, looking for values of $\lambda = \lambda_i$ for which, at given parameters t, c, a, there are solutions that are in L_2 on the positive real axis, meaning that these are reasonable quantum mechanical solutions of (5.5.2), (5.5.3) taken as a Schrödinger equation. The solution of this problem runs along the lines displayed in §4.2.3.

Jaffé Ansatz

$$y_d(z) = \exp\left(\alpha_{1\infty} z - \frac{\alpha_{10}}{z}\right) z^{\alpha_{00}} (z+1)^{\mu} w(z)$$

whereby the singularity parameters are given by

$$\alpha_{10}^2 + G_{-2}\,\alpha_{10} + D_{-4} = 0 \rightsquigarrow \alpha_{10} = \alpha_{101,2}$$

[6] The reason for this choice is that the differential equation in this case may be transformed into its canonical form

$$z^2 \frac{dy}{dz} + \left(-z^2 + c\,z + t\right) \frac{dy}{dz} + (-a\,z + \lambda)\,y = 0,$$

meaning that one of the two characteristic exponents at the finite singularity is zero, viz. $D_0 = D_{-3} = D_{-4} = 0$ in the general form (4.2.44) of the differential equation. Here, a, c are singularity parameters, defining the behaviour of the solutions at the irregular singularities located at $z = 0, \infty$; t is a scaling parameter and λ is the accessory parameter of the differential equation.

thus

$$\alpha_{101} = -\frac{t}{2},$$
$$\alpha_{102} = \frac{t}{2},$$
$$\alpha_{00} = -\frac{D_{-3} + \alpha_{10}\,(G_{-1} - 2)}{G_{-2} + 2\,\alpha_{10}} = 1 - \frac{D_{-3}}{2\,\alpha_{10}}$$

and

$$\alpha_{001} = 1 + \frac{t}{c} - \frac{t}{2},$$
$$\alpha_{002} = 2 - \frac{c}{2},$$
$$\alpha_{1\infty}^2 + G_0\,\alpha_{1\infty} + D_0 = 0 \rightsquigarrow \alpha_{1\infty} = \alpha_{1\infty 1,2}.$$

Therefore

$$\alpha_{1\infty 1} = \frac{t}{2},$$
$$\alpha_{1\infty 2} = -\frac{t}{2},$$
$$\mu = -\frac{D_{-1} + G_0\,\alpha_{00} + \alpha_{1\infty}\,(G_{-1} + 2\,\alpha_{00})}{G_0 + 2\,\alpha_{1\infty}} = -\frac{D_{-1} + 2\,\alpha_{1\infty}\,\alpha_{00}}{2\,\alpha_{1\infty}},$$

with

$$\mu_1 = 2 - a,$$
$$\mu_2 = -a.$$

In order to get the solution being square integrable on the relevant interval, thus on $[0, \infty[$, we choose

$$\alpha_{10} = \alpha_{102} = \frac{t}{2},$$
$$\alpha_{00} = \alpha_{002} = \frac{c}{2},$$
$$\alpha_{1\infty} = \alpha_{1\infty 2} = -\frac{t}{2},$$
$$\mu = \mu_2 = -a.$$

The result of this choice is the *intermediary differential equation*

$$z^2\,\frac{d^2w}{dz^2} + \left[g_2\,z^2 + g_1\,z + g_0 + \frac{g_{-1}}{z+1}\right]\frac{dw}{dz} + \left[d_0 + \frac{d_{-1}}{z+1} + \frac{d_{-2}}{(z+1)^2}\right]w(z) = 0$$

with

$$\begin{aligned}
g_2 &= 2\,\alpha_{1\infty 2} = -t,\\
g_1 &= 2\,(\alpha_{002} + \mu_2) = c - 2\,a,\\
g_0 &= 2\,(\alpha_{102} - \mu_2) = -t + 2\,a,\\
g_{-1} &= 2\,\mu_2 = -2\,a,\\
d_0 &= D_{-2} + 2\,\alpha_{1\infty 2}\,\alpha_{102} + \alpha_{1\infty}^2 - \alpha_{002} + \mu\,[-2\,\alpha_{1\infty 2} + 2\,\alpha_{002} + \mu_2 - 1]\\
&= t^2\left(\frac{1}{4} - \lambda\right) - \frac{c}{2} - a\,(t + c - a + 1),\\
d_{-1} &= \mu_2\,[2\,\alpha_{1\infty 2} - 2\,\alpha_{002} - 2\,\mu_2 + 2 + 2\,\alpha_{102}],\\
d_{-2} &= \mu_2\,(\mu_2 - 1) = a\,(a + 1).
\end{aligned} \tag{5.5.4}$$

The action of the *Jaffé transformation*

$$x = \frac{z-1}{z+1}$$

is

$$\begin{array}{ccccc}
z: & +\infty & 1 & 0 & -1\;, \\
\downarrow & \downarrow & \downarrow & \downarrow & \downarrow \\
x: & +1 & 0 & -1 & -\infty
\end{array}$$

yielding the *final differential equation*

$$(x^2-1)^2\,\frac{\mathrm{d}^2 w}{\mathrm{d}x^2} + \sum_{i=0}^{3} \Gamma_i\, x^i\, \frac{\mathrm{d}w}{\mathrm{d}x} + \sum_{i=0}^{2} \Delta_i\, x^i\, w(x) = 0$$

with

$$\begin{aligned}
\Gamma_3 &= -g_{-1} + 2,\\
\Gamma_2 &= 2\,g_2 - 2\,g_1 + 2\,g_0 + 3\,g_{-1} + 2,\\
\Gamma_1 &= 4\,g_2 - 4\,g_{0d} - 3\,g_{-1} - 2,\\
\Gamma_0 &= 2\,g_{2d} + 2\,g_1 + 2\,g_{0d} + g_{-1d} - 2,\\
\Delta_2 &= d_{-2},\\
\Delta_1 &= -2\,d_{-2} - 2\,d_{-1},\\
\Delta_0 &= d_{-2d} + 2\,d_{-1} + 4\,d_0.
\end{aligned} \tag{5.5.5}$$

A power series

$$w(x) = \sum_{x=0}^{\infty} a_n\, x^n$$

results in the *difference equation*

$$\begin{aligned}
&a_0,\ a_1 \text{ arbitrary},\\
&2\,a_2 + \Gamma_0\,a_1 + \Delta_0\,a_0 = 0,\\
&6\,a_3 + 2\,\Gamma_0\,a_2 + (\Gamma_1 + \Delta_0)\,a_1 + \Delta_1\,a_0 = 0,
\end{aligned}$$

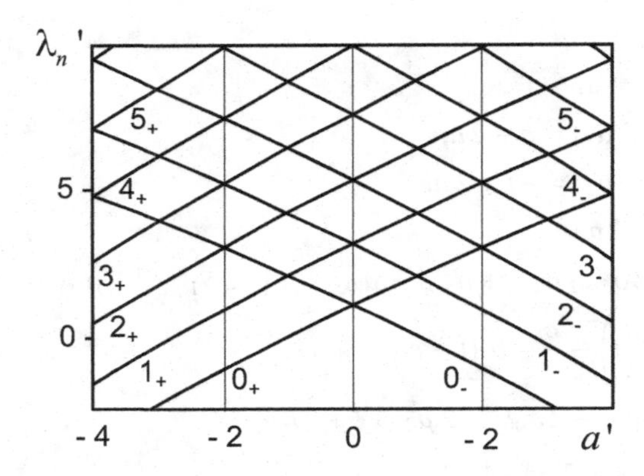

Figure 5.32 Suppression of quantum tunnelling I. Reproduced from Bay et al. (1998),

$$\begin{aligned}\left(1+\frac{\alpha_2}{n}+\frac{\beta_2}{n^2}\right)a_{n+2}&+\left(\frac{\alpha_1}{n}+\frac{\beta_1}{n^2}\right)a_{n+1}\\&+\left(-2+\frac{\alpha_0}{n}+\frac{\beta_0}{n^2}\right)a_n+\left(\frac{\alpha_{-1}}{n}+\frac{\beta_{-1}}{n^2}\right)a_{n-1}\\&\qquad+\left(1+\frac{\alpha_{-2}}{n}+\frac{\beta_{-2}}{n^2}\right)a_{n-2}=0,\ n\geq 2\end{aligned}\tag{5.5.6}$$

with

$$\begin{aligned}&\alpha_2=3,\quad \beta_2=2,\\&\alpha_1=\beta_1=\Gamma_0,\\&\alpha_0=\Gamma_1+2,\quad \beta_0=\Delta_0,\\&\alpha_{-1}=\Gamma_2,\quad \beta_{-1}=\Delta_1-\Gamma_2,\\&\alpha_{-2}=\Gamma_3-5,\quad \beta_{-2}=\Delta_2-2\Gamma_3+6.\end{aligned}$$

Writing

$$\lambda'(z)=\lambda-\frac{c\,(c-2)}{4t},\quad a'=a-1,\quad c'=\frac{c}{2}-1,$$

the numerical result is given in Figures 5.32–5.34.

It is clear that we may consider two relevant intervals of the original equation in z, namely the positive (denoted by +) and the negative (denoted by −) real half-axis. We get two sorts of eigenvalue curves in the λ'–a' coordinate systems for fixed values of t, c' having the λ' coordinate as their symmetry axis.

Interpreting the differential equation as a Schrödinger one, its potential has the form of a double well, the two wells of which are separated by an irregular singularity and thus is the simplest potential that models the suppression of tunnelling fluxes from one well to the neighbouring one. The parameter a in this case governs the asymmetry between the two wells. If the value of the parameter a exceeds a certain threshold (dependent on the other parameters), then there appear eigenvalues lying lower than

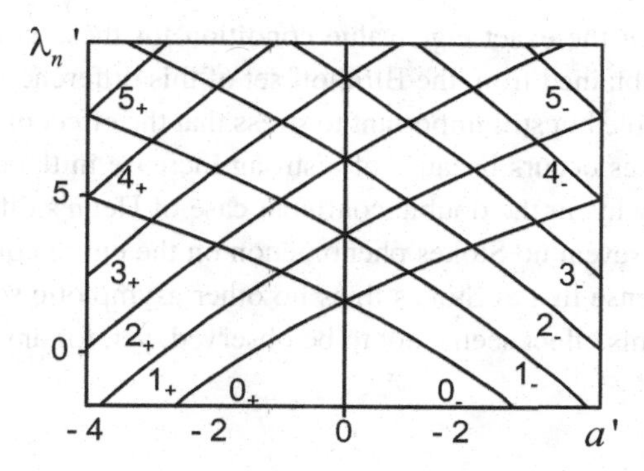

Figure 5.33 Suppression of quantum tunnelling II. Reproduced from Bay et al. (1998),

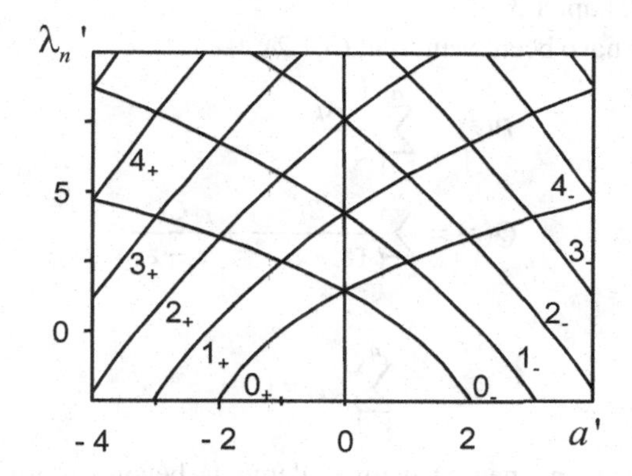

Figure 5.34 Suppression of quantum tunnelling III. Reproduced from Bay et al. (1998),

the minimum of the higher well. It should be mentioned that the corresponding eigenfunctions are generalised polynomials.

In Figures 5.32–5.34 I exhibit the six lowest-lying eigenvalue curves λ'–a' for $c = 0$ and for $t = 1$, $t = 3$, $t = 10$. The ground state is denoted 0 and the excited states are counted according to their number n. The central two-point connection problem on the negative half-axis is denoted by $-$ and on the positive half-axis by $+$.

The double confluent case of Heun's differential equation exhibits several peculiarities: the differential equation has two irregular singularities, the s-ranks of which are 2. When being placed at the origin and at infinity, the differential equation becomes symmetrical with respect to inversion at certain restrictions on the parameters. The generalised Jaffé transformation creates an additional regular singularity at infinity. Such a form is appropriate for solving the central two-point connection problems on the positive as well as on the negative half-axis. The coefficients of the Jaffé expansions obey an irregular fourth-order difference equation of the Poincaré–Perron type.

We have shown that the exact eigenvalue condition for these boundary eigenvalue problems may be obtained from the Birkhoff set of this difference equation.

It is understandable but still important to stress that the effect of suppressing quantum tunnelling fluxes occurs because of a strong increase in the quantum potential. This is the reason why, in the double confluent case of Heun's differential equation, the eigenfunctions reveal no Stokes phenomenon on the entire complex plane of the argument, in the sense that at Stokes lines no other asymptotic solution is added to the existing one. This effect seems not to be observed, yet, for any other equation.

5.6 The Land Beyond Heun

Fuchsian differential equations with more than four singularities may be constructed along the lines of Chapter 3.

The coefficients have been written in (3.1.2):

$$P(z) = \sum_{k=1}^{n} \frac{A_k}{z - z_k},$$

$$Q(z) = \sum_{k=1}^{n} \frac{B_k}{(z - z_k)^2} + \frac{C_k}{z - z_k}$$

with

$$\sum_{k=1}^{n} C_k = 0.$$

However, there are no new structural elements beyond those occurring in the Fuchsian equation with at most four singularities:

- There are three singularities that may always be placed at $z = 0, 1, \infty$.
- One of the two characteristic exponents at each of the finite singularities may be chosen to be zero.
- There are three sorts of parameters: ones that characterise the locations of the singularities, the characteristic exponents and the accessory parameters.
- The number of accessory parameters is $n-3$, where n is the number of singularities of the Fuchsian differential equation.

In a sense, it might be said that the location parameters as well as the characteristic exponents are local or *singularity parameters* characterising features of the singularities of the differential equations and their solutions, respectively, while the accessory parameters are global parameters concerning the differential equation as a whole.

The local solutions about ordinary points may always be written as a power series, while local solutions about regular singularities of the differential equation are Frobenius solutions, the coefficients of which may be solved recursively by means of a linear difference equation. The radius of convergence of the power series in both of these cases ranges to the neigbouring singularity, yielding either a holomorphic or a multiplicative solution.

Last but not least, it should be mentioned that if the power series terminates, the solution becomes a global one, either a holomorphic solution of the whole complex plane or a global solution that becomes multiplicative on surrounding the singularity at hand.

5.6.1 Ellipsoidal Wave Equation: Single Confluent Arscott

Ellipsoidal coordinates are the most general ones within the frame of confocal conic sections. The separation of the Helmholtz equation, which describes vibrations of three-dimensional media within the framework of classical physics

$$\frac{\partial^2 \phi}{\partial x_1^2} + \frac{\partial^2 \phi}{\partial x_2^2} + \frac{\partial^2 \phi}{\partial x_3^2} = \lambda\, \phi \tag{5.6.1}$$

in ellipsoidal coordinates for the angular functions, results in spherical harmonics. For the radial function $R(r)$, however, the differential equation is the so-called *ellipsoidal wave equation*:

$$P_0(r)\,\frac{\mathrm{d}^2 R}{\mathrm{d}r^2} + P_1(r)\,\frac{\mathrm{d}R}{\mathrm{d}r} + P_2(r)\,R = 0$$

with

$$\begin{aligned}
&0 \le r < \infty,\\
&P_0(r) = r\,(r-1)\,(r-c),\\
&P_1(r) = \frac{1}{2}\left[3\,r^2 - 2\,(c+1)\,r + c\right],\\
&P_2(r) = \lambda + \mu\,r + \gamma\,r^2,
\end{aligned}$$

where the ranges of the real-valued parameters are $c \ge 1$, λ, μ, $\gamma \in \mathbb{R}$. Dividing this equation by $P_0(r)$ yields the form

$$\frac{\mathrm{d}^2 R}{\mathrm{d}r^2} + P(r)\,\frac{\mathrm{d}R}{\mathrm{d}r} + Q(r)\,R(r) = 0 \tag{5.6.2}$$

with

$$\begin{aligned}
P(r) &= \frac{\frac{1}{2}}{r} + \frac{\frac{1}{2}}{r-1} + \frac{\frac{1}{2}}{r-c},\\
Q(r) &= \frac{\frac{\lambda}{c}}{r} + \frac{\frac{\lambda+\mu+\gamma}{1-c}}{r-1} + \frac{\frac{\lambda+c\,\mu+c^2\,\gamma}{c\,(c-1)}}{r-c}.
\end{aligned} \tag{5.6.3}$$

The differential equation (5.6.2), (5.6.3) is a specific *single confluent case of the Fuchsian differential equation with five singularities*, called the *ellipsoidal wave equation* or *Lamé wave equation*.

In §3.5.2 I referred to the Fuchsian differential equation with five singularities, the *Arscott differential equation*. Thus, the differential equation (5.6.2), (5.6.3) is the standard form of the single confluent case of the Arscott differential equation.

The inverse of (5.6.2), (5.6.3) is given by

$$\varrho = \frac{1}{r},$$

$$P(\varrho) = \frac{\frac{1}{2}}{\frac{1}{\rho}} + \frac{\frac{1}{2}}{\frac{1}{\rho} - 1} + \frac{\frac{1}{2}}{\frac{1}{\rho} - c},$$

$$Q(\varrho) = \frac{\frac{\lambda}{c}}{\frac{1}{\rho}} + \frac{\frac{\lambda+\mu+\gamma}{1-c}}{\frac{1}{\rho} - 1} + \frac{\frac{\lambda+c\,\mu+c^2\gamma}{c\,(c-1)}}{\frac{1}{\rho} - c},$$

resulting in

$$\frac{dy}{d\varrho} + \tilde{P}(\varrho)\,\frac{dy}{d\varrho} + \tilde{Q}\,w = 0$$

where

$$\tilde{P}(\varrho) = \frac{2}{\rho} - \frac{P(\varrho)}{\rho^2} = \frac{\frac{1}{2}}{\rho} + \frac{\frac{1}{2}}{\rho - \frac{1}{c}} + \frac{\frac{1}{2}}{\rho - 1},$$

$$\tilde{Q}(\varrho) = \frac{Q(\varrho)}{\rho^4}$$

$$= \frac{\gamma}{\rho^3} + \frac{c\,\gamma + \gamma + \mu}{\rho^2} + \frac{c^2\,\gamma + c\,(\gamma + \mu) + \gamma + \lambda + \mu}{\rho} + \frac{\frac{c\,(c^2\gamma + c\,\mu + \lambda)}{1-c}}{\rho - \frac{1}{c}} + \frac{\frac{\gamma+\lambda+\mu}{c-1}}{\rho - 1}.$$

Here, because of the term $\frac{\gamma}{\rho^3}$, it is clearly seen that the singularity of (5.6.2), (5.6.3) at infinity is an irregular one, the s-rank of which is $s = \frac{3}{2}$.

Thus, the differential equation (5.6.2), (5.6.3) is no longer of Heun type, but has the s-rank symbol

$$\left\{\frac{1}{2}, \frac{1}{2}, \frac{1}{2}; \frac{3}{2}\right\}.$$

As already remarked, I refer to the Fuchsian differential equation with five singularities as the *Arscott differential equation*. Therefore, (5.6.2), (5.6.3) denotes a specific *single confluent case of the Arscott differential equation*; indeed, it was actually Felix Medland Arscott in Arscott et al. (1983) who investigated this equation and calculated its discrete spectrum.

In the following we display an important application of a CTCP formulated on the positive real axis of this differential equation that is not only representative in itself but exercises the handling of two eigenvalue parameters.

The characteristic exponents of the finite singularities are given by [cf. equation (2.3.14)]

$$\alpha_{01} = 0, \quad \alpha_{02} = \frac{1}{2},$$
$$\alpha_{11} = 0, \quad \alpha_{12} = \frac{1}{2},$$
$$\alpha_{c1} = 0, \quad \alpha_{c2} = \frac{1}{2}.$$

The Thomé solution at infinity is

$$y(z) = \exp\left(\alpha_\infty \sqrt{r}\right) r^{\alpha_{0i}} \left(1 + \sum_{n=0}^{\infty} \frac{C_n}{r^n}\right)$$

with [cf. (2.3.28)–(2.3.30)]

$$\alpha_{\infty 1} = +2\sqrt{-\frac{\lambda}{c}}, \quad \alpha_{0i1} = \frac{\alpha_{\infty 1}}{4\,\alpha_{\infty 1} - 1},$$
$$\alpha_{\infty 2} = -2\sqrt{-\frac{\lambda}{c}}, \quad \alpha_{0i2} = -\frac{\alpha_{\infty 2}}{4\,\alpha_{\infty 2} - 1}.$$

In the following we formulate a singular boundary eigenvalue problem on the differential equation (5.6.2), (5.6.3) and calculate some of the parameters for which this equation meets the boundary conditions. These are holomorphic at each of the three finite singularities of the Lamé wave equation (5.6.2), (5.6.3). Thus, the eigenvalue condition is looking for solutions that are holomorphic at all of the finite singularities, located at $r = 0$, $r = 1$ and $r = c$, viz. looking for entire solutions. These solutions are called *ellipsoidal wave functions*.

The problem has four parameters: γ, c, λ, μ. The first has a physical meaning, namely the frequency of the vibrating system; the second places one of the three finite singularities of the differential equation and thus is a geometrical parameter, while the last two arise from the process of separating the Helmholtz equation (5.6.1) in ellipsoidal coordinates. For given values of c and γ, it looks for values λ and μ such that the solution $R(r)$ of (5.6.2), (5.6.3) obeys the boundary conditions.

The ansatz

$$y(z) = \sum_{n=0}^{\infty} a_n\, r^n \tag{5.6.4}$$

yield a regular third-order difference equation of Poincaré–Perron type, inclusive of the initial condition

$$
\begin{aligned}
F(\lambda, \mu, \gamma) &= \lambda\, a_0 + \frac{1}{2}\, c\, a_1 = 0, \\
G(\lambda, \mu, \gamma) &= \mu\, a_0 + [\lambda - (1 + c)]\, a_1 + 3\, c\, a_2 = 0, \\
&\gamma\, a_n + \left[\mu + \left(n + \frac{3}{2}\right)(n + 1)\right] a_{n+1} \\
&+ \left[\lambda - (1 + c)(n + 2)^2\right] a_{n+2} + c\left(n + \frac{5}{2}\right)(n + 3)\, a_{n+3} = 0, \quad n = 0, 1, 2, 3, \ldots .
\end{aligned}
\tag{5.6.5}
$$

This difference equation may be written in the form

$$
\begin{aligned}
\frac{1}{2}\, c\, a_1 + \lambda\, a_0 &= 0, \\
3\, c\, a_2 + [\lambda - (1 + c)]\, a_1 + \mu\, a_0 &= 0, \\
\left(c + \frac{\frac{3}{2} c}{n} + \frac{\frac{c}{2}}{n^2}\right) a_{n+1} - \left[(1 + c) - \frac{\lambda}{n^2}\right] a_n & \\
+ \left[1 - \frac{\frac{3}{2}}{n} + \frac{\frac{2\mu+1}{2}}{n^2}\right] a_{n-1} + \frac{\mu}{n^2}\, a_{n-2} &= 0, \quad n = 2, 3, 4, \ldots .
\end{aligned}
$$

The characteristic equation

$$
t^3 - \frac{c + 1}{c}\, t^2 + \frac{1}{c}\, t = 0 \tag{5.6.6}
$$

yields the three solutions

$$
\begin{aligned}
t_1 &= 1, \\
t_2 &= \frac{1}{c}, \\
t_3 &\sim -\frac{\gamma}{n^2} \to 0 \text{ as } n \to \infty.
\end{aligned}
$$

In order to get the particular solution of the difference equation that relates to the third solution t_3 of the characteristic equation (5.6.6), the solution $\{a_n\}$ has to tend to zero $a_n \to 0$ as $n \to \infty$. The consequence is that the radius of convergence of (5.6.4) tends to infinity, yielding an entire function. This function obeys the condition to be holomorphic at each of the three finite singularities, being located at $r = 0$, $r = 1$ and $r = c$.

Because the CTCP is two-parametric, there has to be a more sophisticated numerical procedure in order to calculate the eigenvalues. This is discussed in the following. The main difference from what is discussed above is that the regular difference equation of Poincaré–Perron type (5.6.5) is numerically solved by means of a backward recurrence relation defining the quantities x_n by

$$
\begin{aligned}
x_{N+2} &= x_{N+1} = 0, \\
x_n &= 1, \\
x_n &= -\frac{1}{\gamma}\left\{\left[\mu + \left(n+\frac{3}{2}\right)(n+1)\right] x_{n+1}\right. \\
&\quad \left. + \left[\lambda - (1+c)(n+2)^2\right] x_{n+2} + c\left(n+\frac{5}{2}\right)(n+3)\, x_{n+3}\right\}, \\
&\qquad n = N-1, N-2, N-3, \ldots, 0
\end{aligned}
$$

for sufficiently large values of N. Then

$$
y_n = \frac{x_n}{x_0}, \quad n = 0, 1, 2, 3, \ldots
$$

are estimates of a_n in the sense that

$$
a_n = \lim_{N \to \infty} y_n.
$$

The eigenvalue condition under given values for c and λ for the two eigenvalue parameters μ and γ are the two equations

$$
\begin{aligned}
F(\lambda, \mu, \gamma) &= \lambda\, a_0 + \frac{1}{2}\, c\, a_1 = 0, \\
G(\lambda, \mu, \gamma) &= \mu\, a_0 + [\lambda - (1+c)]\, a_1 + 3\, c\, a_2 = 0,
\end{aligned}
$$

which have to be met by means of a two-dimensional iterative numerical process for this system of two algebraic equations (of the two unknown variables λ and μ).

The details of this numerical procedure have been described in Arscott et al. (1983, p. 371).

The resulting eigenvalues (or eigensolutions) may be numbered such that for each value n, $n = 0, 1, 2, \ldots$, there exist a number m, $m = 0, 1, 2, \ldots, n$ eigenvalues (or eigenfunctions). The numerical starting point may be put at $\gamma = 0$. When $\gamma = 0$, the eigenvalues are given by

$$
\mu = -n\left(n + \frac{1}{2}\right).
$$

The eigenvalues form a pair of eigenvalue curves $\lambda = \lambda(\gamma)$, $\mu = \mu(\gamma)$ for a given value of c. Figures 5.35 and 5.36 sketch the curves $\lambda = \lambda(\gamma)$ and $\mu = \mu(\gamma)$ for $c = 2$, which are just the historic curves Felix Arscott and his collaborators calculated.

5.6.2 Ince Equation

In §3.5.2 it was suggested to call the Fuchsian differential equation with six singularities the *Ince differential equation*. In the following I give an example of this differential equation in mathematical physics, namely in the field of space-dependent classical tracer diffusion under high-density irradiation induction in crystalline solids.

Seeger and Lay (1990, p. 162) showed that irradiation-induced diffusion of tracer atoms on vacancies in crystalline materials (e.g., metals) is described by the classical diffusion equation

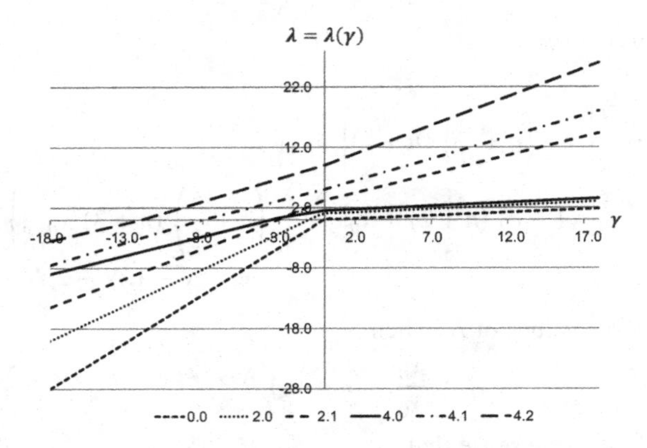

Figure 5.35 Eigenvalue curves $\lambda = \lambda(\gamma)$ for $c = 2$ of the ellipsoidal wave equation. Republished with permission of the American Mathematical Society from Arscott et al. (1983); permission conveyed through Copyright Clearance Center, Inc.

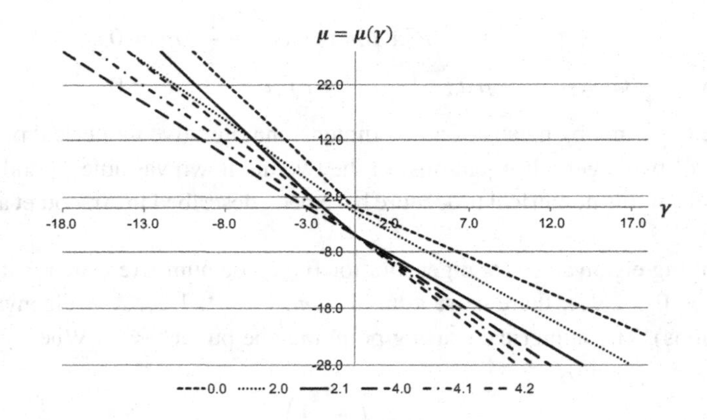

Figure 5.36 Eigenvalue curves $\mu = \mu(\gamma)$ for $c = 2$ of the ellipsoidal wave equation. Republished with permission of the American Mathematical Society from Arscott et al. (1983); permission conveyed through Copyright Clearance Center, Inc.

$$\frac{\mathrm{d}^2 C}{\mathrm{d}z^2} + \left[\frac{1}{z} + \frac{z''}{z'^2}\right] \frac{\mathrm{d}C}{\mathrm{d}z} + \frac{\lambda}{z\,z'^2} C = 0, \tag{5.6.7}$$

where $C = C(z)$ is the concentration of the diffusing particle under the stationary space-dependent diffusion process and λ is the eigenvalue parameter stemming from a separation ansatz. The diffusion problem is governed by space-dependent diffusion coefficients $z(x)$ of the form

$$z(x) = a + b\,\mathrm{sn}^2[c\,x + K(k), k], \tag{5.6.8}$$

where $K(k)$ is the complete elliptic integral of first kind:

$$K(k) = \int_0^1 \frac{\mathrm{d}x}{\sqrt{(1-x^2)(1-k^2 x^2)}}.$$

This quantity $K(k)$, $0 \leq k \leq 1$ determines half of the period of the sinus amplitudinis sn in (5.6.8). There is an alternative to (5.6.8), the diffusion coefficient $z(x)$ of which is given by a pure sinus amplitudinis

$$z(x) = \mathrm{sn}[x, k]. \tag{5.6.9}$$

For the sake of calculatory simplicity, $a = 0,\ b = c = 1$ is taken here, and there is no shift along the x-axis.

Those values λ_i are looked for, which the differential equation (5.6.7) admits physically reasonable particular solutions. The diffusion differential equation (5.6.7) and the formalism of §4.2 are applied; thus the function $z(x)$ in (5.6.8) is to be derived twice. The derivative of the sinus amplitudinis $\mathrm{sn}(x, k)$ is given by (cf. Abramowitz and Stegun, 1970, p. 574)

$$\frac{\mathrm{d}\,\mathrm{sn}(x,k)}{\mathrm{d}x} = \mathrm{cn}(x,k)\,\mathrm{dn}(x,k) = \sqrt{1-\mathrm{sn}^2(x,k)}\,\sqrt{1-k^2\,\mathrm{sn}^2(x,k)}.$$

This derivative may be written in dependence on z exclusively, yielding

$$\begin{aligned} z' = \frac{\mathrm{d}z}{\mathrm{d}x} &= \mathrm{cn}[x,k]\,\mathrm{dn}[x,k] \\ &= \sqrt{1-\mathrm{sn}^2[x,k]}\,\sqrt{1-k^2\,\mathrm{sn}^2[x,k]} \\ &= \sqrt{(1-z^2)\,(1-k^2 z^2)} = \sqrt{g(z)} \end{aligned}$$

with

$$g(z) = (1-z^2)(1-k^2 z^2). \tag{5.6.10}$$

Thus

$$z'^2 = \left(\frac{\mathrm{d}z}{\mathrm{d}x}\right)^2 = g(z). \tag{5.6.11}$$

The second derivative is

$$\begin{aligned} z'' = \frac{\mathrm{d}^2 z}{\mathrm{d}x^2} = \frac{\mathrm{d}\left(\frac{\mathrm{d}z}{\mathrm{d}x}\right)}{\mathrm{d}x} &= \frac{\mathrm{d}}{\mathrm{d}z}\left[\frac{\mathrm{d}z}{\mathrm{d}x}\right] \cdot \frac{\mathrm{d}z}{\mathrm{d}x} \\ &= \frac{\mathrm{d}\sqrt{g(z)}}{\mathrm{d}z}\sqrt{g(z)} \\ &= \frac{1}{2}\frac{\mathrm{d}g}{\mathrm{d}z} = \frac{1}{2}g'. \end{aligned}$$

Thus, the coefficients of (5.6.7) become

$$\begin{aligned}
\tilde{P}(z) &= \frac{1}{z} + \frac{z''}{z'^2} = \frac{1}{z} + \frac{1}{2}\frac{g'}{g} \\
&= \frac{A_{-\frac{1}{k}}}{z+\frac{1}{k}} + \frac{A_{-1}}{z+1} + \frac{A_0}{z} + \frac{A_1}{z-1} + \frac{A_{\frac{1}{k}}}{z-\frac{1}{k}}, \\
\tilde{Q}(z) &= \frac{\lambda}{z\,z'^2} = \frac{\lambda}{z\,g(z)} \\
&= \frac{C_{-\frac{1}{k}}}{z+\frac{1}{k}} + \frac{C_{-1}}{z+1} + \frac{C_0}{z} + \frac{C_1}{z-1} + \frac{C_{\frac{1}{k}}}{z-\frac{1}{k}}
\end{aligned} \tag{5.6.12}$$

with

$$\begin{aligned}
A_{-\frac{1}{k}} &= \frac{1}{2}, \\
A_{-1} &= \frac{1}{2}, \\
A_0 &= 1, \\
A_1 &= \frac{1}{2}, \\
A_{\frac{1}{k}} &= \frac{1}{2}, \\
C_{-\frac{1}{k}} &= \frac{k^2\,\lambda}{2\,(1-k^2)}, \\
C_{-1} &= \frac{\lambda}{2\,(k^2-1)}, \\
C_0 &= \lambda, \\
C_1 &= \frac{\lambda}{2\,(k^2-1)} = C_{-1}, \\
C_{\frac{1}{k}} &= \frac{k^2\,\lambda}{2\,(1-k^2)} = C_{-\frac{1}{k}},
\end{aligned}$$

whereby

$$C_{-\frac{1}{k}} + C_{-1} + C_0 + C_1 + C_{\frac{1}{k}} = \frac{k^2\,\lambda}{1-k^2} + \frac{\lambda}{k^2-1} + \lambda = 0,$$

from which it is seen that (5.6.7) with (5.6.8) is a Fuchsian differential equation. As a result, the s-rank symbol of (5.6.8), (5.6.12) is

$$\left\{\frac{1}{2}, \frac{1}{2}, 1, \frac{1}{2}, \frac{1}{2}; 1\right\},$$

thus equation (5.6.8), (5.6.12) is a Fuchsian differential equation having six singularities. I refer to it as the Ince differential equation.

On this differential equation is formulated a CTCP in order to solve the diffusion problem, the relevant interval of which is given by

$$[0, 1].$$

The characteristic exponents at $x = 0$ are

$$\alpha_{01} = 1,$$
$$\alpha_{02} = 0,$$

while the characteristic exponents at $x = 1$ are

$$\alpha_{11} = \frac{1}{2},$$
$$\alpha_{12} = 0.$$

This shows that the eigensolutions decompose into even and odd functions with respect to $x = K(k)$, resp. $z = 1$. Therefore, in the following, these solutions have to be dealt with separately. As a first step, we consider the even eigensolutions.

Solutions are looked for that are holomorphic at $z = 0$ as well as at $z = 1$. Equation (5.6.7), (5.6.12) may be written in the form

$$P_0(z)\frac{\mathrm{d}^2 y}{\mathrm{d}z^2} + P_1(z)\frac{\mathrm{d}y}{\mathrm{d}z} + P_2(z)\, y = 0 \tag{5.6.13}$$

with

$$P_0(z) = \kappa_5\, z^5 + \kappa_3\, z^3 + \kappa_1\, z,$$
$$P_1(z) = \Gamma_4\, z^4 + \Gamma_3\, z^3 + \Gamma_2\, z^2 + \Gamma_1\, z + \Gamma_0,$$
$$P_2(z) = \Delta_3\, z^3 + \Delta_2\, z^2 + \Delta_1\, z + \Delta_0,$$

whereby the coefficients of the terms in these three equations are given by

$$\kappa_5 = 1,$$
$$\kappa_3 = -\frac{k^2 + 1}{k^2},$$
$$\kappa_1 = \frac{1}{k^2},$$
$$\Gamma_4 = A_{-\frac{1}{k}} + A_{-1} + A_0 + A_1 + A_{\frac{1}{k}},$$
$$\Gamma_3 = -\frac{A_{-\frac{1}{k}} + k\, A_{-1} - k\, A_1 - A_{\frac{1}{k}}}{k},$$
$$\Gamma_2 = -\frac{k^2\, A_{-\frac{1}{k}} + A_{-1} + (k^2 + 1)\, A_0 + A_1 + k^2\, A_{\frac{1}{k}}}{k^2},$$
$$\Gamma_1 = \frac{k\, A_{-\frac{1}{k}} + A_{-1} - A_1 - A_{\frac{1}{k}}}{k^2},$$
$$\Gamma_0 = \frac{A_0}{k^2},$$

$$\begin{aligned}
\Delta_3 &= -\frac{C_{-\frac{1}{k}} + k\,C_{-1} - k\,C_1 - C_{\frac{1}{k}}}{k},\\
\Delta_2 &= -\frac{k^2\,C_{-\frac{1}{k}} + C_{-1} + (k^2+1)\,C_0 + C_1 + k^2\,C_{\frac{1}{k}}}{k^2},\\
\Delta_1 &= \frac{k\,C_{-\frac{1}{k}} + C_{-1} - C_1 - C_{\frac{1}{k}}}{k^2},\\
\Delta_0 &= \frac{C_0}{k^2}.
\end{aligned} \tag{5.6.14}$$

On this differential equation is formulated a central two-point connection problem. Look for parameters $\lambda = \lambda_i$ in dependence on k such that the solution of (5.6.2) behaves holomorphically, simultaneously at $z = -1$ and at $z = +1$. There may be an ansatz of the form

$$y(z) = z^{\alpha} \sum_{n=0}^{\infty} a_n\, z^n, \tag{5.6.15}$$

where the characteristic exponent α is either $\frac{1}{2}$ or 0. The former case produces the odd eigensolutions, the latter one the even eigensolutions.

The ansatz (5.6.15) leads to the difference equation of Poincaré–Perron type

$$\begin{aligned}
&[\kappa_1\,(1+\alpha)(1+\alpha-1) + \Gamma_0\,(1+\alpha)]\;a_1 + [\Gamma_1\,(0+\alpha)]\;a_0 = 0,\\
&[\kappa_1\,(2+\alpha)(2+\alpha-1) + \Gamma_0\,(2+\alpha)]\;a_2 + \Gamma_2\,(1+\alpha)\,[\Gamma_1\,(1+\alpha)+]\;a_1\\
&\qquad + [\kappa_3\,\alpha\,(\alpha-1) + \Gamma_2\,(0+\alpha) + \Delta_0]\;a_0 = 0,\\
&[\kappa_1\,(3+\alpha)(3+\alpha-1) + \Gamma_0\,(3+\alpha)]\;a_3 + [\Gamma_1\,(2+\alpha)]\;a_2\\
&\qquad + [\kappa_3\,(1+\alpha)(1+\alpha-1) + \Delta_0]\;a_1 + [\Gamma_2\,(0+\alpha) + \Delta_1]\;a_0 = 0,\\
&[\kappa_1\,(4+\alpha)(4+\alpha-1) + \Gamma_0\,(4+\alpha)]\;a_4 + [\Gamma_1\,(3+\alpha)]\;a_3\\
&\qquad + [\kappa_3\,(2+\alpha)(2+\alpha-1) + \Gamma_2\,(2+\alpha) + \Delta_0]\;a_2\\
&\qquad\qquad + [\Gamma_3\,(1+\alpha) + \Delta_1]\;a_1 + [\kappa_5\,\alpha\,(\alpha-1) + \Gamma_4\,(0+\alpha) + \Delta_2]\;a_0 = 0,
\end{aligned}$$

$$\begin{aligned}
&\left(\kappa_1 + \frac{\alpha_2}{n} + \frac{\beta_2}{n^2}\right) a_{n+2}\\
&\quad + \left(\frac{\alpha_1}{n} + \frac{\beta_1}{n^2}\right) a_{n+1}\\
&\quad + \left(\kappa_1 + \frac{\alpha_0}{n} + \frac{\beta_0}{n^2}\right) a_n
\end{aligned}$$

$$
\begin{aligned}
&+\left(\frac{\alpha_{-1}}{n}+\frac{\beta_{-1}}{n^2}\right) a_{n-1} \\
&+\left(\kappa_1+\frac{\alpha_{-2}}{n}+\frac{\beta_{-2}}{n^2}\right) a_{n-2} \\
&+\left(\frac{\alpha_{-3}}{n}+\frac{\beta_{-3}}{n^2}\right) a_{n-3}=0, \quad n=3,4,\ldots,
\end{aligned}
\tag{5.6.16}
$$

with

$$
\begin{aligned}
&\alpha_2 = 2\,\alpha\,\kappa_1 + 3\,\kappa_1 + \Gamma_0, \quad \beta_2 = (2+\alpha)\,[(1+\alpha)\,\kappa_1 + \Gamma_0], \\
&\alpha_1 = \Gamma_1, \quad \beta_2 = (1+\alpha)\,\Gamma_1, \\
&\alpha_0 = (2\,\alpha - 1)\,\kappa_3 + \Gamma_2, \quad \beta_0 = \alpha\,(\alpha-1)\,\kappa_3 + \alpha\,\Gamma_2 + \Delta_0, \\
&\alpha_{-1} = \Gamma_3, \quad \beta_{-1} = (\alpha-1)\,\Gamma_3 + \Delta_1, \\
&\alpha_{-2} = (2\,\alpha - 5)\,\kappa_5 + \Gamma_4, \quad \beta_{-2} = (\alpha-2)\,[(\alpha-3)\,\kappa_5 + \Gamma_4], \\
&\alpha_{-3} = 0, \quad \beta_{-3} = \Delta_3.
\end{aligned}
$$

The radius of convergence r of the series in (5.6.15) is generally unity, $r = 1$. However, there may be certain specific values λ_i, $i = 0, 1, 2, \ldots$, of the eigenvalue parameter λ for which this radius of convergence is enlarged to $r = \frac{1}{k}$. These specific values λ_i, $i = 0, 1, 2, \ldots$, are then the eigenvalues of the above-formulated CTCP.

The odd case is dealt with by means of the s-homotopic transformation

$$
y(z) = f(z)\,w(z)
$$

with[7]

$$
f(z) = (z+1)^{\frac{1}{2}}\,(z-1)^{\frac{1}{2}}. \tag{5.6.17}
$$

I will refrain from an explicit presentation of this result, as well as a numerical calculation of the eigenvalues at this point, and leave this to the inclined reader as an exercise at the end of the book.

Summary

The physical problem of calculating eigenvalues of diffusion processes with space-dependent diffusion coefficients has been dealt with in this subsection. In dependence on the concrete functional behaviour of the diffusion coefficient, there appears the type of differential equation to be solved. In order to get an overview of the resulting differential equations in relation to the functional forms of the diffusion coefficients, in the following a tabular is given, in which several situations are listed in a nutshell. It should be mentioned that the physical background is a one-dimensional diffusion

[7] Although only $f(z) = (z-1)^{\frac{1}{2}}$ is necessary, for the sake of symmetry we take (5.6.17).

process in a plate; the diffusion coefficient vanishes at the surfaces of the plate and takes its maximum in the middle of the plate:

Diffusion coefficient	s-Rank symbol/type
$e\,x^2 + f$	$\{1, 1; 1\}$, Gauss
$\mathrm{sn}^2(x\|k)$	$\{1, 1, 1; 1\}$, Heun
$b + c\,\mathrm{th}^2(d\,x + e)$	$\{1, \frac{1}{2}, 1; 1\}$, Heun
$b + c\,\sin^2(d\,x + e)$	$\{1, \frac{1}{2}, \frac{1}{2}; 1\}$, Heun
$b + c\,\mathrm{sn}^2(d\,x + K(k)\|k)$	$\{1, \frac{1}{2}, \frac{1}{2}, 1; 1\}$, Arscott
$\mathrm{sn}(x\|k)$	$\{\frac{1}{2}, \frac{1}{2}, 1, \frac{1}{2}, \frac{1}{2}; 1\}$, Ince

6
Afterword

It has been pointed out several times in this book that the methods presented here are not limited to the equations and problems treated concretely, and also that this is a characteristic feature of the Jaffé approach. The differential equations whose singular boundary eigenvalue problems produce the classical special functions are limited to Fuchsian differential equations with at most three singularities and their confluent and reduced cases, and the method of infinite continued fractions for generating the eigenvalue condition, so far the only transcendental eigenvalue condition, is limited to second-order difference equations.

The explicit presentation of the methods in this book is based on Heun's differential equation and its confluent and reduced cases, mainly for didactic reasons. This is the simplest differential equation, whose singular boundary eigenvalue problems lead to higher special functions. Accordingly, it gives rise to the simplest formulae that lead out of the known terrain, which not only makes the computational effort seem manageable, but also preserves the comprehensibility of the results.

In order to give the reader an impression of the paths through which the calculations for the general case of Fuchsian differential equations with any number of singularities and their confluent and reduced cases lead to the desired results, the fundamental proof of the theory for the general case is sketched in the appendix below, for the irregular second-order difference equation of Poincaré–Perron type. As can easily be seen, this proof for the difference equation of arbitrary high order runs along the same line of argument as that in §1.2.4, which should underline the generalisability of the method. Thus, no fundamental new insights arise here.

The central two-point connection problem as dealt with in this book may be distinguished according to two basic characteristics: if at the two endpoints of the relevant interval the singularities of the underlying differential equation are a regular and an irregular one; if one of the two endpoints of the relevant interval is an ordinary point of the underlying differential equation and the other is an irregular singularity. The former may be denoted the *strong central two-point connection problem*, while the latter may be denoted the *weak central two-point connection problem*. In both cases a singular boundary eigenvalue problem may be formulated, as was shown in §1.2.4.

The question occurs of whether the calculations made above show a pattern that allows us to generalise to the case when the s-rank of the irregular singularity of the underlying differential equation is a general number, say $s \in \mathbb{N}$, $s \geq 2$. Indeed, there is such a pattern; some words on it are in order in the following.

The form of the underlying differential equation is

$$\frac{d^2y}{dz^2} + P(z)\frac{dy}{dz} + Q(z)\,y(z) = 0 \tag{6.0.1}$$

with

$$\begin{aligned} P(z) &= \frac{A}{z} + \sum_{i=0}^{s-2} G_i\, z^i, \\ Q(z) &= \frac{B}{z} + \frac{C}{z^2} + \sum_{i=0}^{2s-4} D_i\, z^i \end{aligned} \tag{6.0.2}$$

on the relevant interval

$$0 \leq z < \infty$$

whereby A, B, C vanish in the case of the weak central two-point connection problem. The following applies for the characteristic exponents of the highest order $\alpha_{s-1,1}$, $\alpha_{s-1,2}$ of the infinite singularity:

$$\alpha_{s-1,1} = \frac{1}{2}\left(-G_{s-2} + \sqrt{G_{s-2}^2 - 4\,D_{2s-4}}\right),$$
$$\alpha_{s-1,2} = \frac{1}{2}\left(-G_{s-2} - \sqrt{G_{s-2}^2 - 4\,D_{2s-4}}\right),$$

and thus

$$\alpha_{s-1,1} - \alpha_{s-1,2} = \sqrt{G_{s-2}^2 - 4\,D_{2s-4}}. \tag{6.0.3}$$

The differential equation (6.0.1), (6.0.2) is encountered by the Jaffé ansatz

$$y(z) = \exp\left(\sum_{n=0}^{s-1} \frac{\alpha_i}{i}\, z^i\right) z^{\alpha_{0r}}\,(z+1)^{\alpha_{0i}-\alpha_{0r}}\, w(z) \tag{6.0.4}$$

with $\alpha_{s-1} = \alpha_{s-1,2}$ yielding an intermediate differential equation of the form

$$\frac{d^2y}{dz^2} + \tilde{P}(z)\frac{dy}{dz} + \tilde{Q}(z)\,y(z) = 0.$$

A Jaffé transformation

$$x = \frac{z}{z+1}$$

eventually yields

$$x\,(x-1)^s\,\frac{d^2w}{dx^2} + \sum_{i=0}^{s} \Gamma_i\,\frac{dw}{dx} + \sum_{i=0}^{s-1} \Delta_i\, w(x) = 0, \tag{6.0.5}$$

which has a solution that is holomorphic at the origin, $z = x = 0$, and thus may be expanded into a power series

$$w(x) = \sum_{n=0}^{\infty} a_n \, x^n .$$

There is a relation between the characteristic exponents α_i of the infinite singularity and the differential equation (6.0.5):

$$\begin{aligned} -\sum_{i=1}^{s} \Gamma_i &= -(G_{s-2} + 2\,\alpha_{s-1,2}) = -\left(G_{s-2} - G_{s-2} - \sqrt{G_{s-2}^2 - 4\,D_{2s-4}}\right) \\ &= \sqrt{G_{s-2}^2 - 4\,D_{2s-4}}, \end{aligned} \tag{6.0.6}$$

which becomes important in the following.

The coefficients a_n, $n = 0, 1, 2, 3, \ldots$, result in an irregular difference equation of Poincaré–Perron type of order s, accompanied by an initial condition. Thus, after having fixed the first term a_0, this initial condition determines the starting values $a_1, a_2, \ldots, a_{n+m}$. The consequence of such an initial condition, therefore, is that the difference equation becomes a linear recurrence relation

$$\begin{aligned} &\left(\kappa_m + \frac{\alpha_m}{n} + \frac{\beta_m}{n^2}\right) a_{n+m} \\ &\quad + \left(-\kappa_{m-1} + \frac{\alpha_{m-1}}{n} + \frac{\beta_{m-1}}{n^2}\right) a_{n+m-1} \\ &\quad + \left(\kappa_{m-2} + \frac{\alpha_{m-2}}{n} + \frac{\beta_{m-2}}{n^2}\right) a_{n+m-2} \\ &\quad \ldots \\ &\quad + \left(-\kappa_{n_0+1} + \frac{\alpha_{n_0+1}}{n} + \frac{\beta_{n_0+1}}{n^2}\right) a_{n-n_0+1} \\ &\quad + \left(\kappa_{n_0} + \frac{\alpha_{n_0}}{n} + \frac{\beta_{n_0}}{n^2}\right) a_{n-n_0} = 0, \\ &\quad n = s - j, s - j + 1, s - j + 2, \ldots, \end{aligned} \tag{6.0.7}$$

whereby the following definitions hold. If the s-rank s is an even number, then

$$s = 2j, \quad j = 1, 2, 3, \ldots .$$

In this case, it is defined as two quantities: $m = j$ and $n_0 = j$.

If the s-rank s is an odd number, then

$$s = 2j + 1, \quad j = 1, 2, 3, \ldots,$$

and it is defined as $m = j$ and $n_0 = j + 1$. Moreover, $s \in \mathbb{N}, s \geq 2$ and the absolute terms $\kappa_{n_0}, \ldots, \kappa_m$ of the coefficients of this difference equation are from *Pascal's triangle*:

$$
\begin{array}{ccccccccccccc}
 & & & & & & 1 & & & & & & \\
 & & & & & 1 & & 1 & & & & & \\
 & & & & 1 & & 2 & & 1 & & & & \\
 & & & 1 & & 3 & & 3 & & 1 & & & \\
 & & 1 & & 4 & & 6 & & 4 & & 1 & & \\
 & 1 & & 5 & & 10 & & 10 & & 5 & & 1 & \\
1 & & 6 & & 15 & & 20 & & 15 & & 6 & & 1 \\
 & & & & & & \cdots & & & & & &
\end{array}
$$

Choosing a_0 in the case of a strong central two-point connection problem and a_0 as well as a_1 in the case of a weak central two-point connection problem converts the difference equation (6.0.7) into an $s + 1$-term recurrence relation that allows the recursive calculation of as many coefficients a_n, $n = 1, 2, 3, \ldots$, as necessary.

There also exists a fundamental relation between the difference equation (6.0.7) and the differential equation (6.0.5):

$$\sum_{i=0}^{s} \Gamma_i = \sum_{i=0}^{s} \alpha_i.$$

Thus, because of (6.0.6), it holds that

$$-\sum_{i=1}^{s} \alpha_i = \sqrt{G_{s-2}^2 - 4\, D_{2s-4}}. \tag{6.0.8}$$

The *mathematical principle of unmasking recessive solutions* is a quite general one. It is important to recognise that the eigenvalue conditions of the CTCP are in no cases holomorphicity requirements. It is a characteristic feature of the CTCP that it may be solved via the mathematical principle of unmasking recessive solutions, both of the resulting difference as well as of the underlying differential equation. This principle works even in cases where the order of the resulting difference equation for the coefficients of the infinite series in the Jaffé ansatz is higher than the order of the underlying differential equation. Thus, as a conclusion, this principle is the basis of the CTCP for linear differential equations in a quite general sense.

The structure of the Jaffé ansatz is

$$y(z) = f(z)\, g[x(z)],$$

where $f(z)$ is an explicit asymptotic factor as $z \to \infty$ and $g[x(z)]$ is an infinite series of power type, the coefficients of which are denoted a_n, $n = 0, 1, 2, 3, \ldots$. This ansatz does have the characteristic feature that its application onto a linear, ordinary second-order differential equation yields once again a linear, ordinary difference equation for the coefficients a_n, $n = 0, 1, 2, 3, \ldots$, however, not necessarily of second order. The order I of the difference equation, being at least one, does not depend on the order of the underlying differential equation but on the number, position and s-ranks of its singularities. The general solution $y^{(g)}(z)$ of the underlying differential as well as the difference equation $a_n^{(g)}$ is a linear vector space, the dimension of which is equal to the order of the underlying equation. Since the order of the differential equation is two, its general solution may be written in the form

$$y^{(g)}(z) = c_1\, y_1(z) + c_2\, y_2(z),$$

where $y_1(z)$ and $y_2(z)$ are linearly independent fundamental solutions and c_1, c_2 are any arbitrary complex-valued constants in z. The general solution of the related difference equation for the a_n (the order of which is supposed to be larger than three) may be written in the form

$$a_n^{(g)} = L_1\, a_{n1} + \sum_{i=2}^{I} L_i\, a_{ni},$$

where a_{ni}, $i = 1, 2, \ldots, I$, are linearly independent fundamental solutions of the underlying difference equation and L_i, $i = 1, 2, \ldots, I$, are any arbitrary complex-valued constants in n.[1] A problem occurs in this situation, as soon as the order of the difference equation exceeds two, since then there is no simple assignment possible any more between the particular solutions of the difference and its underlying differential equation. Nevertheless, there is and must be such an assignment, since it is obvious that the asymptotic behaviour for $n \to \infty$ of all the particular solutions a_{ni}, $i = 2, 3, 4, \ldots, I$, of the difference equation determines the asymptotic behaviour of the infinite series of $g[x(z)]$ of the ansatz $y(z)$ as an eigensolution of the CTCP for the differential equation as $z \to \infty$. This assignment is ruled by the *mathematical principle of unmasking recessive solutions* and is to be displayed in the following. Suppose that $y_1(z)$ and a_{n1} are the maximum solutions of the differential and of the difference equation, respectively, as $z \to \infty$ and $n \to \infty$. Then, the mathematical principle of unmasking recessive solutions states the following:

- *If $L_1 \neq 0$, then $c_1 \neq 0$ and the asymptotic behaviour of $y_1(z)$ as $z \to \infty$ is governed by the asymptotic behaviour of a_{n1} as $n \to \infty$.*
- *If $L_1 = 0$, then $c_1 = 0$ and the asymptotic behaviour of $y(z)$ is governed by the asymptotic factor $f(z)$ as $z \to \infty$.*

Conclusions

- The asymptotic behaviour of the series

$$g[x(z)] = \sum_{n=0}^{\infty} a_n \left(\frac{z}{z+1}\right)^n$$

[1] Even in the case where all the parameters of the differential (and thus of the difference) equation are real-valued, the particular solutions a_n may be complex-valued. However, in this case, these particular solutions always occur as complex-conjugated pairs. This reduces the number of independent coefficients L_i.

as $z \longrightarrow \infty$, determined by the asymptotic behaviour of the coefficients a_n as $n \longrightarrow \infty$, in the former case may be written explicitly.

- The asymptotic behaviour of

$$a_n = \sum_{i=2}^{I} L_i \, a_{ni}$$

as $n \longrightarrow \infty$ in the latter case is such that it does not touch the asymptotic behaviour of $f(z)$ as $z \longrightarrow \infty$.

- This means that in the case $L_1 = 0$, the asymptotic behaviour of $g[x(z)]$ as $z \longrightarrow \infty$ is weaker than $f(z)$ in this limit.

The notion of 'unmasking' means that if there is a constellation of parameters of the differential equation such that $L_1 = 0$, then the recessive particular solutions of the differential equation $y_2(z)$ as well as of the difference equation

$$a_n^{(g)} = \sum_{i=2}^{I} L_i \, a_n^{(i)}$$

appear numerically. This recessive particular solution $y_2(z)$ of the differential equation is by definition the eigensolution of the CTCP, and the underlying constellation of parameters of the differential equation is the eigenvalues.

This principle may serve as a basis for exact solutions of singular boundary eigenvalue problems of linear differential equations, as done in this book. Moreover, the methods are characterised by their straight and simple numerical practicability.

Example

If

$$f(z) = \exp(\alpha_1 \, z)$$

as $z \longrightarrow \infty$, then

$$g[x(z)] \sim \exp(\alpha_2 \, z^{\epsilon})$$

as $z \longrightarrow \infty$ with $\epsilon < 1$ and α_2 arbitrary.

I guess that the mathematical principle of unmasking recessive solutions is quite a general one, constituting and ruling the basis of singular boundary eigenvalue problems of linear equations.

Appendix A
Standard Central Two-Point Connection Problem

This book makes a claim to generality. The exact solution of the central two-point connection problem with the help of the presented method, which I refer to as the Jaffé method in honour of George Cecil Jaffé, is not restricted to a certain s-rank of the irregular singularity. To emphasise this claim to generality, I will present below the central proof for this general case, which establishes the relation between the index-asymptotic behaviour of the particular solutions of the difference equation arising from the Jaffé method and the variable-asymptotic behaviour of the particular solution of the underlying differential equation. In its simplest form, this proof was presented in §1.2.4.

In order to be able to do this, a few preliminary remarks are necessary. We assume (for reasons of computational effort) that there is a regular singularity at the origin and an irregular singularity of s-rank s at infinity, and that certain particular solutions (namely decreasing, i.e., the recessive ones) are to be connected with each other.

It is noteworthy that the above-noted relation between the index-asymptotic behaviour of the particular solutions $\{a_n\}$ of the difference equation coming out of the Jaffé method on the one side and the variable-asymptotic behaviour of the particular solutions of the underlying differential equation on the other side in the general case is largely analogous to the presentation in §1.2.4. The mathematical mechanism behind this consists essentially of replacing the term $\{1/2\}$ in the proof of §1.2.4 by the term $\frac{s-1}{s}$, $s = 2, 3, 4, \ldots$, in the corresponding exponents. The crucial thing is that the value of this fraction remains below the value of one, which indeed is the case for all values of s:

$$\frac{s-1}{s} < 1 \quad \text{for } s = 2, 3, 4, \ldots .$$

The procedure starts with the differential equation

$$\frac{\mathrm{d}^2 y}{\mathrm{d}z^2} + P(z)\,\frac{\mathrm{d}y}{\mathrm{d}z} + Q(z)\,y = 0 \tag{A.1}$$

with

$$P(z) = \frac{A}{z} + G_0 + G_1\,z + \cdots + G_{s-2}\,z^{s-2},$$
$$Q(z) = \frac{B}{z^2} + \frac{C}{z} + D_0 + D_1\,z + \cdots + D_{2\,(s-2)}\,z^{2\,(s-2)}, \tag{A.2}$$

the s-rank symbol of which looks like

$$\{1; s\}.$$

The pair of characteristic coefficients of the regular singularity at the origin is given by [cf. (1.2.12)]

$$\alpha_{0r1} = \frac{1 - A + \sqrt{(1-A)^2 - 4\,B}}{2},$$
$$\alpha_{0r2} = \frac{1 - A - \sqrt{(1-A)^2 - 4\,B}}{2},$$

while from the $s + 1$ pairs of characteristic coefficients of the singularity at infinity, only the pair of highest order is of interest:

$$\alpha_{s1} = -\frac{1}{2}\left(G_{s-2} + \sqrt{G_{s-2}^2 - 4\,D_{2\,(s-2)}}\right),$$
$$\alpha_{s2} = -\frac{1}{2}\left(G_{s-2} - \sqrt{G_{s-2}^2 - 4\,D_{2\,(s-2)}}\right).$$

The decisive thing now is that in each case just one particular solution of the differential equation at the origin decreases, and all the others increase when approaching the singularity radially along the positive real axis. At infinity, there is just one particular solution that decreases exponentially, while all the others increase exponentially. This means that

$$\alpha_{0r1} > 0,\ \alpha_{0r2} < 0,\ \alpha_{s1} > 0,\ \alpha_{s2} < 0$$

is supposed to hold.

In solving the central two-point connection problem it is, therefore, a matter of finding the condition for a particular solution of the differential equation (A.1), (A.2), which decreases along the positive real axis when approaching the singularity along the positive real axis simultaneously at the origin and at infinity.

To achieve this, we make the Jaffé ansatz

$$y(z)$$
$$= \exp\left(\frac{\alpha_{s2}}{s}\,z^s + \frac{\alpha_{(s-1)2}}{s-1}\,z^{s-1} + \cdots + \frac{\alpha_{22}}{2}\,z^2 + \alpha_{12}\,z\right) z^{\alpha_{0r2}}\,(z+1)^{\alpha_{0i2}-\alpha_{0r2}}\,v(z)$$

and eventually carry out the Jaffé transformation

$$x = \frac{z}{z-1}$$

for the resulting differential equation for $v(z)$. Here, α_{0i2} is the first-order characteristic exponent of the Thomé solution for the irregular singularity at infinity, which results from α_{s2}. Finally, the function $v[x(z)]$ is holomorphic at the origin $z = x = 0$ and can, therefore, be represented as a Taylor series:

$$v[x(z)] = \sum_{n=0}^{\infty} a_n \, x^n. \tag{A.3}$$

Eventually, we get the coefficients a_n of the infinite power series in (A.3) in the form of an irregular difference equation of Poincaré–Perron type, the order of which is s.

If the s-rank of the irregular singularity is an even (natural) number, then this difference equation has the form[1]

$$\begin{aligned}
&\left(\kappa_{\frac{s}{2}} + \frac{\alpha_{\frac{s}{2}}}{n} + \frac{\beta_{\frac{s}{2}}}{n^2}\right) a_{n+\frac{s}{2}} \\
&\quad + \left(-\kappa_{\frac{s}{2}-1} + \frac{\alpha_{\frac{s}{2}-1}}{n} + \frac{\beta_{\frac{s}{2}-1}}{n^2}\right) a_{n+\frac{s}{2}-1} \\
&\quad\quad + \left(-\kappa_{\frac{s}{2}-2} + \frac{\alpha_{\frac{s}{2}-2}}{n} + \frac{\beta_{\frac{s}{2}-2}}{n^2}\right) a_{n+\frac{s}{2}-2} \\
&\quad\quad\quad + \cdots \\
&\quad\quad\quad\quad + \left(-\kappa_{-\frac{s}{2}+1} + \frac{\alpha_{-\frac{s}{2}+1}}{n} + \frac{\beta_{-\frac{s}{2}+1}}{n^2}\right) a_{n-\frac{s}{2}+1} \\
&\quad\quad\quad\quad\quad + \left(-\kappa_{-\frac{s}{2}+1} + \frac{\alpha_{-\frac{s}{2}+1}}{n} + \frac{\beta_{-\frac{s}{2}+1}}{n^2}\right) a_{n-\frac{s}{2}+1} \\
&\quad\quad\quad\quad\quad\quad + \left(\kappa_{-\frac{s}{2}} + \frac{\alpha_{-\frac{s}{2}}}{n} + \frac{\beta_{-\frac{s}{2}}}{n^2}\right) a_{n-\frac{s}{2}} \\
&\quad\quad\quad\quad\quad\quad = 0, \qquad n = s-j, s-j+1, s-j+2, \ldots,
\end{aligned} \tag{A.4}$$

with $j = 1, 2, 3, \ldots$.

If the s-rank of the irregular singularity is an odd (natural) number, then this difference equation has the form

[1] This difference equation is accompanied by initial conditions such that after having determined a_0, the difference equation turns into a recurrence relation and all the subsequent coefficients $\{a_n\}$, $n = 1, 2, 3, \ldots$, may be calculated step by step.

$$
\begin{aligned}
&\left(\kappa_{\frac{s-1}{2}} + \frac{\alpha_{\frac{s-1}{2}}}{n} + \frac{\beta_{\frac{s-1}{2}}}{n^2}\right) a_{n+\frac{s-1}{2}} \\
&\quad + \left(-\kappa_{\frac{s-1}{2}-1} + \frac{\alpha_{\frac{s-1}{2}-1}}{n} + \frac{\beta_{\frac{s-1}{2}-1}}{n^2}\right) a_{n+\frac{s-1}{2}-1} \\
&\quad + \left(-\kappa_{\frac{s-1}{2}-2} + \frac{\alpha_{\frac{s-1}{2}-2}}{n} + \frac{\beta_{\frac{s-1}{2}-2}}{n^2}\right) a_{n+\frac{s-1}{2}-2} \\
&\quad + \cdots \\
&\quad + \left(-\kappa_{-\frac{s-1}{2}+1} + \frac{\alpha_{-\frac{s-1}{2}+1}}{n} + \frac{\beta_{-\frac{s-1}{2}+1}}{n^2}\right) a_{n-\frac{s-1}{2}+1} \\
&\quad + \left(-\kappa_{-\frac{s-1}{2}+1} + \frac{\alpha_{-\frac{s-1}{2}+1}}{n} + \frac{\beta_{-\frac{s-1}{2}+1}}{n^2}\right) a_{n-\frac{s-1}{2}+1} \\
&\quad + \left(\kappa_{-\frac{s-1}{2}} + \frac{\alpha_{-\frac{s-1}{2}}}{n} + \frac{\beta_{-\frac{s-1}{2}}}{n^2}\right) a_{n-\frac{s-1}{2}} \\
&\quad = 0, \quad n = s-j, s-j+1, s-j+2, \ldots,
\end{aligned}
\tag{A.5}
$$

with $j = 1, 2, 3, \ldots$.

The values α_i and β_i result from the parameters of the differential equation (A.1), (A.2) and the values κ_i result from the sth row of Pascal's triangle.

As can be seen, for $s = 2$ the difference equation (A.4) for the central two-point connection problem results in the second-order difference equation of the single confluent case of Heun's differential equation and for $s = 4$ the difference equation (A.4) for the central two-point connection problem results in the fourth-order difference equation of the triconfluent case of Heun's differential equation.

As can be seen, for $s = 3$ the difference equation (A.5) results in the third-order difference equation of the biconfluent case of Heun's differential equation.

It is of central importance that the Birkhoff set of this difference equation shows that there is exactly one particular solution which exhibits the strongest exponentially increasing asymptotic behaviour. This particular solution alone determines whether the function represented by the convergent series (A.3) decays exponentially towards zero for $n \longrightarrow \infty$ or increases exponentially.

Considering the first-order characteristic equation of the difference equation (A.4), (A.5):

$$(t-1)^s = 0,$$

one recognises that the ratio of two successive terms of all particular solutions of the difference equation (A.4), (A.5) approaches unity:

$$t_n = \frac{a_{n+1}}{a_n} \sim 1$$

as $n \longrightarrow \infty$.

The second-order characteristic equation, moreover, shows how this asymptotics looks in detail. This results, from the difference equation (A.4) for even values of s, to

$$\begin{aligned}
&\left(\kappa_{\frac{s}{2}} + \frac{\alpha_{\frac{s}{2}}}{n}\right) t^{s} \\
&\quad + \left(-\kappa_{\frac{s}{2}-1} + \frac{\alpha_{\frac{s}{2}-1}}{n}\right) t^{s-1} \\
&\quad\quad + \left(\kappa_{\frac{s}{2}-2} + \frac{\alpha_{\frac{s}{2}-2}}{n}\right) t^{s-2} \\
&\quad\quad\quad + \cdots \\
&\quad\quad\quad\quad + \left(\kappa_{-\frac{s}{2}-1} + \frac{\alpha_{-\frac{s}{2}-1}}{n}\right) t \\
&\quad\quad\quad\quad\quad + \kappa_{-\frac{s}{2}} + \frac{\alpha_{-\frac{s}{2}}}{n} = 0
\end{aligned} \tag{A.6}$$

and for odd values of s, to

$$\begin{aligned}
&\left(\kappa_{\frac{s-1}{2}} + \frac{\alpha_{\frac{s-1}{2}}}{n}\right) t^{s} \\
&\quad + \left(-\kappa_{\frac{s-1}{2}-1} + \frac{\alpha_{\frac{s-1}{2}-1}}{n}\right) t^{s-1} \\
&\quad\quad + \left(\kappa_{\frac{s-1}{2}-2} + \frac{\alpha_{\frac{s-1}{2}-2}}{n}\right) t^{s-2} \\
&\quad\quad\quad + \cdots \\
&\quad\quad\quad\quad + \left(\kappa_{-\frac{s-1}{2}-1} + \frac{\alpha_{-\frac{s-1}{2}-1}}{n}\right) t \\
&\quad\quad\quad\quad\quad + \kappa_{-\frac{s-1}{2}} + \frac{\alpha_{-\frac{s-1}{2}}}{n} = 0,
\end{aligned} \tag{A.7}$$

the solutions of which are given by

$$t_n = \frac{a_{n+1}}{a_n} \sim 1 + \gamma \frac{s-1}{s\, n^{\frac{1}{s}}} + O\left(\frac{1}{n^{\frac{1}{2s}}}\right) \tag{A.8}$$

as $n \longrightarrow \infty$ with

$$\gamma = \sqrt[s]{-\sum_{i} \alpha_i}. \tag{A.9}$$

This algebraic equation has s solutions, which can be illustrated as shown in Figures A.1 and A.2, indicating particular solutions of the underlying difference equation. All of these solutions (A.8) of the equation (A.9) indicate exponential behaviours of the difference equations (A.6) and (A.7) as $n \longrightarrow \infty$.

Now, we can prove the following auxiliary theorem:

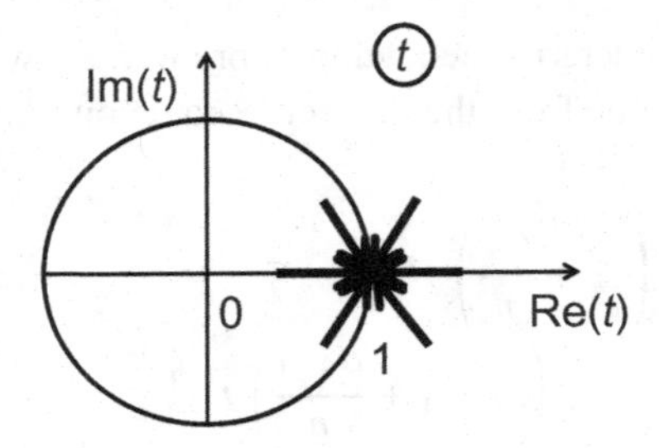

Figure A.1 First-order asymptotics for even s-ranks of the irregular singularity.

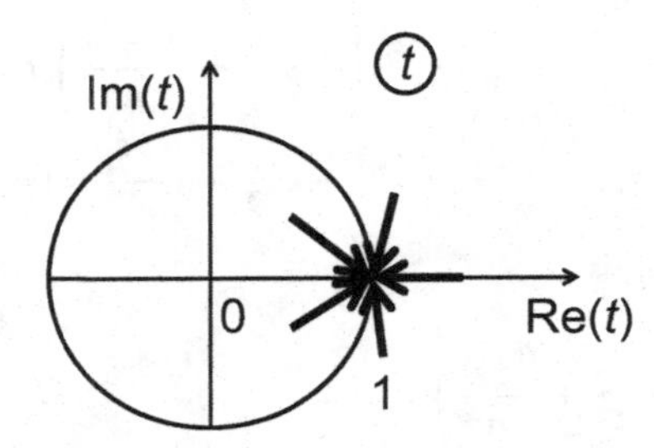

Figure A.2 First-order asymptotics for odd s-ranks of the irregular singularity.

Theorem (Auxiliary). *Suppose a sequence a_n, $n = 1, 2, 3, \ldots$, of numbers behaves like*

$$a_n \sim \exp\left(\gamma\, n^{\frac{s-1}{s}}\right), \qquad s = 2, 3, 4, \ldots, \tag{A.10}$$

as $n \longrightarrow \infty$, where γ is any complex-valued parameter. Then the ratio of two consecutive terms a_{n+1} and a_n is

$$\frac{a_{n+1}}{a_n} \sim 1 + \gamma \frac{s-1}{s} \frac{1}{n^{\frac{1}{s}}} + O\left(\frac{1}{n^{\frac{2}{s}}}\right) \tag{A.11}$$

as $n \longrightarrow \infty$.

Proof of Auxiliary Theorem:

$$
\begin{aligned}
\frac{a_{n+1}}{a_n} &\sim \frac{\exp\left[\gamma\,(n+1)^{\frac{s-1}{s}}\right]}{\exp\left(\gamma\, n^{\frac{s-1}{s}}\right)} \\
&\sim \frac{\exp\left[\gamma\, n^{\frac{s-1}{s}}\left(1+\frac{s-1}{s\,n}+\cdots\right)\right]}{\exp\left[\gamma\, n^{\frac{s-1}{s}}\right]} \\
&\sim \frac{\exp\left[\gamma\, n^{\frac{s-1}{s}}+\frac{\gamma\,(s-1)}{s}\,n^{-\frac{1}{s}}+\cdots\right]}{\exp\left[\gamma\, n^{\frac{1}{s}}\right]} \\
&\sim \frac{\exp\left[\gamma\, n^{\frac{1}{s}}\right]\exp\left[\frac{(s-1)\gamma}{s\,n^{\frac{1}{s}}}+\cdots\right]}{\exp\left[\gamma\, n^{\frac{1}{s}}\right]} \\
&\sim 1+\frac{s-1}{s}\,\frac{\gamma}{n^{\frac{1}{s}}}+O\left(\frac{1}{n^{\frac{2}{s}}}\right)
\end{aligned}
\tag{A.12}
$$

as $n \longrightarrow \infty$ •

Formula (A.10) is an asymptotic solution of the difference equation (A.4) as $n \longrightarrow \infty$, from which follows (A.11) as the ratio of two consecutive terms of this function. On the other hand, this ratio is also given by the solution (A.8) of the characteristic equation (A.6) of the difference equation (A.4) (which in its part is a result of the difference equation).

In the following it is shown that the maximum solution of the difference equation (A.4) generates the maximum particular solution of the differential equation (A.1), (A.2). This maximum solution is indicated by the arrow lying on the real axis, pointing from $t > 1$ to the value $t = 1$ in Figures A.1, A.2. Such a particular solution for the difference equation always exists irrespective of whether the s-rank s of the irregular singularity of the underlying differential equation at infinity is an even or an odd number. For this represents exactly the particular solution that causes the strongest exponential increase, while all other $s - 1$ particular solutions cause a weaker asymptotic behaviour as $n \longrightarrow \infty$.

So, it is a matter of determining the functional behaviour of precisely these particular solutions of the difference equation (A.4), (A.5) on the one hand, and the differential equation (A.1), (A.2) on the other hand, and their relationship to each other. Or, to put it briefly: if the coefficients a_n [and thus the particular solution of the difference equation (A.4), (A.2)] for $n \longrightarrow \infty$ increase exponentially, so does the function $v[x(z)]$ for $z \longrightarrow \infty$, which is represented by the Taylor series (A.3). This is shown in the following.

Proposition. *If the coefficients a_n in the Taylor series* (A.3)

$$v[x(z)] = \sum_{n=0}^{\infty} a_n \left(\frac{z}{z+1}\right)^n = \sum_{n=0}^{\infty} a_n \exp\left[n \ln\left(\frac{z}{z+1}\right)\right]$$

behave like

$$a_n \sim \exp\left(\gamma_1 n^{\frac{s-1}{s}}\right)$$

as $n \longrightarrow \infty$, then the function $v[x(z)]$ behaves like

$$v[x(z)] \sim \exp\left\{\left[\frac{(s-1)\gamma_1}{s}\right]^s \frac{z^{s-1}}{s-1}\right\}$$

as $z \longrightarrow \infty$.

Proof Consider

$$\begin{aligned}
\sum_{n=0}^{\infty} a_n x^n = \sum_{n=0}^{\infty} a_n \left(\frac{z}{z+1}\right)^n &\sim \sum_{n=0}^{\infty} \exp\left(\gamma_1 n^{\frac{s-1}{s}}\right) \left(\frac{z}{z+1}\right)^n \\
&= \sum_{n=0}^{\infty} \exp\left(\gamma_1 n^{\frac{s-1}{s}}\right) \exp\left[n \ln\left(\frac{z}{z+1}\right)\right] \\
&= \sum_{n=0}^{\infty} \exp\left[\gamma_1 n^{\frac{s-1}{s}} + \ln\left(\frac{z}{z+1}\right) n\right] \\
&= \sum_{n=0}^{\infty} \exp\left[h(z,n)\right] = \sum_{n=0}^{\infty} u(z,n)
\end{aligned}$$

with

$$u(z,n) = \exp\left[h(z,n)\right] \overset{\text{Auxiliary I}}{\sim} \exp\left[\gamma_1 n^{\frac{s-1}{s}} - \frac{n}{z}\right]$$

as n, $z \longrightarrow \infty$. Here, use is made of Auxiliary I in §1.2.4.

On the basis of the Euler–MacLaurin sum formula (cf., e.g., deBruijn, 1961; Abramowitz and Stegun, 1970), we have the calculation

$$\begin{aligned}
v[x(z)] &= \int_{n=0}^{\infty} u(z,n)\, \mathrm{d}n \\
&= \int_{n=0}^{\infty} \exp\left[h(z,n)\right] \mathrm{d}n \\
&= \int_{n=0}^{\infty} \exp\left[\gamma_1 n^{\frac{s-1}{s}} - \frac{n}{z}\right] \mathrm{d}n. \qquad \text{(A.13)}
\end{aligned}$$

It is important to understand that for $\gamma_1 > 0$ or for $\Re(\gamma_1) > 0$, the function $u(z,n) = u_z(n)$ always has a hump that has two properties as $n \to \infty$, $z = \text{const.}$:

- The apex of the hump tends to infinity as $n \longrightarrow \infty$.
- The hump becomes smaller and taller, viz. a peak, a delta function in the limit $n \longrightarrow \infty$.

These two properties are the conditions for applying the Lagrange integration method, which is done in the following.

According to this method, the function $u(z,n) = u_z(n)$ is approximated by a second-order polynomial in the limit $n \longrightarrow \infty$, with z a constant (cf., e.g., deBruijn, 1961):

$$u_z(n) \sim g_1(n) = a_1 \; [n - n_0]^2 + b_1 \quad \text{as } n \longrightarrow \infty \tag{A.14}$$

with

$$\begin{aligned} a_1 &= a_1(n_0, \gamma_1, z), \\ b_1 &= b_1(n_0, \gamma_1, z), \\ n_0 &= n_0(\gamma_1, z), \end{aligned} \tag{A.15}$$

where n_0 is the location of the hump on the n-axis. In order to calculate an approximation of the integral function in (1.2.98), this function $g_1(n)$ may be integrated from the lower zero n_{11} of $g_1(n)$ to the upper one n_{12}, yielding

$$\begin{aligned} \int_0^\infty g_1(z,n)\, \mathrm{d}n &\sim \int_{n_{11}}^{n_{12}} g_1(z,n)\, \mathrm{d}n \\ &\sim v(z) \end{aligned}$$

as $n \longrightarrow \infty$ and for fixed values of the independent variable z.

According to the Lagrange integration method, there are three determining conditions for the three parameters a_1, b_1, n_0:

$$\begin{aligned} g_1(n = n_0) &= u_z(n = n_0), \\ \left.\frac{\mathrm{d}g_1(n)}{\mathrm{d}n}\right|_{n=n_0} &= \left.\frac{\mathrm{d}u_z(n)}{\mathrm{d}n}\right|_{n=n_0} = 0, \\ \left.\frac{\mathrm{d}^2 g_1(n)}{\mathrm{d}n^2}\right|_{n=n_0} &= \left.\frac{\mathrm{d}^2 u_z(n)}{\mathrm{d}n^2}\right|_{n=n_0}. \end{aligned}$$

In the following, these three conditions are evaluated.

- The first condition yields

$$\boxed{\exp\left[\gamma_1\, n_0^{\frac{s-1}{s}} - \frac{n_0}{z}\right] = a_1\, (n_0 - n_0)^2 + b_1 = b_1.}$$

- The second condition yields

$$\frac{\mathrm{d}u}{\mathrm{d}n} = \left[\frac{(s-1)\,\gamma_1}{s\, n^{\frac{1}{s}}} - \frac{1}{z}\right] \exp\left(\gamma_1\, n^{\frac{s-1}{s}} - \frac{n}{z}\right)$$

thus

$$\left.\frac{\mathrm{d}u}{\mathrm{d}n}\right|_{n=n_0} = \left[\frac{(s-1)\,\gamma_1}{s\,n_0^{\frac{1}{s}}} - \frac{1}{z}\right] \exp\left[\gamma_1\, n_0^{\frac{s-1}{s}} - \frac{n_0}{z}\right] = 0,$$

resulting in

$$\frac{(s-1)\,\gamma_1}{s\,n_0^{\frac{1}{s}}} - \frac{1}{z} = 0$$

or

$$\boxed{n_0(\gamma_1, z) = \left[\frac{(s-1)\,\gamma_1}{s}\,z\right]^s.}$$

On the other hand [cf. (A.14)],

$$\frac{\mathrm{d}g_1(n)}{\mathrm{d}n} = 2\,a_1\,(n - n_0)$$

from which follows

$$\left.\frac{\mathrm{d}g_1(n)}{\mathrm{d}n}\right|_{n=n_0} = 2\,a_1\,(n_0 - n_0) = 0,$$

as is shown to be correct.

- The third condition requires the carrying out of the second derivative

$$\frac{\mathrm{d}^2 u_z}{\mathrm{d}n^2} = u_z(n)\left[\left(\frac{\mathrm{d}h}{\mathrm{d}n}\right)^2 + \frac{\mathrm{d}^2 h}{\mathrm{d}n^2}\right].$$

If $h = h(n)$ is an algebraic function in n – as is the case here – then

$$\frac{\mathrm{d}^2 u_z}{\mathrm{d}n^2} \sim \exp\left(\gamma_1\, n^{\frac{s-1}{s}} - \frac{n}{z}\right)$$

as $n \to \infty$, thus

$$\left.\frac{\mathrm{d}^2 u_z}{\mathrm{d}n^2}\right|_{n=n_0} \sim \exp\left(\gamma_1\, n_0^{\frac{s-1}{s}} - \frac{n_0}{z}\right)$$

as $n_0 \to \infty$ and

$$\frac{\mathrm{d}^2 g_1}{\mathrm{d}n^2} = 2\,a_1,$$

from which follows

$$*a_1 = \frac{1}{2}\left.\frac{\mathrm{d}^2 u_z}{\mathrm{d}n^2}\right|_{n=n_0} = \frac{1}{2}\,u_z(n)\left.\left[\left(\frac{\mathrm{d}h}{\mathrm{d}n}\right)^2 + \frac{\mathrm{d}^2 h}{\mathrm{d}n^2}\right]\right|_{n=n_0}$$

$$\sim \frac{1}{2}\exp\left(\gamma_1\, n_0^{\frac{s-1}{s}} - \frac{n_0}{z}\right)$$

as $n_0 \longrightarrow \infty$, meaning $z \longrightarrow \infty$. Then

$$n_0 = \left[\frac{(s-1)\,\gamma_1}{s}\,z\right]^s,$$

$$n_0^{\frac{s-1}{s}} = \left[\frac{(s-1)\,\gamma_1}{s}\,z\right]^{s-1},$$

thus

$$a_1(n_0) \sim \exp\left\{\left[\frac{(s-1)\,\gamma_1}{s}\right]^s \frac{z^{s-1}}{s-1}\right\}$$

as $n_0 \longrightarrow \infty$ or $z \longrightarrow \infty$.

Summarising the calculations, note that the function

$$u_z(n) = \exp\left(\gamma_1\, n^{\frac{s-1}{s}} - \frac{n}{z}\right)$$

in the limit $n \longrightarrow \infty$ is approximated by the quadratic parabola

$$g_1(n) = a_1\,[n - n_0]^2 + b_1$$

with[2]

$$n_0(z) = \left[\frac{(s-1)\,\gamma_1}{s}\,z\right]^s,$$

$$a_1(n_0, z) \sim \frac{1}{2}\exp\left[\gamma_1\, n_0^{\frac{s-1}{s}} - \frac{n_0}{z}\right] \quad \text{as } n_0 \longrightarrow \infty,$$

$$b_1(n_0\, z) = \exp\left[\gamma_1\, n_0^{\frac{s-1}{s}} - \frac{n_0}{z}\right].$$

Incidentally

$$\frac{b_1}{a_1} = 2.$$

The calculation of the zeros of $g_1(n)$ is done by

$$a_1\,[n_1 - n_0]^2 + b_1 = 0$$

or

$$a_1\,n_1^2 - 2\,a_1\,n_0\,n_1 + a_1\,n_0^2 + b_1 = 0.$$

Thus

$$n_{11,2} = \frac{1}{2\,a_1}\left[2\,a_1\,n_0 \pm \sqrt{4a_1^2\,n_0^2 - 4\,a_1\left(a_1\,n_0^2 + b_1\right)}\right]$$

$$= n_0 \pm \sqrt{-\frac{b_1}{a_1}} = \left(\frac{(s-1)\,\gamma_1}{s}\,z\right)^s \pm \gamma_1^{\frac{s-1}{s}}\, z^{\frac{s-1}{s}}.$$

[2] Thus, for $z \rightarrow \infty$, $a_1 \rightarrow -b_1$.

The final result is summarised by stating

$$\boxed{\begin{aligned} v[x(z)] &= \sum_{n=0}^{\infty} a_n\, x^n \\ &\sim \sum_{n=0}^{\infty} \exp\left(\gamma_1\, n^{\frac{s-1}{s}}\right) x^n \text{ as } n \to \infty \\ &\sim \int_{n_{11}}^{n_{12}} \left[a_1\, (n-n_0)^2 + b_1\right] \mathrm{d}n \\ &= \left.\frac{a_1}{3}\,(n-n_0)^3 + b_1\, n\right|_{n_{11}}^{n_{12}} \\ &\sim \exp\left\{\left[\tfrac{(s-1)\,\gamma_1}{s}\right]^s \tfrac{z^{s-1}}{s}\right\} \end{aligned}}$$

or

$$\boxed{v[x(z)] \sim \exp\left\{\left[\frac{(s-1)\,\gamma_1}{s}\right]^s \frac{z^{s-1}}{s-1}\right\}}$$

as $z \to \infty$, which is to be shown. □

Appendix B
Curriculum Vitae of George Cecil Jaffé

This book is based on an idea of George Cecil Jaffé, which he published in the *Zeitschrift für Physik* in 1933. As a Jewish professor he had to abandon his position in the same year. I do not know whether he would have traced his idea if he could have stayed. Nevertheless, honour to whom honour is due! In the following I give a short curriculum vitae from the German Wikipedia of this marvellous scientist, who – and whose idea – deserves to be remembered and maintained. That is why this book has been written against oblivion, against the oblivion of the discriminated life of George Cecil Jaffé and against the oblivion of his marvellous mathematical idea.

George Cecil Jaffé[1] was born in 1880 in Moscow as a German citizen. He was the son of the merchant Ludwig Jaffé (1845–1923) from Hamburg and his wife Henriette, née Marks (1853–1929), an American from New Orleans. For the sake of a German school education for their children, the family returned to Hamburg in 1888. There, George attended the 'Volksschule' (elementary school), the 'Realgymnasium' and, eventually, a humanistic grammar school for eight years. After his school-leaving examination in 1898 he first studied mathematics, physics and chemistry at the Ludwig Maximilian University of Munich, emphasising chemistry. Thereafter, he continued to study at the Universität Leipzig. Under the supervision of his teacher, **Wilhelm Ostwald**, a later Nobel Prize winner, Jaffé wrote a dissertation thesis about saturated solutions, which obtained him a doctoral degree. Moreover, in Leipzig, the famous physicist **Ludwig Boltzmann** had great influence on him.

Endowed with a recommendation from Boltzmann, Jaffé continued his studies at Cambridge in 1903/04. Once there, the French Nobel Prize winner **Pièrre Curie** visited Cambridge and asked Jaffé to work with him and his wife, **Marie Curie**,[2] in their laboratory in Paris. Jaffé accepted. After a 1-year stay, Jaffé went on to the United States. There, he attended several universities on the East Coast, as well as the National Bureau of Standards and Technology (NIST). Eventually he returned to Leipzig, where he habilitated in 1908 with a dissertation on the electrical conductivity

[1] This curriculum vitae is transcribed, with slight modifications, from the text of the German Wikipedia, published under 'George Jaffé'. See also *Neue Deutsche Biographie*, Volume 10.

[2] Marie Skłodowska Curie was the first person – and the only woman – to get the Nobel Prize twice. Moreover, she was the only mother whose daughter also got a Nobel Prize.

of pure hexane. Subsequently, he became a scientific assistant and gave lectures at the Universität Leipzig as a 'Privatdozent'.

In 1911 and 1912 he worked once again in the laboratory of the Curies in Paris, this time as a Carnegie Scholar. In 1916 he was appointed 'außerordentlicher Professor' of the Universität Leipzig.

The academic career of George Jaffé was interrupted by the First World War. After he became a soldier in 1915, he was promoted to the rank of lieutenant and got several high-level military decorations. In particular, he earned special recognition for his deciphering skills. In 1919 he returned to Leipzig from military service.

In 1923 George Jaffé was appointed ordinary full professor of theoretical physics. He worked on the ionisation of gases, light absorption in metals and insulators, hydrodynamics, high vacuum discharge, the theory of relativity, anisotropic radiation fields and statistical mechanics.

From 1926 to 1933 George Jaffé worked as a full professor of theoretical physics at the Universität Gießen (located in the northern part of Hesse, Germany). There, he concerned himself with the conductivity of ions and with kinetic gas theory. In 1932 he held the office of Dean of the Faculty of Philosophy.

At the Universität Gießen, George Jaffé could teach and research only for a period of 6 years. Since he was of Jewish descent, it was suggested by the chancellor of the university, im Interesse des ungestörten Lehrbetriebs den Beginn seiner Vorlesungen hinauszuschieben (i.e., to postpone the start of his lectures in the interests of an undisturbed teaching programme). On 6 September 1933 he was sent a dismissal notice by the Hessian State Ministry (Hessisches Staatsministerium), on the basis of §4 Reichsgesetz from 7 April 1933. On behalf of his military earnings in the First World War, this ejection was mitigated into a forced retirement in March 1934. In 1938 George Jaffé left the Deutsche Physikalische Gesellschaft at the instigation of the Dutch Nobel Prize winner **Peter Debye**.

After this politicaly motivated dismissal, George Jaffé worked on private scientific projects in Freiburg/Breisgau. In 1939 he emigrated to the United States where, at the Louisiana State University in Baton Rouge, he was a visiting lecturer up to 1942; thereafter, he became an associate professor and in 1946 he was appointed to a full professorship.

His main scientific interest was electrical conductivity in liquids and related problems. In 1950 he retired, aged 70. As an emeritus professor, he continued his scientific research on electrical conductivity of semiconductors and on the diffusion of neutrons.

In 1912 George Jaffé married the pianist Paula Hegner (1889–1943), with whom he had a son.

His final years were spent partially in Germany; he died in March 1965 in Göppingen (Baden-Württemberg, Germany). His body was transported to the United States, where he was buried at the Hebrew Rest Cemetery in New Orleans.

According to his will, his inheritance is kept at the library of the Universität Gießen. It encompasses thirteen volumes of private notes, showing that George Cecil Jaffé was a man of extraordinarily broad education.

Table B.1 *The academic career of George Cecil Jaffé*

Date	Academic Steps	Location	Remark
1898	School-Leaving Examination	Hamburg	
1898	Studies of Mathematics, Physics and Chemistry	Universität München	Professor: Wilhelm Ostwald
	Graduation	Universität Leipzig	Influenced by Ludwig Boltzmann
1903/04	Research Fellow	Cambridge University	Recommended by Ludwig Boltzmann
1904/05	Post-Doc Research Work	Laboratory of École municipale de physique et de chimie industrielles (EPCI, today: ESPCI)	With Pierre and Marie Curie
1908	Habilitation	Universität Leipzig	
1911/12	Carnegie Grant	Université Sorbonne, Paris	With Marie Curie
1916	Ausserordentlicher Professor (Extraordinary Professor)	Universität Leipzig	
1915–1919	Interruption by First World War	Military service at several places	
1919–1923	Research and Lecturing	Universität Leipzig	
1923	Ordentlicher Professor (Ordinary Professor)	Universität Leipzig	
1923–1926	Research and Lecturing	Universität Leipzig	
1926	Ordinarius	Universität Gießen	
1926–1933	Research and Lecturing	Universität Gießen	
1932	Dean of the Faculty of Philosophy	Universität Gießen	
1934	Forced Retirement	Universität Gießen	
1938	Discharge from the Deutsche Physikalische Gesellschaft		
1934–1939	Private Research Projects	Freiburg/Breisgau	
1939	Emigration to the United States	Baton Rouge, LA	
1939–1942	Visiting Lecturer	Louisiana State University in Baton Rouge, LA	
1943–1946	Associate Professor	Louisiana State University in Baton Rouge, LA	
1946–1950	Full Professor	Louisiana State University in Baton Rouge, LA	
1950	Retirement	Louisiana State University in Baton Rouge, LA	
1950–1965	Research Activities as an Emeritus Professor		

References

Abramowitz, M. and Stegun, I. (1970). *Handbook of Mathematical Functions*, 9th ed. Dover Publications.

Adams, R. (1928). On the irregular cases of the linear ordinary difference equation. *Trans. Amer. Math. Soc.*, **30**(3), 507–541.

Arscott, F. M., Taylor, P. J. and Zahar, R. V. M. (1983). On the numerical construction of ellipsoidal wave functions. *Math. Comput.*, **40**(161), 367–380.

Aulbach, B., Elaydi, S. and Ladas, G. (eds). (2004). *Proceedings of the Sixth International Conference on Difference Equations: New Progress in Difference Equations*. Chapman & Hall/CRC Press.

Bay, K., Lay, W. and Akopyan, A. M. (1997). Avoided crossings of the quartic ascillator. *J. Phys. A: Math. Gen.*, **30**, 3057–3067.

Bay, K., Lay, W. and Slavyanov, S. Yu. (1998). Asymptotic and numeric study of eigenvalues of the double confluent Heun equation. *J. Phys. A: Math. Gen.*, **31**, 8521–8531.

Behnke, H. and Sommer, F. (1976). *Theorie der analytischen Funktionen einer komplexen Veränderlichen.* Springer.

Bieberbach, L. (1965). *Theorie der gewöhnlichen Differentialgleichungen,* 2nd ed. Springer.

Birkhoff, G. D. and Trjitzinsky, W. J. (1933). Analytic theory of singular difference equations. *Acta. Math.*, **60**, 1–89.

Bronstein, I. N., Semendjajew, K. D., Musiol, G. and Mühlig, H. (2001). *Taschenbuch der Mathematik*, 5. überarbeitete und erweiterte Auflage. Harri Deutsch.

Byrd, P. F. and Friedman, M. D. (1971). *Handbook of Elliptic Integrals for Engineers and Scientists*, 2nd ed. Springer.

Coddington, E. A. and Levinson, N. (1955). *Theory of Ordinary Differential Equations.* McGraw-Hill. Reprinted 1984, Krieger.

Courant, R. and Hilbert, D. (1968). *Methoden der Mathematischen Physik I,* 3rd ed. Springer.

deBruijn, N. G. (1961). *Asymptotic Methods in Analysis*, 2nd ed. North-Holland.

Erdélyi, A. (1956). *Asymptotic Expansions.* Dover.

Esslinger, J. (1990). *Quanteneffekte bei der Diffusion von Kinken auf Versetzungen.* Dissertation, Universität Stuttgart.

Fuchs, L. (1866). Zur Theorie der linearen Differentialgleichungen. *J. Reine und Angew. Math.*, **66**, 121–160.

Gelfand, I. M. and Schilow, G. J. (1964). *Verallgemeinerte Funktionen: Einige Fragen zur Theorie der Differentialgleichungen*, Vol. III. Deutscher Verlag der Wissenschaften.

Gelfand, I. M. and Wilenkin, N. (1964). *Verallgemeinerte Funktionen: Einige Anwendungen der Harmonischen Analyse: Gelfandsche Raumtripel*, Vol. IV. Deutscher Verlag der Wissenschaften.

Gutzwiller, M. (1990). *Chaos in Classical and Quantum Mechanics.* Springer.

Heun, K. (1889). Zur Theorie der Riemannschen Fuctionen zweiter Ordnung mit vier Verzweigungspunkten. *Math. Ann.*, **33**, 161–179.

Hille, E. (1997). *Ordinary Differential Equations in the Complex Domain.* Dover.

Hurwitz, A., Courant, R. and Röhrl, H. (1964). *Allgemeine Funktionentheorie und Elliptische Funktionen.* Springer.

Ince, E. L. (1956). *Ordinary Differential Equations.* Dover.

Jaffé, G. C. (1933). Zur Theorie des Wasserstoffmolekülions. *Z. Phys.*, **87**, 535–544.

Lay, W. (1997). The quartic oscillator. *J. Math. Phys.*, **38**(2), 639–647.

Olver, F. W. J. (1974). *Asymptotics and Special Functions.* Academic Press.

Olver, F. W. J., Lozier, D. W., Boisvert, R. F. and Clark, C. W. (eds.) (2010). *NIST Handbook of Mathematical Functions.* Cambridge University Press.

Nörlund, N. E. (1924). *Vorlesungen über Differenzenrechnung.* Springer.

Perron, O. (1909). Über einen Satz des Herrn Poincaré. *J. Reine Agnew. Math.*, **136**, 17–38.

Perron, O. (1910). Über die Poincarésche lineare Differenzengleichung. *J. Reine Agnew. Math.*, **137**, 6–64.

Perron, O. (1911). Über lineare Differenzengleichungen. *Acta Math.*, **34**, 109–137.

Perron, O. (1913). *Die Lehre von den Kettenbrüchen.* Teubner Verlag 3rd ed., 1954. Volume 1, *Elementare Kettenbrüche*; Volume 2, *Analytische und funktionentheoretische Kettenbrüche.*

Pincherle, S. (1894). Delle funzioni ipergeometriche e di varie questioni ad esse attinenti. *Gion. Mat. Battaglini*, **32**, 209–291.

Poincaré, H. (1885). Sur les équations linéaires aux différentielles ordinaires et aux différences finies. *Amer. J. Math.*, **7**, 203–258. Reprinted in *Oeuvres*, **1**, 226–289.

Poincaré, H. (1886). Sur les intégrales irrégulières des équations linéaires. *Acta Math.*, **8**, 295–344. Reprinted in *Oeuvres*, **1**, 290–332.

Ronveaux, A. (ed.) (1995). *Heun's Differential Equations.* Oxford University Press.

Rubinowicz, A. (1972). *Sommerfeldsche Polynommethode.* Springer.

Schubert, M. and Weber, G. (1980). *Quantentheorie – Grundlagen, Methoden, Anwendung Teil I.* VEB Deutscher Verlag der Wissenschaften.

Seeger, A. and Lay, W. (eds.) (1990). *Proceedings of the Centennial Workshop on Heun's Equation – Theory and Applications.* 3–8 September 1989, Schloß Ringberg (Rottach-Egern). Max-Planck-Institut für Metallforschung, Institut für Physik, Stuttgart.

Seeger, A. and Schiller, P. (1966). Kinks in dislocation lines and their effects on the internal friction in crystals. In W. P. Mason (ed.), *Physical Acoustics*, Vol. III, pp. 361ff. Academic Press.

Slavyanov, S. Yu. (1996). *Asymptotic Solutions of the One-Dimensional Schrödinger Equation.* (Translations of Mathematical Monographs, **151**). American Mathematical Society.

Slavyanov, S. Yu. and Lay, W. (2000). *Special Functions. A Unified Theory Based on Singularities.* Oxford University Press.

Slavyanov, S. Yu. and Veshev, N. A. (1997). Structure of avoided crossings for eigenvalues related to equations of Heun's class. *J. Phys. A: Math. Gen.*, **30**, 673–687.

Whittaker, E. T. and Watson, G. N. (1927). *A Course of Modern Analysis,* 4th ed. Cambridge University Press.

Index

Printed in the United States
by Baker & Taylor Publisher Services

Printed in the United States
by Baker & Taylor Publisher Services